AF389421

Microgrids: Planning, Protection and Control

Microgrids: Planning, Protection and Control

Editor

Hossam A. Gabbar

MDPI • Basel • Beijing • Wuhan • Barcelona • Belgrade • Manchester • Tokyo • Cluj • Tianjin

Editor
Hossam A. Gabbar
Ontario Tech University
Canada

Editorial Office
MDPI
St. Alban-Anlage 66
4052 Basel, Switzerland

This is a reprint of articles from the Special Issue published online in the open access journal *Energies* (ISSN 1996-1073) (available at: https://www.mdpi.com/journal/energies/special_issues/ microgrids_planning_protection_control).

For citation purposes, cite each article independently as indicated on the article page online and as indicated below:

LastName, A.A.; LastName, B.B.; LastName, C.C. Article Title. *Journal Name* **Year**, *Volume Number*, Page Range.

ISBN 978-3-0365-3661-3 (Hbk)
ISBN 978-3-0365-3662-0 (PDF)

Contents

Tri Ardriani, Pekik Argo Dahono, Arwindra Rizqiawan, Erna Garnia, Pungky Dwi Sastya, Ahmad Husnan Arofat and Muhammad Ridwan
A DC Microgrid System for Powering Remote Areas [†]
Reprinted from: *Energies* **2021**, *14*, 493, doi:10.3390/en14020493 . . . . **1**

Michael Stadler, Zack Pecenak, Patrick Mathiesen, Kelsey Fahy and Jan Kleissl
Performance Comparison between Two Established Microgrid Planning MILP Methodologies Tested On 13 Microgrid Projects
Reprinted from: *Energies* **2020**, *13*, 4460, doi:10.3390/en13174460 . . . . **17**

Juan Carlos Oviedo Cepeda, German Osma-Pinto, Robin Roche, Cesar Duarte, Javier Solano and Daniel Hissel
Design of a Methodology to Evaluate the Impact of Demand-Side Management in the Planning of Isolated/Islanded Microgrids
Reprinted from: *Energies* **2020**, *13*, 3459, doi:10.3390/en13133459 . . . . **39**

Giovanni Artale, Giuseppe Caravello, Antonio Cataliotti, Valentina Cosentino, Dario Di Cara, Salvatore Guaiana, Ninh Nguyen Quang, Marco Palmeri, Nicola Panzavecchia and Giovanni Tinè
A Virtual Tool for Load Flow Analysis in a Micro-Grid
Reprinted from: *Energies* **2020**, *13*, 3173, doi:10.3390/en13123173 . . . . **63**

Andres Arias Londoño, Oscar Danilo Montoya, Luis Grisales-Noreña
A Chronological Literature Review of Electric Vehicle Interactions with Power Distribution Systems
Reprinted from: *Energies* **2020**, *13*, 3016, doi:10.3390/en13113016 . . . . **89**

Oscar Danilo Montoya, Walter Gil-González and Edwin Rivas-Trujillo
Optimal Location-Reallocation of Battery Energy Storage Systems in DC Microgrids
Reprinted from: *Energies* **2020**, *13*, 2289, doi:10.3390/en13092289 . . . . **113**

Yuichiro Yoshida and Hooman Farzaneh
Optimal Design of a Stand-Alone Residential Hybrid Microgrid System for Enhancing Renewable Energy Deployment in Japan
Reprinted from: *Energies* **2020**, *13*, 1737, doi:10.3390/en13071737 . . . . **133**

Saher Javaid, Mineo Kaneko, Yasuo Tan
Structural Condition for Controllable Power Flow System Containing Controllable and Fluctuating Power Devices
Reprinted from: *Energies* **2020**, *13*, 1627, doi:10.3390/en13071627 . . . . **151**

Kishan Veerashekar, Halil Askan and Matthias Luther
Qualitative and Quantitative Transient Stability Assessment of Stand-Alone Hybrid Microgrids in a Cluster Environment
Reprinted from: *Energies* **2020**, *13*, 1286, doi:10.3390/en13051286 . . . . **171**

Lennart Petersen, Florin Iov and German Claudio Tarnowski
A Model-Based Design Approach for Stability Assessment, Control Tuning and Verification in Off-Grid Hybrid Power Plants
Reprinted from: *Energies* **2020**, *13*, 49, doi:10.3390/en13010049 . . . . **215**

Article

A DC Microgrid System for Powering Remote Areas [†]

Tri Ardriani [1], Pekik Argo Dahono [1,*], Arwindra Rizqiawan [1], Erna Garnia [2], Pungky Dwi Sastya [3], Ahmad Husnan Arofat [3] and Muhammad Ridwan [3]

[1] Institut Teknologi Bandung, School of Electrical Engineering and Informatics, Jl. Ganesha 10, Bandung 40111, Indonesia; ardriani.t@gmail.com (T.A.); windra@std.stei.itb.ac.id (A.R.)

[2] Faculty of Economics, Universitas Sangga Buana, Jl. PHH Mustofa (Suci) 68, Bandung 40111, Indonesia; erna.garnia@usbypkp.ac.id

[3] Technology Development Division, PT. Len Industri (Persero), Jl. Soekarno-Hatta 442, Bandung 40111, Indonesia; pungkydwisastya@gmail.com (P.D.S.); ahmad.husnan@len.co.id (A.H.A.); muhammad.ridwan@len.co.id (M.R.)

* Correspondence: pekik@konversi.ee.itb.ac.id

[†] This paper is an extended version of our paper published in the Proceeding of 2018 Conference on Power Engineering and Renewable Energy (ICPERE), 29–31 October 2018, Solo, Indonesia.

Abstract: DC microgrid has been gaining popularity as solution as a more efficient and simpler power system especially for remote areas, where the main grid has yet to be built. This paper proposes a DC microgrid system based on renewable energy sources that employs decentralized control and without communication between one grid point and another. It can be deployed as an individual isolated unit or to form an expandable DC microgrid through DC bus for better reliability and efficiency. The key element of the proposed system is the power conditioner system (PCS) that works as an interface between energy sources, storage system, and load. PCS consists of modular power electronics devices and a power management unit, which controls power delivery to the AC load and the grid as well as the storage system charging and discharging sequence. Prototypes with 3 kWp solar PV and 13.8 kWh energy storage were developed and adopt a pole-mounted structure for ease of transportation and installation that are important in remote areas. This paper presents measurement results under several conditions of the developed prototypes. The evaluation shows promising results and a solid basis for electrification in remote areas.

Keywords: DC microgrid; power conditioner system; renewable energy; scalable microgrid

Citation: Ardriani, T.; Dahono, P.A.; Rizqiawan, A.; Garnia, E.; Sastya, P.D.; Arofat, A.H.; Ridwan, M. A DC Microgrid System for Powering Remote Areas . *Energies* **2021**, *14*, 493. https://doi.org/10.3390/en14020493

Received: 20 December 2020
Accepted: 14 January 2021
Published: 18 January 2021

Publisher's Note: MDPI stays neutral with regard to jurisdictional claims in published maps and institutional affiliations.

1. Introduction

As an archipelago with more than 17,000 islands, Indonesia faces a challenge in delivering electricity to all its citizens, particularly to those who live in the remote areas and outer islands. According to the Indonesian Ministry of Energy and Mineral Resources [1], although the total electrification ratio of Indonesia in 2019 has reached almost 99%, there are places that lag behind and around 1 million families still without access to electricity. Additionally, strong grids are only available to the main islands of the country, where the central government and most of the population live. There are also areas where electricity is available only for several hours a day.

Located in the equator, the solar potential in Indonesia is estimated to be around 208 GWp, much higher than other types of RES such as hydro (75 GWp), wind (60 GWp), and geothermal (29.5 GWp) [2]. It is one of the most evenly distributed RES throughout the country. Therefore, a solar-based system is very suitable to accelerate providing electricity to rural villages.

Solar PV has been a popular choice RES because they are getting cheaper by the day and are easy to install. In urban cities, small-scale solar power systems are installed on rooftops, such as described in [3,4], both to provide green energy and to reduce bills. The disadvantage of PV rooftops is that it usually needs the AC grid to run and cannot

1

operate in stand-alone mode. Extending the main grid to remote areas require a lot of time and effort. One viable solution to this problem is to build independent power systems that do not need to rely on the main grid, i.e., a microgrid. These systems tap into RES near the local load, effectively eliminating the cost needed to draw long cables from the main grid and reducing dependency on fossil fuel-based power generation.

There are two types of microgrids, AC and DC microgrids. In AC microgrids, energy sources that produce DC power, such as PV panels and a fuel cell, will obviously need DC-AC conversion to connect to the lines. Interestingly, sources that produce AC power, such as wind, hydro, and geothermal, may require AC-DC-AC conversion for better synchronized connection to the grid [5–7]. Meanwhile, in DC microgrid, both DC-producing and AC-producing sources may require only one conversion, resulting in fewer converters needed, which in turn gives better efficiency. Nowadays, DC microgrid is gaining popularity due to its simplicity and higher power quality than its AC counterpart [7–9]. In DC system, the control is simpler because there is no problem with synchronization and reactive power [5–17].

Many DC microgrid systems have been proposed in the literature. Refs. [18,19] discuss microgrid systems that are reliant on AC utility. These systems are similar to the ones in [3,4] in that they are more suited for urban areas where the network is strong. In [10], a DC microgrid system for rural areas is designed to be able to operate independently in the absence of power network, but it only has stand-alone mode, meaning it does not have power sharing capability.

For rural areas, building centralized microgrids is a poor choice [11–13]. This is due to the socio-environmental conditions of rural areas. Firstly, the geographical terrain is difficult. This calls for a system that can be easily transported and installed. Secondly, communities are formed in clusters where homes, schools, hospital, and other public facilities are built apart from each other, which means a centralized generation system will have a lot of conductor losses. Thirdly, in rural areas, communication infrastructure is lacking. Therefore, the system needs to be able to operate in stand-alone mode. To accommodate future expansion, the equipment that go into the system have to be modular and can be installed in a plug-and-play manner [5].

Distributed DC microgrid systems are proposed in [11–17]. Refs. [11–15] take into account the fact that a lot of home appliances can run on low-voltage DC power. These devices also do not require a lot of power. Because of that, the system can be designed with low specification that costs considerably less than other systems. Unfortunately, its strength can also be its weakness, because at present, AC-powered home appliances are still very common.

The systems described in [16,17] are made of a DC bus and power converters that interface it with the energy source, ESS, and other microgrid elements. Loads will tap into the DC bus directly. This system can be built for large scale; however, loads may be located far away from the source and the distribution losses may be quite high.

In addition to the topology, discussion on power electronics technology and control methods is indispensable. There are a lot of different power converter topologies to choose from, all ranging from basic, uni-directional DC-DC converter, to multilevel and modified converters [14,16,20–23] with unique features that can be harnessed for a plethora of different purposes. Although, a system with many different kinds of power converters, such as in [10,16,24], may be expensive due to the higher design and production cost. Furthermore, replacement units may be limited to the same manufacturer, making it an inflexible system. Ref. [25] proposes a uniform design of multi-purpose converters for microgrids, but presently the technology has only been applied to DC-to-DC conversion.

In the realm of control techniques, one of the central issues is the power sharing method between microgrid elements. Ref. [26] proposes forecasting algorithm for an isolated DC microgrid system to regulate power flow. This algorithm has succeeded in reducing generation costs, but the predictive aspect of the control is prone to uncertainties and complex to implement. Ref. [27] propose droop control method based on state of charge (SoC). These methods are easier to implement, especially because in rural areas, maintenance and operation have to be kept simple. Ref. [28] proposes using peer-to-peer control between microgrids, while [29] opts using switching frequency modulation-based communication, which is a good choice for systems in urban area where communication link can be established fairly easily, but not as feasible in the country and remote islands. Generally, individual microgrid control is preferred because failure in one system may not affect the others [30,31].

This paper proposes a DC microgrid system that has all the requirements mentioned above, as well as the main building block of the system. The key features of this system are as follow: (1) Modular; (2) expandable; (3) independent/without communication; (4) easily transported; and (5) each system covers a small ground area. The last point is crucial for places with land-ownership problems. Figure 1 shows a multi-point DC microgrid of the proposed system, with the primary equipment denoted as PCS. Each system consists of the following four elements: (1) A locally available energy source; (2) ESS; (3) AC loads; and (4) DC bus interconnection. If more points are to be integrated, they can be connected via the DC grid. Each point is capable of both taking and giving energy from and to the grid.

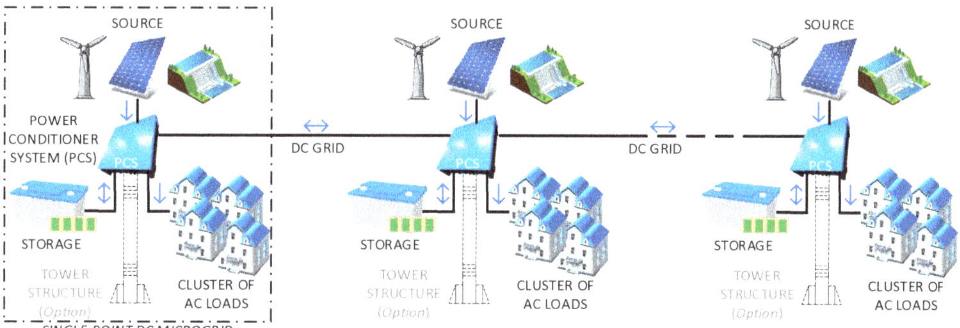

Figure 1. The proposed DC microgrid system.

This paper is structured as follows: PCS as the main element of the proposed system will be explained in Section 2. Comparison with other DC microgrid systems that have been proposed and implemented for rural areas will also be covered in this section. Section 3 discusses the evaluation of PCS to show the feasibility of implementing this system. Section 4 touches on the business and investment side of the proposed DC microgrid installation for a remote area in Indonesia. Finally, Section 5 contains the conclusion to this paper.

2. A Modular, Independent, and Expandable DC Microgrid System

2.1. Power Conditioner System

PCS is the primary element in the DC microgrid system proposed in this paper. It interfaces an energy source such as solar PV, wind, and hydro, an ESS, AC loads, and the DC grid. PCS consists of power converters to harvest energy from the source, charge and discharge the ESS, and regulate power to loads and grid. Figure 2a shows the inner configuration of a PCS. It may be deployed to form an isolated, independent system, but it can also connect to other PCS and share power between each other. Each PCS has its own management unit that keeps track the energy stored in its ESS, and export or import power to and from its neighboring points according to this data [32].

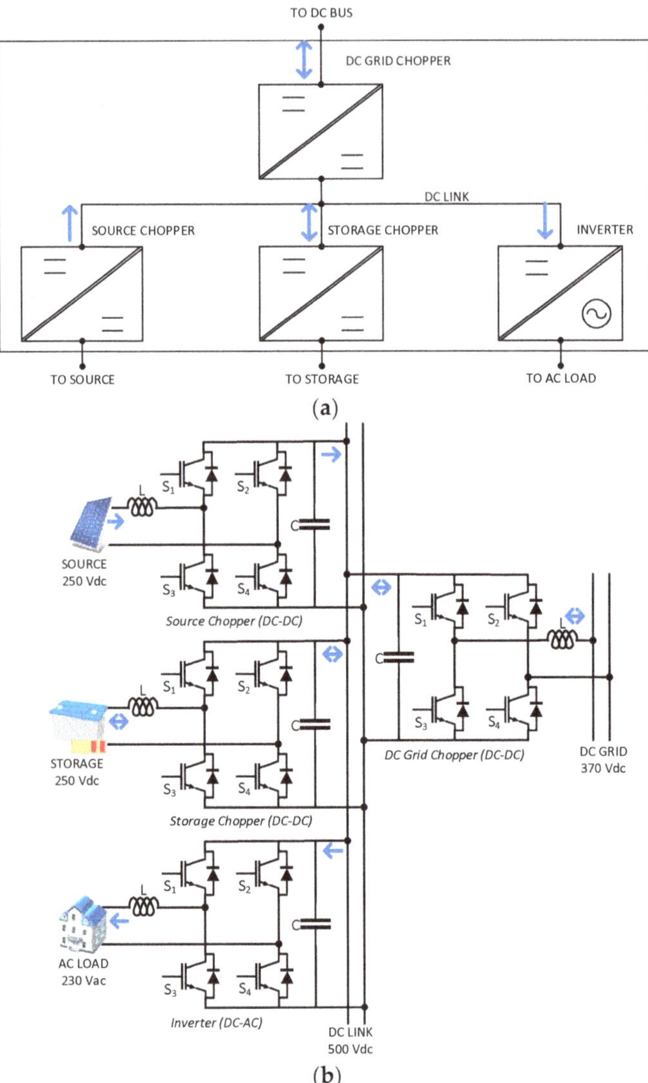

Figure 2. (**a**) Block diagram of a power conditioner system (PCS). (**b**) Converters topology employed in PCS.

Figure 2b shows the converter topology that is being used for this system. It can be seen that the hardware has the same bridge topology that can be used either as DC-DC, DC-AC, or AC-DC converter. However, at present the system is designed primarily for solar PV. By using this approach, the technology is made simple, which is important for application in remote islands. Modular plug and play feature may be expected of this system.

The power converters inside PCS are connected by an internal DC link bus (Figure 2b), whose voltage is being maintained at 500 V. At the beginning of operation, precharging action takes place as soon as the ESS is plugged into the storage chopper terminal. Once the DC link voltage is established, the management unit then activates the source chopper, inverter, and DC grid chopper depending on the specified SoC limits. Each PCS has the total capacity of 3 kW.

Table 1 lists the specification of the converters in every PCS, whereas Table 2 shows the state of each converter based on SoC. The flowchart in Figure 3 shows the decision-making process of the management unit according to guidelines set in Table 2.

Table 1. Specifications of each converter in a PCS unit.

	Source Chopper	Storage Chopper	DC Grid Chopper	Inverter
Rated Power	3000 W	3000 W	1000 W	3000 VA
DC Link Voltage	DC 500 V	DC 500 V	DC 500 V	DC 500 V
Input/Output Voltage	DC 250 V	DC 250 V	DC 370 V	AC 230 V
Rated Current	DC 12 A	DC 12 A	DC 2.7 A	AC 13 A
Other	MPPT Algorithm	Bidirectional power flow	Controlled DC Bus Voltage Range: DC 350–390 V Bidirectional power flow	Rated Frequency: 50 Hz (1-phase)

Table 2. PCS converter states based on state of charge (SoC).

SoC	Source Chopper *	Storage Chopper	DC Grid Chopper	Inverter
0–10%	ON or OFF	Charging	Charging	OFF
10–30%	ON or OFF	Charging	Charging	ON
30–70%	OFF	Charging	Charging	ON
	ON	Discharging	Discharging	
70–100%	ON or OFF	Discharging	Discharging	ON

* Source chopper operates based on the availability of power generated by PV panels.

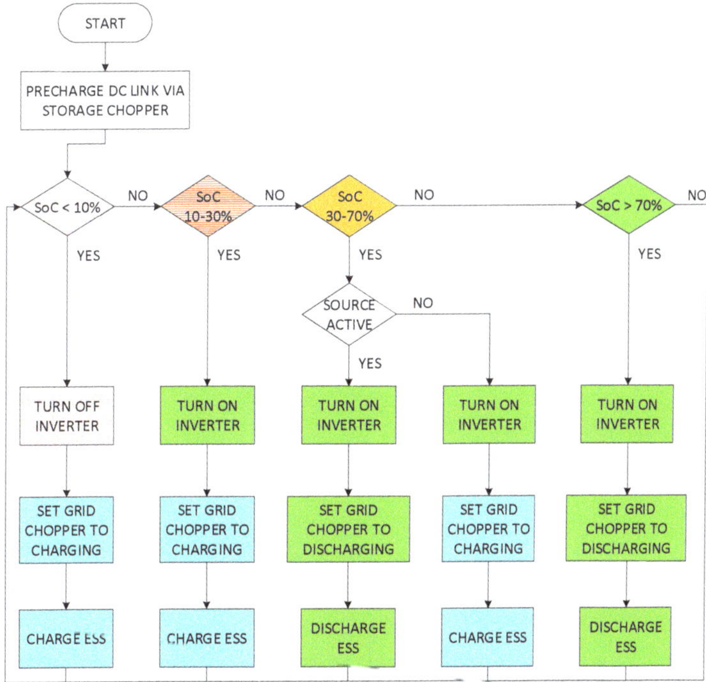

Figure 3. Flowchart of the PCS management unit.

The proposed DC microgrid system has the capability of working both as a single-point and multi-point system to serve single or multiple clusters of loads. Figure 4 shows the multi-point configuration with different kinds of loads. A single-point system converts power from the source and stores it in the battery and/or transfer it to the AC loads. In the multi-point mode, the proposed DC microgrid system works similar to a single point, but power sharing capability is through the DC bus. Each point controls its charge and discharge states with its own management unit the way it does in a single-point system and independent of what the other point does.

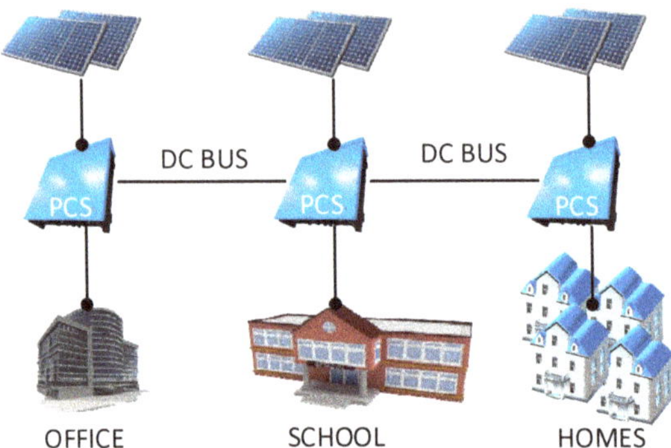

Figure 4. Multi-point DC microgrid configuration.

Figure 5 shows the power flow in the system based on the storage SoC level. The system that has higher energy level, or state of charge (SoC), exports power, whereas the system that lacks energy imports it. Consider the leftmost system. Its SoC is above 70%, so the DC grid chopper works in discharge mode regardless of what state the source chopper and storage chopper are in. Similarly, if the SoC is under 30%, then the management unit will automatically set the DC grid chopper to charging (not shown in the figure).

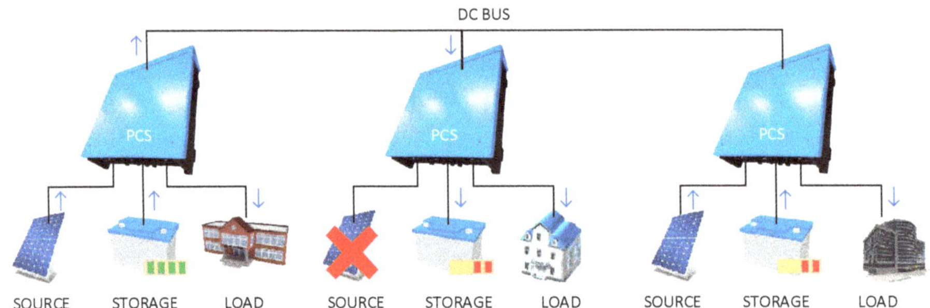

Figure 5. Power flow in multi-point DC microgrid system.

In the middle-level SoC, DC grid chopper operation is based on the status of the source and storage choppers. If the storage chopper is running, then power is supplied from the source and the PCS has the ability to send power out via the DC grid terminal. On the contrary, if the source chopper is not running, then the power of the unit is wholly supplied by battery. This causes the management unit to set the DC grid chopper to charging. Consider the middle and rightmost system in Figure 5. The system in the middle charges its storage system because the source is unavailable, while the system on the right-hand side exports power even though its storage is not in full condition because the source is available. The operational settings that are shown in Table 2 are configurable prior to field installation.

2.2. Implementation of the Proposed DC Microgrid System in Isolated Remote Areas

In rural areas, oftentimes the land to install public facilities is difficult to obtain because of unclear ownership or the land is forbidden to be used as per the local or cultural belief. To minimize the installation area, the proposed DC microgrid system can be configured as pole-mounted along with the other equipment such as PV panels and batteries. Figure 6 shows a solar tower with a set of PV panels, a PCS, batteries, and a control panel. Multi-point implementation of this configuration is shown in Figure 1.

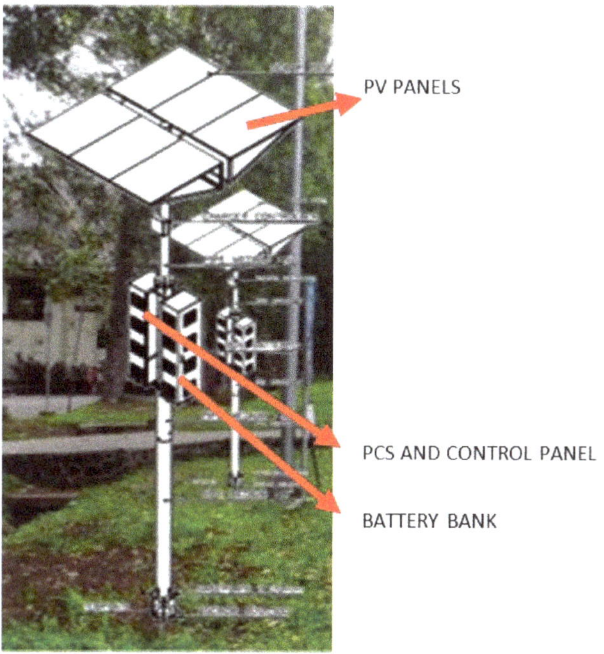

Figure 6. Solar tower: An implementation of PCS.

Houses in remote islands are assumed to use electricity for very basic needs, such as lighting, cooking, and, in places with access to communication facilities, to power up phones. Therefore, in this context, each house may be limited to use around 0.5 kWh each day. An individual system of the proposed DC microgrid has capacity of 3 kWp and 13.8 kWh of ESS per cluster, which is reasonable to serve 0.5 kWh per day for 10 houses. The system such as the one in Figure 6 uses 12 PV panels of 260 Wp each to achieve 3 kWp. The panels are positioned as a canopy shading around 18 m^2 of ground to minimize land utilization.

In the proposed system, every point is designed with higher power rating, where loads can connect directly to the load converter side of PCS. Table 3 compares the key characteristics of proposed method and several DC microgrid systems in the literature. All of these systems have modular components that make scaling up the size of the network easy, and each independent unit is designed for small-power use of rural homes. The main advantage of the proposed DC microgrid system over the others are the AC output into which the loads plug. Even though a lot of home appliances operate on DC power, AC-powered devices are still in abundance and easily obtained. By using a power system that outputs AC power, users are not required to change all of their electronic appliances. It will also be easier to switch to the AC utility grid when it reaches the remote areas.

Table 3. Comparison of several DC microgrid systems.

No.	Source	Description	Topology
1.	Ref [12]	• Distributed generation that can be integrated with centralized power source and storage via DC grid. • Every unit consists of PV panels, battery, and power converters. • DC grid voltage: 380 V. • Load voltage: 48 V. • Power rating per unit: 125 W.	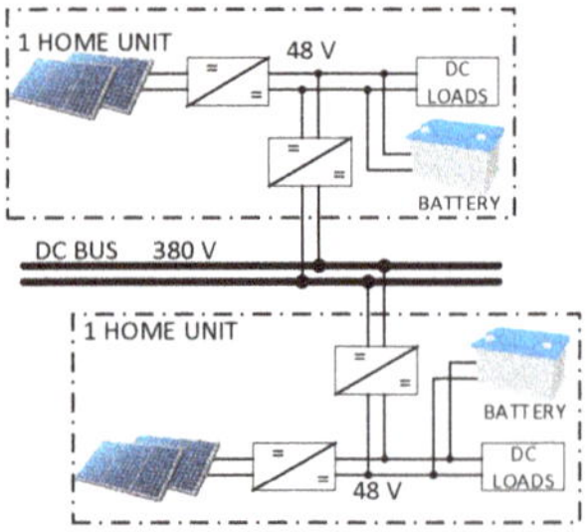 1 HOME UNIT — 48 V — DC LOADS — BATTERY; DC BUS 380 V; 1 HOME UNIT — BATTERY — DC LOADS — 48 V
2.	Ref [13]	• Centralized generation for several clusters of loads, one cluster consists of several households. • Distributed voltage control assisted by digital communication. • DC grid voltage: 360–400 V. • Load voltage: 12 V. • Power rating per unit: 100 W.	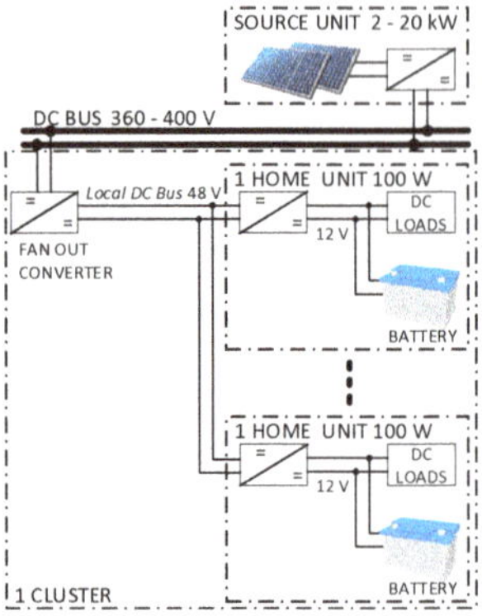 SOURCE UNIT 2 - 20 kW; DC BUS 360 - 400 V; Local DC Bus 48 V; FAN OUT CONVERTER; 1 HOME UNIT 100 W — 12 V — DC LOADS — BATTERY; 1 HOME UNIT 100 W — 12 V — DC LOADS — BATTERY; 1 CLUSTER
3.	Ref [14]	• Distributed generation. • DC grid voltage: 380 V. • Load voltage: 48 V. • Power rating per unit: 100 W. • DC loads are connected directly to 48 V bus to avoid conversion losses.	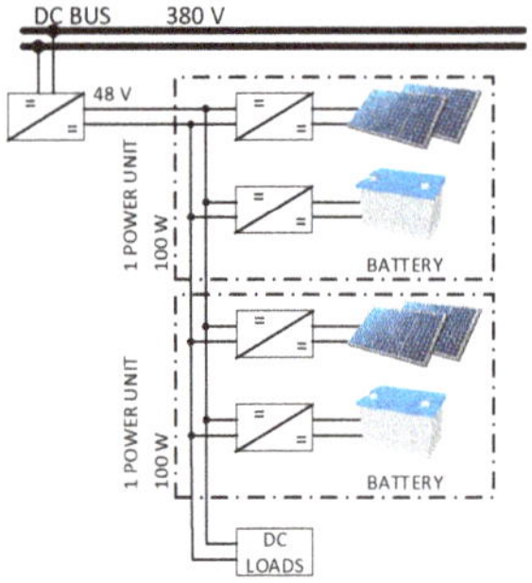 DC BUS 380 V; 48 V; 1 POWER UNIT 100 W — BATTERY; 1 POWER UNIT 100 W — BATTERY; DC LOADS
4.	Ref [15]	• Distributed generation. • DC grid voltage: 380 V. • Load voltage: 48 V. • Power rating per unit: 200 W. • Coordination between units is achieved by the converters' interaction within the units. • Communal loads are supplied by the power generated from the home units.	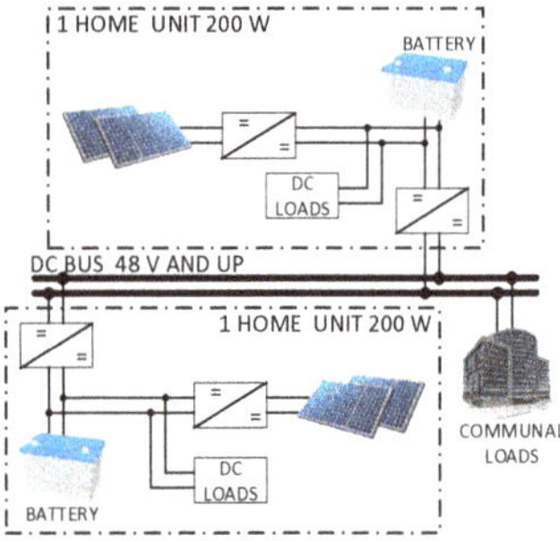 1 HOME UNIT 200 W — BATTERY — DC LOADS; DC BUS 48 V AND UP; 1 HOME UNIT 200 W — DC LOADS — BATTERY; COMMUNAL LOADS

Table 3. *Cont.*

No.	Source	Description	Topology
5	Proposed System	<ul><li>Distributed generation.</li><li>DC grid voltage: 370 V.</li><li>Load voltage: 230 VAC 1-phase</li><li>Power rating per unit: 3 kW.</li><li>Each unit supplies a cluster of houses and manages its own energy consumption individually.</li><li>Outputs 230 VAC to accommodate common home appliances that run on AC power.</li></ul>	

3. Functional and Performance Evaluation of the Proposed DC Microgrid System

3.1. Evaluation Methodology

There are two evaluation sections: Lab testing and field testing. In lab testing, the proposed DC microgrid system are tested as a single-point system and a multi-point system. Dummy SoC data are used to avoid having to charge or discharge the batteries to the desired level as doing so is too time-consuming. Single-point system evaluation mainly focuses on each point's functional specifications. Precharging sequence is done manually using 192-ohm resistor and 40-amps DC circuit breaker. The multi-point system evaluation tests the PCS operation sequence. Their DC grid terminals are being connected and the parameters observed. The units automatically precharge as they get connected to power source and then follow the same procedure as in single system initialization.

The parameters and equipment used for lab testing are as follow: Two sets of batteries: OPZV batteries 240-V 1,000-Ah and LiFePo batteries 256-V 70-Ah; two sets of PV modules: each consists of 40 50-Wp 12-V PV panels in series and parallel to achieve a total 340 V and 2 kWp; and two sets of 3-kW heater as resistive loads.

The field testing is conducted to gather data regarding conductor losses. Conductor losses need to be evaluated because in the target location for implementing this system, the residential area is thinly distributed. Figure 7 shows the field-testing setup. PCS are loaded with LED lights located 100 m away from it, assuming that the houses in rural villages could be this far apart from the energy source. Measurements are done at the AC output terminal of the PCS and at the load connection point. The charging and discharging process are also monitored with changing load currents.

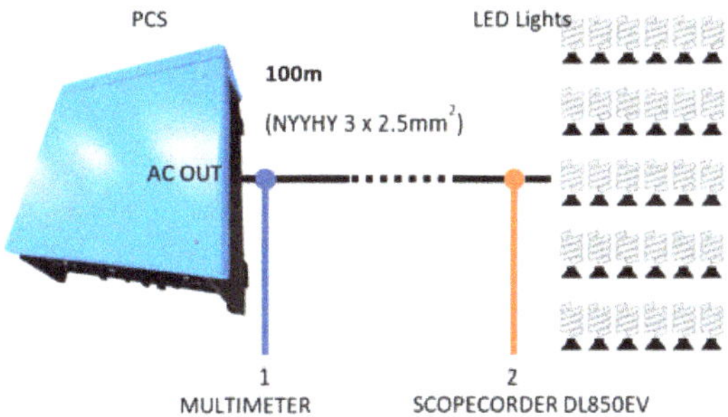

Figure 7. PCS field evaluation with distant loads.

3.2. Evaluation Results

3.2.1. Lab Testing

The storage chopper activates after the management unit (operated manually by using a laptop PC) gives the signal. Current drawn from battery indicates that the battery chopper is working. The measured voltage at DC link is 500 V. The battery chopper charges battery after the PV is connected to the source input terminal and the SoC in the management unit is set to 4%. The battery charging operation indicates that the PV chopper is working. When the inverter is activated and loaded, the current flows from both the PV and battery via their respective choppers.

Performance of the inverter is summarized in Figure 8. During this test, PV is deactivated, and the supply goes solely from the battery. The inverter voltage output is stable at 228 V AC, loaded at maximum 2.4 kW. The amount of DC power that is converted into AC is generally measured by the efficiency that can be calculated by using Equation (1).

$$Efficiency = \frac{AC\ Output}{Battery\ Input} \tag{1}$$

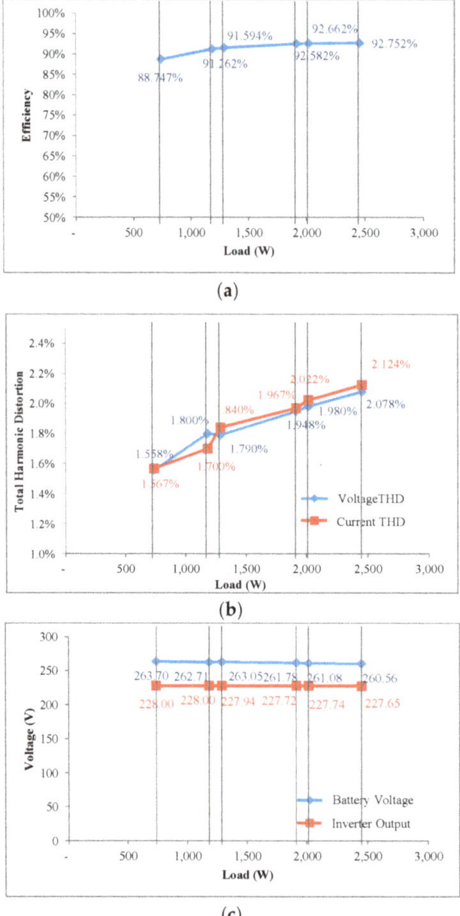

Figure 8. Inverter evaluation: (**a**) Efficiency vs. load, (**b**) voltage and current THD vs. load, and (**c**) battery and inverter voltage vs. load.

The efficiency data are shown in Figure 8a. The efficiency rises together with the load increase, reaching over 90% with load above 1100 W. It can be expected that at full load the efficiency will be higher than 92.75%. The power quality measurement results in Figure 8b shows that the harmonic level is well below standardized limits according to the IEEE Std 519-2014. The distortion levels increase along with the load in an almost linear manner. According to the trend, at full load they are predicted not to exceed the standard limit. The inverter also feeds stable voltage throughout the test, delivering up to 80% of rated output power without any significant drop in voltage, as shown in Figure 8c.

The DC grid chopper is being activated last. During this test, because the SoC is set to 4%, the DC grid chopper runs in charging mode (refer to Figure 4). When it is not connected to the other PCS to draw current from, the measured voltage at the DC grid chopper terminal is 390 V.

During this test, the power consumption of PCS units is also measured. In idle state when no switching occurs, the power consumed is 16–20 W. When switching occurs, the power consumption rises to 100 W.

For the multi-point evaluation, the precharging sequence is done automatically using precharging module within PCS. The precharging is done in 1.5 s and the discharging in 2 s. To make the PCS share power between each other, SoC of one PCS is set to 30% with the PV disconnected, while SoC of the other PCS is set to 70%. The management unit of the first PCS will detect that its battery needs to be charged. Therefore, management unit of the first PCS sets its DC grid chopper to charging mode. Meanwhile, management unit of the second PCS sets its DC grid chopper to discharging mode. Because the first PCS is set to absorb power from the DC bus and the second PCS to inject power, power will naturally flow from the second PCS to the first one.

3.2.2. Field Testing

Figure 9a shows that the inverter output voltage drops at the load point of connection. The proposed DC microgrid system does not use closed-loop control to keep constant voltage in the consumer to keep the simplicity. In addition, the household devices in remote area are not sensitive to constant voltage. At 9 A of load current (70% of rated current), the voltage drop does not exceed 10% of its nominal value, which is still acceptable for common AC-powered household devices. The voltage drops more when the load current is bigger due to the losses at the cable. Figure 9b shows the measured power at both end of the cable. During this experiment, the cable type NYYHY 3×2.5 mm^2 that is readily available in general stores was used. If a better quality is required, a different type of cable with less resistance may solve the problem.

Figure 10 shows the charging and discharging of battery bank with varying load current. This experiment is done by increasing and then decreasing the load demand. When the PV produces power more than the load demand (the charging process at the left-hand side), PCS will charge the battery, signified by the negative battery power. When the load demands power higher from the PV capacity (the discharging process in the middle), the battery enters discharging mode, hence the positive battery power. Battery gives out the necessary power to meet the load demand, and the power it exerts rises along with the increase in load current. PCS then goes back to charging mode when the load demand decreases to under the PV production. This experiment shows that the charging and discharging of the battery bank has worked well to meet the needs of the varying loads.

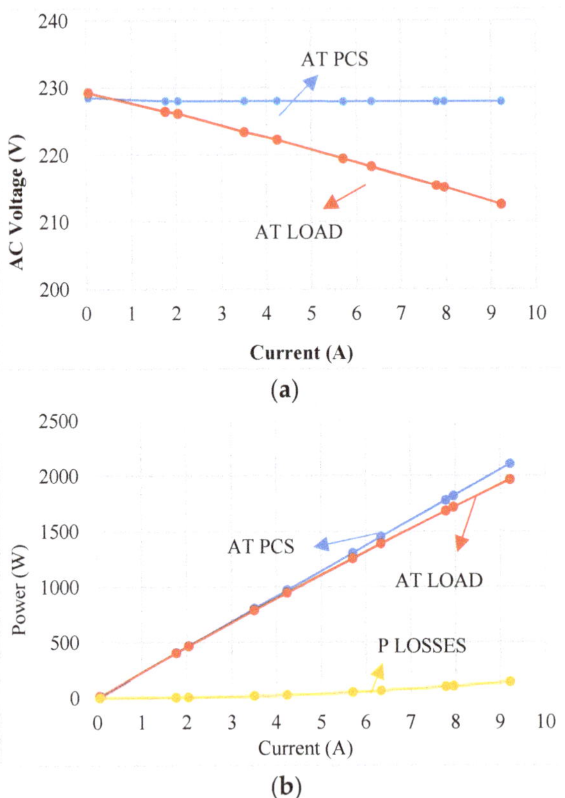

Figure 9. Field testing results: (**a**) Voltage drop on cable, and (**b**) losses on cable.

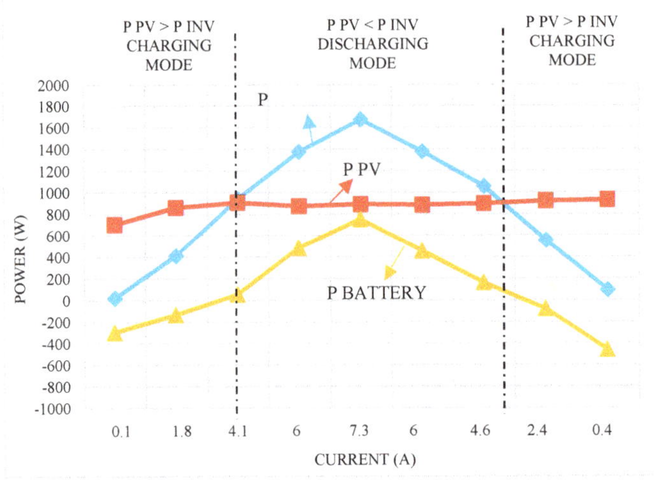

Figure 10. Battery charging and discharging process with varying load current.

4. LCOE and Investment Analysis

Levelized cost of energy (LCOE) is defined as the average total cost to build and operate a power plant per output energy in a certain period of time. LCOE can be calculated Equation (2).

$$LCOE = \frac{NPVCost}{NPVProduce} = \frac{\sum \frac{I(t)+M(t)+F(t)}{(1+r)^t}}{\sum \frac{E(t)}{(1+r)^t}} \tag{2}$$

where:

I = initial investment cost,
M = O&M cost,
F = fuel cost,
E = energy produced by the plant,
r = discount rate,
t = time,
$NPVCost$ = net present value from the total cost spent in the plant's lifetime, and
$NPVProduce$ = net present value of the total energy generated by the plant during its lifetime.

The *LCOE* is computed using the following assumptions: Three systems such as described in Figure 1 will be installed, which amount to total capacity of 9 kWp, each system can be used to serve 10 homes running common home electronic appliances; investment cost of each system is US$23,375; system efficiency is 95%; sun light is accessible for an average 4.5 h/day all year; PV panel output degradation is 0.5%/year; O&M cost is 0.5% from investment cost/year; O&M growth rate is 2%; interest rate is 9%/year, and because 70% of the investment comes from bank loan, the discount rate is 10.8%; and the project lifespan is 20 years.

For 20 years, the total *NPV* for investment and O&M is US$72,975 and the total generated energy is 107.4 kW. Therefore, the $LCOE = \frac{NPVCost}{NPVProduce} = \frac{US\$72,975}{107.4} = US\$0.68/kWh$. If the tariff is the same as *LCOE*, the internal rate of return (IRR) is 10.82% and pay out in 8.26 years. This implies that in order for this project to be economically feasible, the *LCOE* specifies the minimum tariff for this system. US$ 0.68/kWh is inexpensive compared to diesel generator that is generally the main electricity source in remote areas. Adding up the costs of transporting fuels, the total production cost may be well above US$ 1/kWh. Moreover, by using the proposed DC microgrid, we will reduce dependency on the already dwindling fossil fuel reserves.

5. Conclusions

This paper proposes a new DC microgrid system concept that is suitable for accelerating electricity delivery to rural and remote areas that has no access to the main utility grid. Its modularity makes it easily transported and provides flexibility to the overall system. The DC interface between points is also advantageous when they need to combine with other points due to the absence of the need to synchronize as in AC-interfaced systems.

PCS is the key equipment for the proposed microgrid system. It interfaces the microgrid elements such as energy sources, ESS, and load, by using the same converter topologies for those different purposes. It can be deployed as an individual isolated unit and also form a grid through the DC bus for better reliability and efficiency. PCS manages the available power and the load requirements, while able to exchange power between connected points.

Prototypes have been developed to demonstrate the proposed DC microgrid system, with capacity of 3 kWp of solar PV and 13.8 kWh of battery each. One such system can be used to supply a cluster of 10 households. They are implemented in a pole-mounted structure to save space, while the PV panels form a canopy shading 18 m^2 of ground. The system can be expanded to create a DC microgrid network by connecting several single isolated systems through 370 VDC bus.

The proposed DC microgrid system has been tested in laboratory and in the field, both as a single-point and a multi-point system. Each of them runs well both in either modes, reaching 92.75% of efficiency, and confirmed low THD levels, where the voltage THD is 2.078% and current THD is 2.124%. The prototype field testing successfully demonstrated the battery charge and discharge sequences following several load conditions. Voltage drops is also tested to confirm the acceptable level of voltage in the consumer's household devices. Financial analysis based on LCOE concludes that the proposed system is feasible for implementation with the minimum tariff US$0.68/kWh.

Author Contributions: Conceptual development P.A.D., A.R., and T.A.; economic analysis E.G.; implementation and field testing A.H.A., P.D.S., and M.R. All authors have read and agreed to the published version of the manuscript.

Funding: This research is funded by the LPDP, Indonesia. This research is also partially funded by the Indonesian Ministry of Research and Technology/National Agency for Research and Innovation, and Indonesian Ministry of Education and Culture, under World Class University Program managed by Institut Teknologi Bandung.

Institutional Review Board Statement: Not applicable.

Informed Consent Statement: Not applicable.

Data Availability Statement: Not applicable.

Acknowledgments: The authors would also like to thank Sanken Electric Co., Ltd. and PT Len Industri (Persero), for cooperation during this research.

Conflicts of Interest: The authors declare no conflict of interest.

Abbreviations

AC	Alternating Current
DC	Direct Current
GWp	gigawatt-peak
ESS	Energy Storage System
kWh	kilowatt-hour
kWp	kilowatt-peak
LCOE	Levelized Cost of Energy
O&M	Operation and Maintenance
PCS	Power Conditioner System
PV	Photovoltaic
RES	Renewable Energy Sources
SoC	State of Charge
THD	Total Harmonic Distortion

References

1. Kementrian Energi dan Sumber Daya Mineral. Available online: https://www.esdm.go.id/id/media-center/arsip-berita/tumbuh-3-persen-per-tahun-rasio-elektrifikasi-triwulan-iii-capai-9886-persen (accessed on 20 December 2020).
2. Direktorat Jenderal Energi Baru, Terbarukan, dan Konservasi Energi. *Statistik EBTKE 2016*; Direktorat Jenderal Energi Baru, Terbarukan, dan Konservasi Energi: Jakarta, Indonesia, 2016.
3. Hassan, M.U.; Saha, S.; Haque, M. A framework for the performance evaluation of household rooftop solar battery systems. *Int. J. Electr. Power Energy Syst.* **2021**, *125*, 106446. [CrossRef]
4. Alam, M.J.E.; Muttaqi, K.M.; Sutanto, D. A Three-Phase Power Flow Approach for Integrated 3-Wire MV and 4-Wire Multigrounded LV Networks with Rooftop Solar PV. *IEEE Trans. Power Syst.* **2012**, *28*, 1728–1737. [CrossRef]
5. Justo, J.J.; Mwasilu, F.; Lee, J.; Jung, J.-W. AC-microgrids versus DC-microgrids with distributed energy resources: A review. *Renew. Sustain. Energy Rev.* **2013**, *24*, 387–405. [CrossRef]
6. Planas, E.; Andreu, J.; Gárate, J.I.; De Alegría, I.M.; Ibarra, E. AC and DC technology in microgrids: A review. *Renew. Sustain. Energy Rev.* **2015**, *43*, 726–749. [CrossRef]
7. Rodriguez-Diaz, E.; Vasquez, J.C.; Guerrero, J.M. Intelligent DC Homes in Future Sustainable Energy Systems: When efficiency and intelligence work together. *IEEE Consum. Electron. Mag.* **2016**, *5*, 74–80. [CrossRef]

8. Patterson, B.T. DC, Come Home: DC Microgrids and the Birth of the "Enernet". *IEEE Power Energy Mag.* **2012**, *10*, 60–69. [CrossRef]

9. Kumar, D.; Zare, F.; Ghosh, A. DC Microgrid Technology: System Architectures, AC Grid Interfaces, Grounding Schemes, Power Quality, Communication Networks, Applications, and Standardizations Aspects. *IEEE Access* **2017**, *5*, 12230–12256. [CrossRef]

10. Schumacher, D.; Beik, O.; Emadi, A. Standalone Integrated Power Electronics Systems: Applications for Off-Grid Rural Locations. *IEEE Electrif. Mag.* **2018**, *6*, 73–82. [CrossRef]

11. Nasir, M.; Khan, H.A.; Zaffar, N.A.; Vasquez, J.C.; Guerrero, M. Scalable solar dc microgrid: On the Path to Revolutionizing the Electrification Architecture of Developing Communities. *IEEE Electrif. Mag.* **2018**, *6*, 63–72. [CrossRef]

12. Jhunjhunwala, A.; Kaur, P. Solar Energy, dc Distribution, and Microgrids: Ensuring Quality Power in Rural India. *IEEE Electrif. Mag.* **2018**, *6*, 32–39. [CrossRef]

13. Madduri, P.A.; Poon, J.; Rosa, J.; Podolsky, M.; Brewer, E.; Sanders, S.R. Scalable DC Microgrids for Rural Electrification in Emerging Regions. *IEEE J. Emerg. Sel. Top. Power Electron.* **2016**, *4*, 1195–1205. [CrossRef]

14. Li, D.; Ho, C.N.M. A Module-Based Plug-n-Play DC Microgrid with Fully Decentralized Control for IEEE Empower a Billion Lives Competition. *IEEE Trans. Power Electron.* **2020**, *36*, 1764–1776. [CrossRef]

15. Nasir, M.; Jin, Z.; Khan, H.A.; Zaffar, N.A.; Vasquez, J.C.; Guerrero, J.M. A Decentralized Control Architecture Applied to DC Nanogrid Clusters for Rural Electrification in Developing Regions. *IEEE Trans. Power Electron.* **2019**, *34*, 1773–1785. [CrossRef]

16. Dastgeer, F.; Gelani, H.E.; Anees, H.M.; Paracha, Z.J.; Kalam, A. Analyses of efficiency/energy-savings of DC power distribution systems/microgrids: Past, present and future. *Int. J. Electr. Power Energy Syst.* **2019**, *104*, 89–100. [CrossRef]

17. Dragicevic, T.; Lu, X.; Vasquez, J.C.; Guerrero, J.M. DC Microgrids—Part II: A Review of Power Architectures, Applications, and Standardization Issues. *IEEE Trans. Power Electron.* **2016**, *31*, 3528–3549. [CrossRef]

18. Neto, P.J.D.S.; Barros, T.A.; Silveira, J.P.; Filho, E.R.; Vasquez, J.C.; Guerrero, J.M. Power management techniques for grid-connected DC microgrids: A comparative evaluation. *Appl. Energy* **2020**, *269*, 115057. [CrossRef]

19. Mi, Y.; Guo, J.; Yu, S.; Cai, P.; Ji, L.; Wang, Y.; Yue, D.; Fu, Y.; Jin, C. A Power Sharing Strategy for Islanded DC Microgrid with Unmatched Line Impedance and Local Load. *Electr. Power Syst. Res.* **2021**, *192*, 106983. [CrossRef]

20. Adam, G.P.; Vrana, T.K.; Li, R.; Li, P.; Burt, G.; Finney, S. Review of technologies for DC grids—Power conversion, flow control and protection. *IET Power Electron.* **2019**, *12*, 1851–1867. [CrossRef]

21. Cornea, O.; Andreescu, G.-D.; Muntean, N.; Hulea, D. Bidirectional Power Flow Control in a DC Microgrid Through a Switched-Capacitor Cell Hybrid DC–DC Converter. *IEEE Trans. Ind. Electron.* **2017**, *64*, 3012–3022. [CrossRef]

22. Vuyyuru, U.; Maiti, S.; Chakraborty, C. Active Power Flow Control between DC Microgrids. *IEEE Trans. Smart Grid* **2019**, *10*, 5712–5723. [CrossRef]

23. Rathore, A.K.; Patil, D.R.; Srinivasan, D. A Non-Isolated Bidirectional Soft Switching Current fed LCL Resonant DC/DC Converter to Interface Energy Storage in DC Microgrid. *IEEE Trans. Ind. Appl.* **2015**, *52*, 1. [CrossRef]

24. Lu, S.-Y.; Wang, L.; Lo, T.-M.; Prokhorov, A.V. Integration of Wind Power and Wave Power Generation Systems Using a DC Microgrid. *IEEE Trans. Ind. Appl.* **2015**, *51*, 2753–2761. [CrossRef]

25. Cheng, T.; Lu, D.D.-C.; Qin, L. Non-Isolated Single-Inductor DC/DC Converter with Fully Reconfigurable Structure for Renewable Energy Applications. *IEEE Trans. Circuits Syst. II Express Briefs* **2017**, *65*, 351–355. [CrossRef]

26. Sechilariu, M.; Locment, F.; Wang, B. Photovoltaic Electricity for Sustainable Building. Efficiency and Energy Cost Reduction for Isolated DC Microgrid. *Energies* **2015**, *8*, 7945–7967. [CrossRef]

27. Strunz, K.; Abbasi, E.; Huu, D.N. DC Microgrid for Wind and Solar Power Integration. *IEEE J. Emerg. Sel. Top. Power Electron.* **2014**, *2*, 115–126. [CrossRef]

28. Werth, A.; Andre, A.; Kawamoto, D.; Morita, T.; Tajima, S.; Tokoro, M.; Yanagidaira, D.; Tanaka, K. Peer-to-Peer Control System for DC Microgrids. *IEEE Trans. Smart Grid* **2016**, *9*, 3667–3675. [CrossRef]

29. Choi, H.-J.; Jung, J.-H. Enhanced Power Line Communication Strategy for DC Microgrids Using Switching Frequency Modulation of Power Converters. *IEEE Trans. Power Electron.* **2017**, *32*, 4140–4144. [CrossRef]

30. Papadimitriou, C.; Zountouridou, E.; Hatziargyriou, N. Review of hierarchical control in DC microgrids. *Electr. Power Syst. Res.* **2015**, *122*, 159–167. [CrossRef]

31. Sahoo, S.K.; Sinha, A.K.; Kishore, N.K. Control Techniques in AC, DC, and Hybrid AC–DC Microgrid: A Review. *IEEE J. Emerg. Sel. Top. Power Electron.* **2018**, *6*, 738–759. [CrossRef]

32. Ardriani, T.; Sastya, P.D.; Arofat, A.H.; Dahono, P.A. A Novel Power Conditioner System for Isolated DC Microgrid System. In Proceedings of the 2018 Conference on Power Engineering and Renewable Energy (ICPERE), Solo, Indonesia, 29–31 October 2018; pp. 1–5.

Article

Performance Comparison between Two Established Microgrid Planning MILP Methodologies Tested On 13 Microgrid Projects

Michael Stadler [1,2,3,4,*], **Zack Pecenak** [1], **Patrick Mathiesen** [1], **Kelsey Fahy** [1] and **Jan Kleissl** [4]

[1]   Bankable Energy|XENDEE Inc., 6540 Lusk Blvd, San Diego, CA 92121, USA; ZPecenak@xendee.com (Z.P.); pmathiesen@xendee.com (P.M.); fahykt@xendee.com (K.F.)

[2]   Bioenergy and Sustainable Technologies Research GmbH, 3250 Wieselburg, Austria

[3]   Center for Energy and Innovative Technologies (CET), 3681 Hofamt Priel, Austria

[4]   Center for Energy Research, University of California at San Diego, 9500 Gilman Dr., San Diego, CA 92037, USA; jkleissl@eng.ucsd.edu

*   Correspondence: mstadler@xendee.com or mstadler@cet.or.at; Tel.: +1-619-431-1689

Received: 10 August 2020; Accepted: 28 August 2020; Published: 28 August 2020

Abstract: Mixed Integer Linear Programming (MILP) optimization algorithms provide accurate and clear solutions for Microgrid and Distributed Energy Resources projects. Full-scale optimization approaches optimize all time-steps of data sets (e.g., 8760 time-step and higher resolutions), incurring extreme and unpredictable run-times, often prohibiting such approaches for effective Microgrid designs. To reduce run-times down-sampling approaches exist. Given that the literature evaluates the full-scale and down-sampling approaches only for limited numbers of case studies, there is a lack of a more comprehensive study involving multiple Microgrids. This paper closes this gap by comparing results and run-times of a full-scale 8760 h time-series MILP to a peak preserving day-type MILP for 13 real Microgrid projects. The day-type approach reduces the computational time between 85% and almost 100% (from 2 h computational time to less than 1 min). At the same time the day-type approach keeps the objective function (OF) differences below 1.5% for 77% of the Microgrids. The other cases show OF differences between 6% and 13%, which can be reduced to 1.5% or less by applying a two-stage hybrid approach that designs the Microgrid based on down-sampled data and then performs a full-scale dispatch algorithm. This two stage approach results in 20–99% run-time savings.

Keywords: Microgrid; DER; planning; MILP; optimization; run-time; full time-series optimization; data reduction; DER-CAM; XENDEE

1. Introduction

Microgrid deployment is accelerating rapidly and roughly 2300 Microgrids were operational or planned worldwide in 2018 [1]. In the last 6 months of 2018, 240 additional Microgrid projects were added to the Navigant database, demonstrating a steady increase in Microgrid projects. More impressive is the increase in 2019. As of June 2019, Navigant identifies 4475 Microgrid projects worldwide [2]. Microgrid Knowledge [3] estimates that the Microgrid market will reach US$31 billion by the year 2027, underscoring the need for effective, fast Microgrid design and planning tools to keep up with the increasing number of projects.

The research community provides several different methodologies to plan a Microgrid from an economic perspective. All methodologies need to match energy supply with Microgrid demand to determine the annual energy costs, Net Present Value (NPV), or emissions from Microgrid adoption. Investment costs, operation and maintenance costs, subsidies, tax incentives or carbon costs among others are considered in this calculation [4]. The goal of these approaches is to determine the optimal combination of Distributed Energy Resources (DER) and their sizes to meet the demand, subject to constraints and inputs.

In simulation or trial and error approaches, the user changes input data (e.g., investment costs for DER) to analyze the impact on the results (e.g., adopted technologies in a Microgrid) [5]. While simulation approaches are helpful to understand a complex system by running multiple iterations in a manual fashion, simulation approaches do not have built-in mechanisms (i.e., mathematical solvers) to find the best or optimal solution (e.g., optimal DER capacity). Since there are often millions of combinations for technology choices and operational levels, simulation approaches can require enormous numbers of iterations to find the optimal technology combinations. Optimal operational dispatches (e.g., unit commitment for multiple DERs) are also elusive for simulation approaches, and will significantly increase runtimes. Fescioglu-Unver et al. [6] conclude that rule-based, i.e., assumption-based simulation approaches are not viable to guarantee optimal dispatch results; instead optimization techniques should be used to increase the profitability of Microgrids.

Mixed Integer Linear Programming (MILP) optimization algorithms and associated mathematical solvers can overcome the limitations of simulation approaches and deliver optimal economic and/or green house gas solutions in a single run, creating a viable path to identify the best DER portfolio and dispatch.

As indicated in [7] the process of designing a Microgrid, which comprises conceptual design, technical design, electrical analysis, power flow analysis, and implementation, can be very time consuming. Thus, any economic optimization algorithm attempting to deliver the optimal DER portfolio, lowest costs, and optimal dispatch must be fast while maintaining accuracy. Examples for such optimization tools are REopt [8] and DER-CAM [9]. These tools are already actively used in the Microgrid industry for real Microgrid design. Reopt uses a full-scale MILP approach optimizing each hour of the year explicitly while DER-CAM relies on a peak-preserving day-type approach to reduce run-times.

Other examples for economic Microgrid optimization algorithms can be found in [10–12]. A common challenge for these optimization algorithms is run-time, which ranges between 0.1 and 280 h, depending on the considered technologies in a Microgrid and optimization approach (see Figure 1). Each optimization approach using a full annual dataset of 8760 hourly data points exhibits run-times above 2.8 h, rendering such approaches impractical for real-world Microgrid design projects since dozen or even hundreds of sensitivity runs might be needed. [10,11] demonstrate that down-sampling the data to representative days can reduce the run-time below 1 h. However, down-sampling impacts on the objective function and technology adoption need to be analyzed for multiple Microgrid and DER projects. Schütz et al. in [11] perform a comparison for two test cases between an 8760 optimization and different k-means down-sampling approaches. Gabrielli et al. in [10] test different optimization methods to address the issue of discontinuity between representative periods when modeling seasonal storage in energy systems, but for only two test cases.

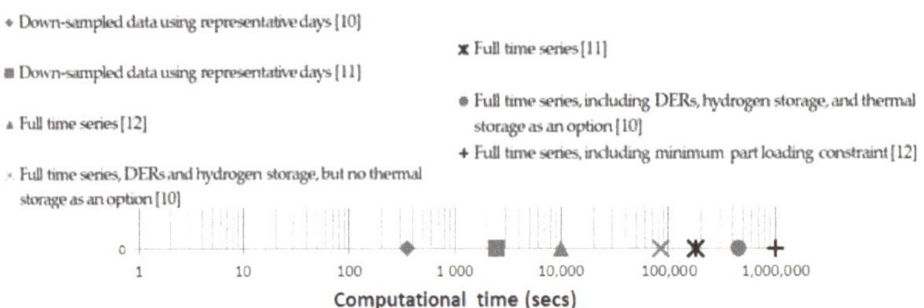

Figure 1. Run-time comparisons for different optimization approaches and use-cases. The different model setups tested in [10–12] result in considerable different run-times. The legend is ordered from small to large computational time. Please note the logarithmic scale.

Fahy et al. [13] demonstrate a peak load preserving down-sampling method and compare the technology selection as well as the Objective Function (OF) to a full-scale time-series optimization approach (FSO). For a single example, the results show OF differences below 1% and no technology adoption difference, but run-time savings of 90%. The authors of [13] also show that clustering with k-means always delivers worse OF results than the peak-preserving day-type approach selected for this study.

The literature review and our research indicate that traditional FSO approaches might be prohibitive for wide-spread Microgrid design, unless special hybrid optimization (HO) approaches are used. Typically, such HO approaches use two stages, in which the first stage optimizes the Microgrid technology adoption based on down-sampled representative day optimization (RO) [14]. Technology adoption results from the first stage are used to inform the second stage for dispatch optimization. Pecenak et al. [15] introduce a new HO approach that applies a minimum DER constraint, derived from the first HO stage, to the second stage. This approach also guarantees robust Microgrid outage modeling solutions by combining the peak preserving day-type approach with a FSO approach.

Down-sampling methodologies and HO approaches show great potential for industry applications, but an extensive performance comparison involving more than two test sites is lacking. Thus, this research compares the peak load preserving down-sampling RO approach from [13] and the FSO approach for 13 real Microgrid projects in the US. We address how the Microgrid setup and input data drive OF differences between the RO and FSO, as well as the impact of DER sizing deviations in the two models. Additionally, we research how the embedded MILP dispatch modeling of the second (dispatch) stage in an HO reduces the OF differences between RO and FSO.

2. Model Description and Used Data Down-Sampling

The mathematical optimization model has been documented in the literature numerous times and is based on the Distributed Energy Resources Customer Adoption Model (DER-CAM) [9]. Several studies have expanded on DER-CAM. Mashayekh et al. [16] added power flow and multi-node capabilities, allowing for optimal placement of DER technologies in a distribution network. To keep run-times low for such a power flow version the RO is needed. Cardoso et al. [17] describe an Ancillary Service market MILP extension for DER-CAM and show how such markets impact the Microgrid design. Milan et al. [18] introduce nonlinear efficiency modeling for CHP systems and describe the MILP in detail. Especially the modelling of nonlinear behavior increases the run-times considerably and call for RO approaches. A DER-CAM version with considerations of passive building measures is established in [19], which allows DER and building technology optimization to create zero carbon solutions. Another version considers electric vehicle (EV) modeling under uncertainty [20] and has been applied to assess the impact of EV interconnections on optimal DER solutions. The authors of [21] consider outage modeling in DER-CAM by adding a particle swarm optimization to determine the optimal investment and operation of DER equipment. Solar variability has been incorporated by [22] and the impact on Microgrid design has been studied. The most recent version of DER-CAM implements also an efficient multi-year optimization [23] based on a RO. In this paper we use DER-CAM, implemented in XENDEE [24], as basis for the peak load preserving down-sampling RO runs for the 13 Microgrid projects.

The process of solving the MILP based on Figure 2 can be very time consuming since the amount (represented as arrow width) of each energy flow is not static over the modeled time horizon, but can change considerably with each time-step, because of, e.g., available solar radiation or changes in electric rates. Solving such a MILP with full time-series data sets, each containing 8760 data points for hourly resolution or 35,040 for 15 min resolution (or even more data points in a multi-year setting), can take hundreds of hours. Thus, down-sampling methodologies are used to reduce the run-time. We refer to this down-sampling representative day optimization as RO.

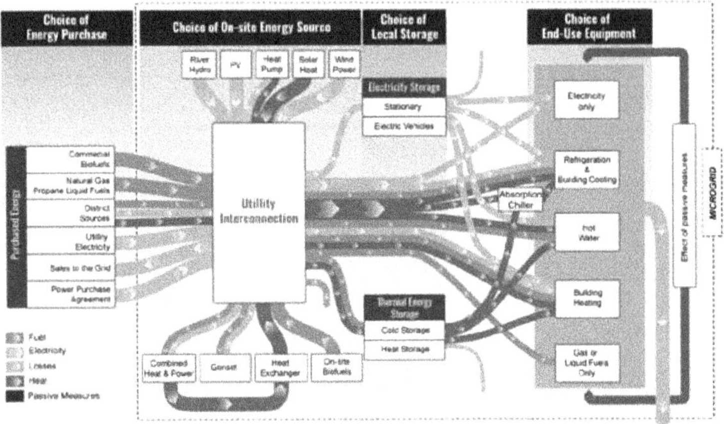

Figure 2. Sankey diagram for the Microgrid DER-CAM/XENDEE MILP. The five energy end-uses on the right hand side need to be supplied with energy at minimized annual energy costs or CO_2 emissions. The MILP analyzes the energy flows (different arrows) in each time-step and decides on the optimal investment capacities and technologies as well as energy flows in each time-step, constituting an optimal dispatch profile.

2.1. Peak Preserving Day-Types Representative Optimization (RO)

The peak preserving RO is a special data reduction method that preserves both total annual energy demand and demand peaks. The peak preserving approach reduces annual hourly demand data into typical weekday and weekend profiles as well as peak demand day profiles. For each month m, one 24 h profile of each day-type is constructed with an hourly resolution. The total annual energy consumption is calculated using multipliers to scale up typical weekday, weekend, and peak demand day-type profiles, with the multipliers ND representing the number of times each day-type d occurs in a given month m ($ND_{m,d}$).

The peak demand day profile for each month is constructed by selecting at each hour the maximum demand across any given day in the month. The resulting peak day profile represents both peak daily consumption and peak monthly demand. An example of the peak demand profile is shown in Figure 3.

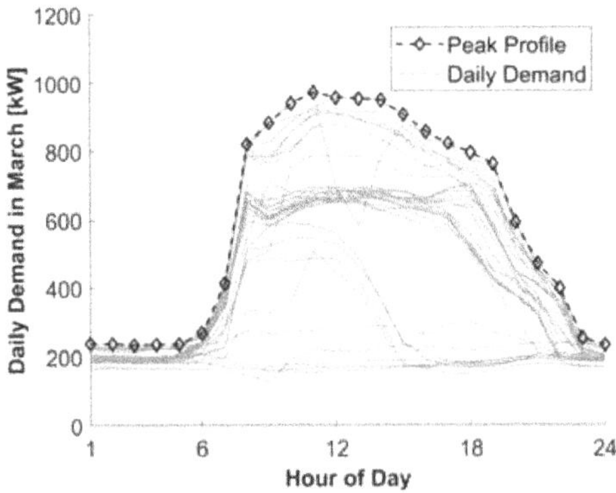

Figure 3. Daily demand for each day in an example March month and peak demand day profile constructed from selecting maximum hourly demand across all days.

The representative weekday and weekend demand profiles must both represent the average weekday and weekend demand behavior, while also maintaining the monthly total energy consumption.

Therefore, the average weekday and weekend profiles are adjusted to account for the energy contained in the peak demand profiles. The monthly demand data are separated into sets of weekday and weekend data. The sets are summed to calculate the total weekday demand and weekend demand for each hour in the day. The total weekday demand data for month m is modified by subtracting the peak demand, multiplied by the number of peak days expected to occur in the month, from the weekday consumption at each hour the peak occurred on a weekday. As indicated by [13] the number of peak days is not crucial and just using one peak day profile in the optimization is sufficient. The same approach is used to modify the weekend consumption data, based on peaks occurring on weekends. The modified total weekday and weekend demand data sets are averaged into 24-h representative weekday and weekend demand profiles.

RO MILP

The energy end-uses (u) are grouped into three characteristic groups d: weekdays, weekend days, and peak days for each month. The MILP approach is solved for the entire year, resulting in 36 daily profiles and 864 hourly data points. A brief overview of the MILP is given in what follows and additional selected constraints are given in Figure 4.

The objective function minimizes the total costs C

$$
\begin{aligned}
= & \sum_{m} \mathrm{MFix_m} + \sum_{u,m,d,h} u\!\sim_{u,m,d,h} \cdot C_{u,m,d,h} \cdot \mathrm{ND_{m,d}} \\
& + \sum_{u,m,p} \mathrm{max}u\!\sim\sim_{u,p,m} \cdot D_{u,p,m} \\
& + \sum_{g} num_g \cdot \mathrm{IFix_g} \cdot \mathrm{ANN_g} + \sum_{cUs} (pur_{cUs} \cdot \mathrm{IFix_{cUs}} + cap_{cUs} \cdot \mathrm{IVar_{cUs}}) \cdot \mathrm{ANN_{cUs}} \\
& + \sum_{j,u,m,d,h} \frac{gen_{j,u,m,d,h}}{\eta_j} \cdot \mathrm{GENC_{j,u,m,d,h}} \cdot \mathrm{ND_{m,d}} + \sum_{u,m,d,h} dr_{u,m,d,h} \cdot \mathrm{DRC_{u,m,d,h}} \cdot \mathrm{ND_{m,d}} \\
& - \sum_{i,m,d,h} sell_{i,m,d,h} \cdot \mathrm{S_{m,d,h}} \cdot \mathrm{ND_{m,d}}
\end{aligned}
\tag{1}
$$

Major constraint Energy Balance :

$$
\begin{aligned}
\sum_{u,m,d,h} \mathrm{LOAD_{u,m,d,h}} + \sum_{i,m,d,h} sell_{i,m,d,h} + \sum_{s,m,d,h} \sin_{s,m,d,h} + \sum_{u,m,d,h} dr_{u,m,d,h} \\
= \sum_{u,m,d,h} u\!\sim_{u,m,d,h} + \sum_{i,u,m,d,h} gen_{i,u,m,d,h} + \sum_{s,u,m,d,h} sout_{s,u,m,d,h}
\end{aligned}
\tag{2}
$$

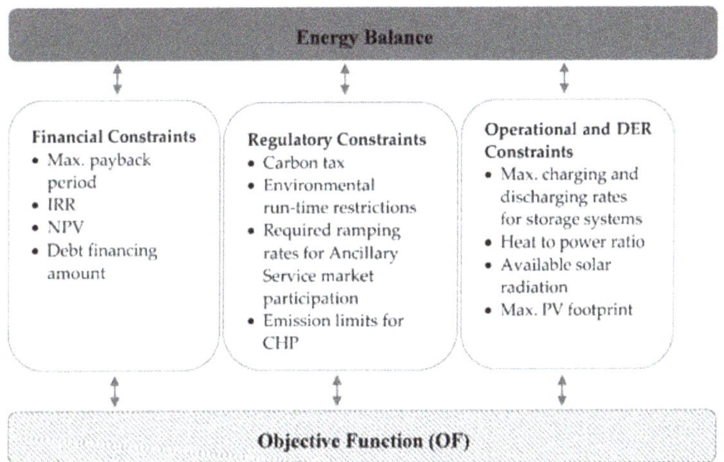

Figure 4. Selected constraints of the MILP.

Indices

c	continuous generation technologies (assumed to be available in any size), $c \in \mathbf{C} =$ {photovoltaic panels, solar thermal panels, and absorption chillers}
d	day-types, $d \in \mathbf{D} =$ {week, peak, weekend}
g	discrete generation technologies (explicitly modeled in discrete sizes), internal combustion engines (ICE), micro turbines (MT), fuel cells (FC), and gas turbines (GT), with and without heat exchangers (HX), $g \in \mathbf{G} =$ {ICE, ICEHX, MT, MTHX, FC, FCHX, GT, GTHX}. All discrete technologies without HX are referred to as DG, DG with HX as CHP
h	hours in a day $h \in \mathbf{H} = \{1, 2, \ldots, 24\}$
i	DER technologies, $i \in \mathbf{I} = \mathbf{J} \cup \mathbf{S}$
j	generation technologies, $j \in \mathbf{J} = \mathbf{G} \cup \mathbf{C}$
m	months in a year, $m \in \mathbf{M} = \{1, 2, \ldots, 12\}$
p	utility demand periods, $p \in \mathbf{P} =$ {coincident, on peak, mid peak, off peak}
s	energy storage technologies, stationary storage and heat storage, $s \in \mathbf{S} =$ {electric energy storage systems, heat storage}
u	energy end-uses for each day-type (d), including electricity-only (eo), cooling (cl), space heating (sh), water heating (wh), and natural gas loads (ng), $u \in \mathbf{U} =$ {eo, cl, sh, wh, ng}

Parameters

ANN_i	annuity rate of investing in DER technology i
$ND_{m,d}$	number of days of type d in month m
$C_{u,m,d,h}$	volumetric electricity charges
$D_{u,p,m}$	charges applied to peak power demand for end-use u during period p, and month m
$DRC_{u,m,d,h}$	volumetric demand response costs
$GENC_{j,u,m,d,h}$	fuel costs, maintenance costs
$IFix_i$	fixed investment cost of DER technology i
$IVar_{c \cup s}$	variable investment cost of continuous energy conversion technology c, or storage technology s
$LOAD_{u,m,d,h}$	Microgrid energy demand for end-use u, in month m, day-type d, and hour h
$MFix_m$	fixed monthly utility charges/contract demand charges
$S_{m,d,h}$	electricity sales price in month m, day-type d, and hour h
η_i	energy conversion efficiency for i

Decision Variables

$cap_{c \cup s}$	installed capacity of continuous generation technology c, or storage technology s
$dr_{u,m,d,h}$	energy demand of end-use u removed by demand response measures in month m, day d, and hour h
$gen_{j,u,m,d,h}$	useful (e.g., electric output) energy provided by generation technology j for end-use u in month m, day-type d, and hour h
num_g	number of installed units of discrete generation technology g
$pur_{c \cup s}$	binary purchase decision for continuous generation technology c, or storage technology s
$sell_{i,u,m,d,h}$	energy sales from technology i that is exported in month m, day-type d, and hour h
$sin_{s,m,d,h}$	energy input to storage technology s, in month m, day-type d, and hour h
$sout_{s,u,m,d,h}$	energy output from storage technology s for end-use u, in month m, day-type d, and hour h
$u\!\sim_{u,m,d,h}$	utility purchase for end-use u, during month m, day-type d, and hour h

2.2. Full-Scale Time-Series Optimization (FSO)

Since DER-CAM was programmed as a RO model, the FSO requires some adjustment of the day-type framework to emulate an FSO. The FSO MILP model is derived from the RO model by modifying $ND_{m,d}$ to represent the real number of days in a month instead of the number of representative day-types. Thus, instead of e.g., using 22 representative weekdays, eight weekend days, and one peak profile for the RO, we convert $ND_{m,d}$ into a binary matrix containing ones to identify the real days observed in each month. In the case of January 2020, the matrix consists of ones from 1 to 31. For February 2020 it consists of ones from 1 to 29 and zeros for 30 and 31, etc. Days must be linked in time to allow energy to shift between consecutive days, creating a real seasonal model. The authors of [15] describe the changes needed to create an FSO model in detail.

Additionally, we link the RO and FSO to create a Hybrid Optimization (HO) approach. In such an HO approach the sizing (e.g., DER capacity) solution from the down-sampled RO will be used as fixed input for the FSO. In other words, the FSO just optimizes the dispatch of the RO-designed Microgrid using the full time-series data which preserves short run-times. To differentiate between a real FSO, which also sizes DERs, and the full time-series dispatch optimization within the HO, we call the latter TSO. The second part of this paper will compare the OF and run-time results of a simple RO with those of an HO, utilizing a TSO as a second stage.

3. Microgrid Projects

3.1. General Description of Microgrid Projects

Table 1 presents an overview of the Microgrid projects. Cases were selected to represent a diverse variety of host types, geographic locations, tariff characteristics, and total load consumption. These Microgrid projects have been modeled by the authors in detail. All the projects are optimized using the RO, FSO, and HO models, and results are assessed from an economic perspective.

All Microgrids are grid connected without grid outages considered, except for Mil2, for which we modeled a 24-h outage on the day with the highest electric peak demand, which in this case constitutes also the highest daily energy consumption. None of the Microgrids are allowed to sell electricity to the utility, except for Un1, which is on a net-metering tariff and can export surplus electricity to the utility. To analyze effects of electricity sales we will show hypothetical sensitivity runs for selected sites in Section 4.2.

The tariffs and technology data are summarized in the Appendices A and B.

3.2. Electric Load Data

For each case except Un4, hourly metered load data for one year was used in this analysis (8760 data points per case for FSO and HO). In the absence of metered load data, for Un4, data from the Commercial and Residential Hourly Load Profile database for a hospital was used since the modeled site is a medical University [25]. Segments of the time-series electric load data for Un3 and Un4 are provided in Figure 5.

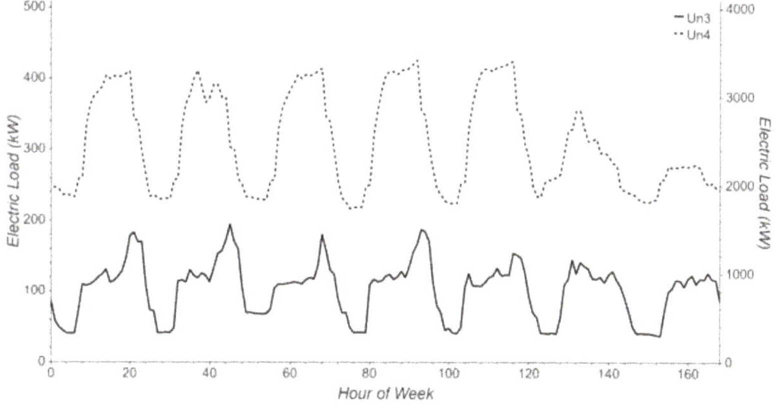

Figure 5. Example time-series electric load data for Un3 (solid line, left *y*-axis) and Un4 (dashed line, right *y*-axis). Hours 120–168 are weekend days.

Table 1. Overview of considered Microgrid projects for this research. FER: Flat energy rate, FSER: Flat seasonal energy rate, FER-winter: Flat energy rate just for winter months, TOUER: Time-Of-Use energy rate, TOUER-summer: Time-Of-Use energy rate just for summer months, NCDC: Non-coincident demand charge, PDC: Peak demand charge, MPDC: Mid peak demand charge. PV: Photovoltaics, EES: Electric Energy Storage, CHP: Combined Heat and Power, DG: Distributed Generation as natural gas or diesel fired backup systems. All Microgrid projects are in the US, due to confidentiality reasons the exact locations cannot be revealed.

Case	Type	State/Territory	Techn. Modeled	Tariff Characteristics	Annual Electrical Cons. (MWh)	Annual Heating Cons. (MWh)	Electric Peak Load (MW)
Ind	Industrial/Pharmaceutical	Puerto Rico	PV, EES	FER, NCDC, PDC, MPDC	22,642	n/a	3.96
Res	Residential/Public	Connecticut	PV, EES, DG	TOUER, NCDC	1640	n/a	0.37
Man	Industrial/Materials	Puerto Rico	PV, EES, CHP	FER, NCDC	78,400	41,854	12.48
Com	Commercial/Public	Washington State	PV, EES, CHP	FER, NCDC	4263	667	0.93
Un1	University	Colorado	PV, EES, DG	TOUER-summer, FER-winter, PDC	12,076	n/a	2.85
Un2	University	Hawai'i	PV, EES	FER, NCDC	3338	n/a	0.97
Un3	University	California	PV, EES, DG	TOUER, NCDC, PDC	825	n/a	0.20
Un4	University	Vermont	PV, EES, DG	FER, NCDC	26,713	6817	5.00
Mil1	Military	Texas	PV, EES, DG	TOUER, NCDC	330,648	n/a	67.61
Mil2	Military	New Mexico	PV, EES, DG	TOUER, NCDC	78,878	n/a	15.99
Mil3	Military	Maryland	PV, EES, DG	FSER, NCDC	187,645	n/a	33.96
Mil4	Military	California	PV, EES, DG	TOUER, NCDC	86,349	n/a	15.00
Mil5	Military	Massachusetts	PV, EES, DG	TOUER, NCDC	16,564	n/a	3.41

Figure 6 summarizes the electric load data statistics. Total annual electric load is the sum of the hourly electric load profile over the entire year. Annual load variability is the sum of the absolute value of hourly energy ramp rates (ERR_n) normalized by the total annual electric load to ensure that load volatilities can be compared between sites (Equation (3)).

$$Annual\ Load\ Variability = \frac{\Sigma(abs(ERR_n))}{\Sigma_{m,d,h} load_{m,d,h}} = \sum_{m,d,h} \frac{load_{m,d,h} - load_{m,d,h-1}}{\Sigma_{m,d,h} load_{m,d,h}} \tag{3}$$

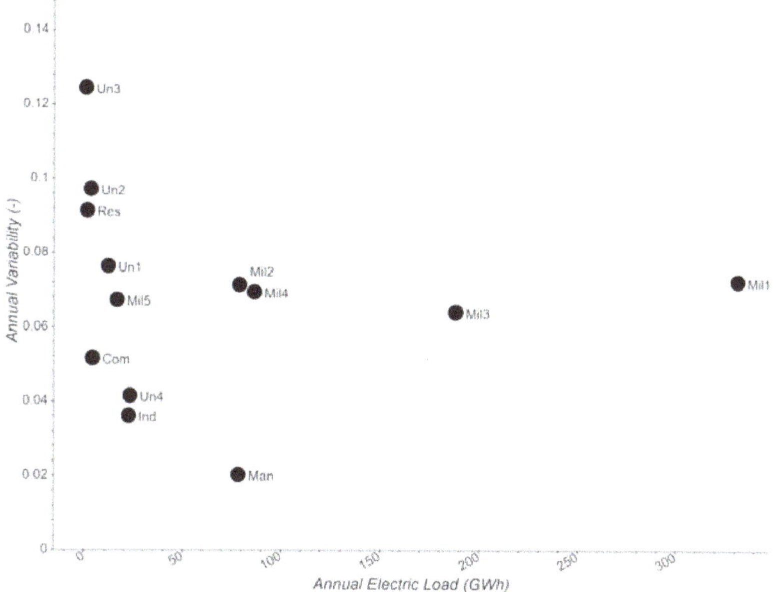

Figure 6. Summary of electric load data for all cases as a function of annual electric load (GWh) and load variability (-) expressed as the sum of the absolute values of all 1 h power changes normalized by the total annual load.

For example, the Un3 metered data is more volatile, with severe late afternoon ramps and a total variability equal to 12.4% of its annual load. Un4, on the other hand, is relatively smooth, and therefore, shows low variability numbers of 4.2% in Figure 6, which also can be attributed to the load modeling. Overall, the two industrial sites (Ind and Man) have the lowest variability. The military sites Mil1 and Mil3 have significantly larger annual loads than the other cases.

3.3. Solar Radiation Data

Several sources of solar radiation data were used for the different Microgrid projects as input for the RO, FSO, and TSO MILP: real measurements of solar radiation (two projects), Helioscope [26] data (six projects), and PVWatts data based on the NREL National Solar Radiation Database from satellite data (NSRDB, five projects) [27]. While the FSO and TSO use the 8760 PV Watts output directly, the RO uses an average daily profile for each month constructed from the 8760 time-series.

Similar to Figure 6, Figure 7 compares the solar production data across all sites via the capacity factors and total variability. Similar to the demand, total solar variability is calculated by summing over the absolute values of solar production ramp rates (SRR_n) and normalizing them by the total energy production (Equation (4)).

$$Annual\ Solar\ Variability = \frac{\Sigma(abs(SRR_n))}{\Sigma_{m,d,h}PV_{m,d,h}} = \sum_{m,d,h}\frac{PV_{m,d,h} - PV_{m,d,h-1}}{\Sigma_{m,d,h}PV_{m,d,h}} \tag{4}$$

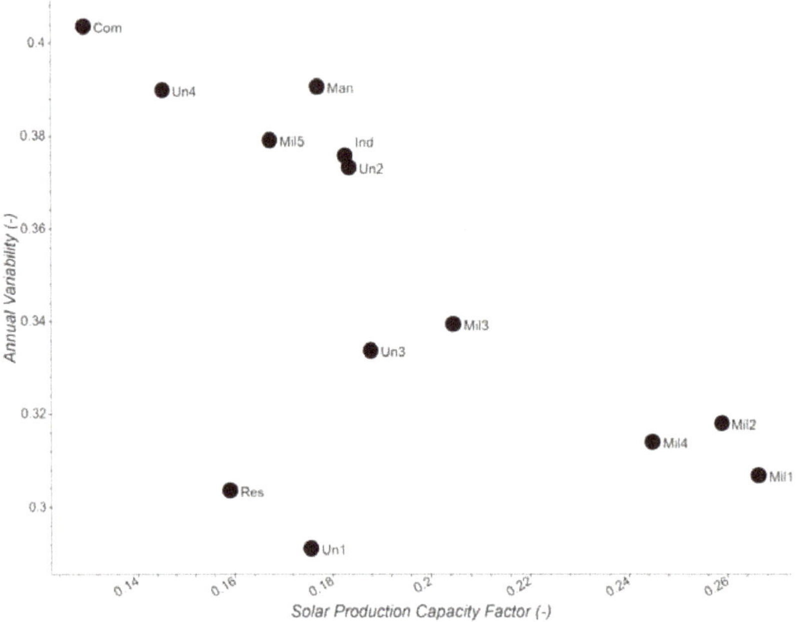

Figure 7. Solar production summary for all cases expressed through the annual capacity factor and the normalized annual solar variability.

Figure 7 is useful to infer case-by-case differences in the solar resource. For example, Mil1, Mil2, and Mil4 are located in sunny locations with little variability due to clouds. As such, variability is low and total solar production and capacity factors are high. Conversely, Com is in a region with low total solar production and high variability, indicating frequent cloud cover and ramp events.

4. Results

4.1. Representative Optimization (RO) versus Full-Scale Time-Series Optimization (FSO)

Table 2, Figures 8 and 9 present the high level results for all 13 Microgrid cases, comparing the objective functions of RO and FSO, technology adoptions, and run-times. The run-time savings for the analyzed cases can range between almost 100% and 85%.

In total, 10 cases out of the 13 show OF differences below 1.5%. Un4, with a very high solar variability (see Figure 7) and small load variability shows the highest OF difference. This could explain the 97.4% difference in PV adoption between the RO and FSO. Some cases show very similar OFs for the RO and FSO models despite significant changes in the technology adoption, which is expected for MILP approaches. Examples are the Ind case with a −32% difference in EES adoption, but only a −0.5% OF difference or the Res case with a 56.9% difference in EES adoption, but only a −0.3% difference in the OF (see also Figure 9).

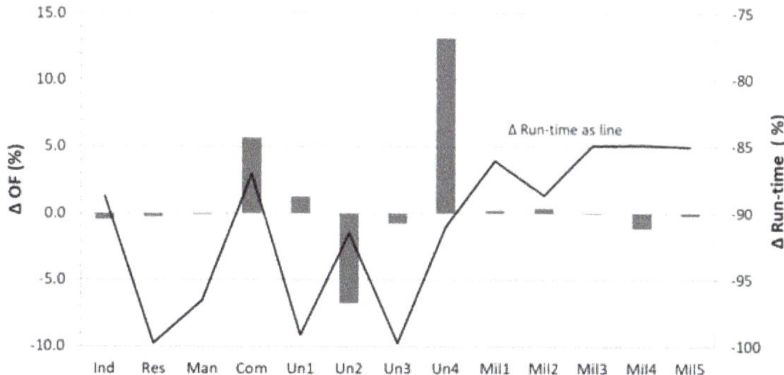

Figure 8. OF differences (bars) and run-time savings (line) for the RO approach compared to the FSO. Negative run-time numbers represent savings.

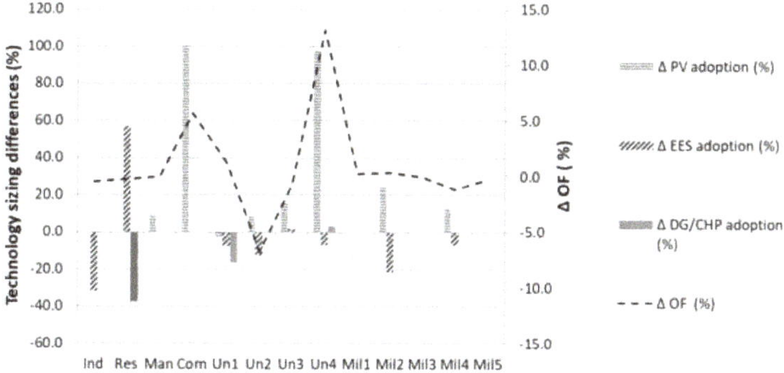

Figure 9. Variations in RO technology adoption compared to the FSO. The OF difference compared to FSO is shown as a dashed line.

It is worth noting that the Ind and Res cases experience opposite solar and load volatilities: Ind has a relative high solar variability compared to the Res case (see Figure 7) and a small load variability compared to the Res case (see Figure 6). Note that in the Ind and Res case the available PV space is fully utilized, explaining the exact same PV sizes in both optimization models.

Un3 with the highest load variability experiences one of the highest run-time savings between the two MILP models (99.8%), a very low OF difference of −0.8%, moderate PV difference (16%), very small EES differences (2%), and no difference in fuel fired DG adoption.

The Mil2 includes a 24 h outage and the possibility to install DG units. Since the RO model preserves the peak loads from the full time-series load data and the DG units contribute to the worst case outage modeling (highest power demand and electric energy consumption in a day in this case), the installed fuel fired DG units in the FSO and RO model match. The differences in the PV (24.3%) and EES (−21.4%) adoptions are influenced by the different granularity of the solar radiation modeling in the RO and FSO—the RO uses an average monthly solar production profile, while the FSO uses the full-scale time-series. However, the OF (e.g., project cost) differs by only 0.4%.

Com and Un4 show significant OF differences with higher costs in the RO model (i.e., ΔOF >0), which creates a budget cushion for these projects when modeled with RO, but could also render these projects economically unattractive. Their technology selection in the RO is higher except for the smaller EES numbers in the RO model for Un4 (Figure 9). Un2, on the other hand, shows significant lower costs in the RO model, indicating that there is no clear trend on whether the RO is over- or underestimating OFs.

Figure 9 also shows that if sizing of several technologies differ for a Microgrid (seven cases), then RO oversizing of one technology is usually balanced by undersizing of another. In four of those seven cases RO oversizes PV and undersizes storage compared to the FSO, indicating that the technologies could be to some extent interchangeable.

Among these 13 Microgrid designs, there is no clear relationship between the solar variability and the deviation in the OF solution (Figure 10). Com, Un4, and Un2 tend to have higher solar variability and are the three cases which exhibit OF deviations greater than 1.5%. However, sites Man, Ind, and Mil5 also have high solar variability and small OF differences. However, for Mil5 no DER investments are optimal and installed. On the other hand, the load variability seems to have no significant impact on the OF differences, indicating that the peak-preserving down-sampling is an effective method to capture load spikes.

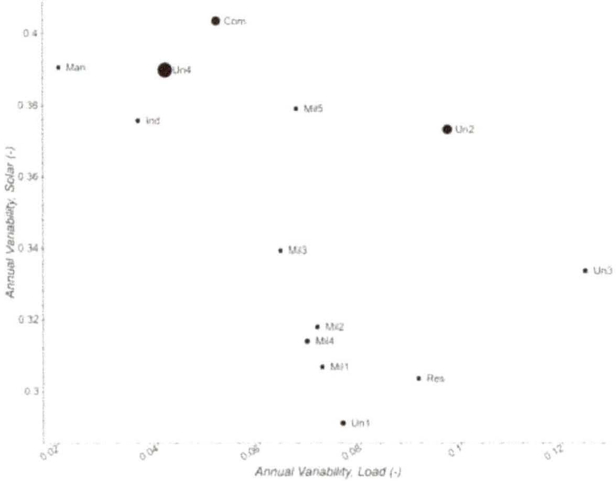

Figure 10. Scatter plot comparing the absolute value of OF deviation of each case as a function of solar variability (*y*-axis) and load variability (*x*-axis). The size of the circle represents the OF deviation of each case.

4.2. Sensitivity to Electricity Sales

A limitation of the Microgrid projects is that only site Un1 is explicitly considering energy sales to the utility, which inherently limits the economic viability and sizing for DERs, especially for solar PV. Thus, sensitivity scenarios are performed to assess the impact of electricity sales on OF differences as well as DER technology selections. For these sensitivities, two cases were selected, one with a minor OF difference of −1.2% (Mil4) and one with a considerable difference of −6.8% (Un2). These sites are representative as universities and military sites are prime Microgrid candidates considering their abundance in the set of real Microgrid projects in Table 1.

The sales prices are assumed to be the same as the energy purchase prices from the utility during the same time period. Capacity bidding or Ancillary Service market participation is not considered.

For the Un2 case, including energy sales the OF difference is reduced to −2.0% (see column 1 in Table 3 for the Un2 cases). In this particular case, the OF deviation reduction may be explained by identical PV capacity investments. When sales are considered, PV is attractive enough that both algorithms invest in PV to its spatial limit. However, the Un2 sales with FSO case shows less EES, which causes the Microgrid to import and export more energy on an annual basis compared to the RO approach (see Table 4 cells Un2 sales/A/B/D/E).

Table 2. Overview results for the 13 modeled Microgrids. OF: Objective Function; R-time: Run-time; RO: Down-sampled representative day-types optimization.

Case	1	2	3	4	5	6	7	8	9	10	11	12	13
	Δ OF (%)	R-Time RO (mins)	R-Time FSO (mins)	Δ R-Time (%)	PV RO (kW)	PV FSO (kW)	Δ PV Compared to FSO (%)	EES RO (kWh)	EES FSO (kWh)	Δ EES Compared to FSO (%)	DG/CHP RO (kW)	DG/CHP FSO (kW)	Δ DG/CHP Compared to FSO (%)
Ind	−0.5	0.2	1.6	−89	1568	1568	0.0	396	582	−32.0	0	0	n/a
Res	−0.3	0.3	121.0	−100	715	715	0.0	1048	668	56.9	100	160	−37.5
Man	0.0	0.3	7.4	−97	358	328	9.1	0	0	n/a	9975	9975	0
Com	5.7	0.2	1.7	−87	182	0	100.0 *)	0	0	n/a	0	0	n/a
Un1	1.3	1.0	121.1	−99	8969	9211	−2.6	8243	8909	−7.5	500	600	−16.7
Un2	−6.8	0.0	0.5	−92	1627	1501	8.4	2242	2573	−12.9	0	0	n/a
Un3	−0.8	0.2	88.5	−100	257	222	15.8	320	314	1.9	100	100	0
Un4	13.2	0.2	2.4	−91	995	504	97.4	2227	2400	−7.2	3000	2900	3.4
Mil1	0.3	0.2	1.5	−86	0	0	n/a	0	0	n/a	0	0	n/a
Mil2	0.4	0.3	2.2	−89	6107	4913	24.3	6600	8400	−21.4	12,000	12,000	0
Mil3	−0.1	0.2	1.4	−85	0	0	n/a	0	0	n/a	0	0	n/a
Mil4	−1.2	0.2	1.2	−85	13,053	11,600	12.5	7800	8400	−7.1	0	0	n/a
Mil5	−0.2	0.2	1.1	−85	0	0	n/a	0	0	n/a	0	0	n/a

*) Differences (Δ) are calculated as (RO data-FSO data)/FSO data, except for Δ PV adoption compared to the FSO approach to avoid an undefined in the Com case. Only in this case we use (RO data-FSO data)/RO data.

Table 3. Overview results for the sensitivity runs considering energy sales to the utility. OF: Objective Function; R-time: Run-time; RO: Representative Optimization approach; FSO: Full Scale Time-Series Optimization.

Case	1	2	3	4	5	6	7	8	9	10	11	12	13
	Δ OF (%)	R-Time RO (mins)	R-Time FSO (mins)	Δ R-Time (%)	PV RO (kW)	PV FSO (kW)	Δ PV Compared to FSO (%)	EES RO (kWh)	EES FSO (kWh)	Δ EES Compared to FSO (%)	DG/CHP RO (kW)	DG/CHP 8FSO (kW)	Δ DG/CHP Compared to FSO (%)
Un2 no sales	−6.8	0.0	0.5	−92	1627	1501	8.4	2242	2573	−12.9	0	0	n/a
Un2 sales	−2.0	0.0	0.6	−92	2994 *)	2994 *)	0	1412	1358	4	0	0	n/a
Mil4 no sales	−1.2	0.2	1.2	−85	13,053	11,600	12.5	7800	8400	−7.1	0	0	n/a
Mil4 sales	−1.1	0.1	0.8	−88	18,945	16,856	12.4	9000	9600	−6.3	0	0	n/a

*) In the Un2 case with sales the PV adoption is reaching the maximum available space at the site.

The Mil4 sales case does not show significant changes compared to the case without sales. The OF differences change from −1.2% to −1.1% and the PV/EES adoptions show slightly smaller differences compared to the Mil4 no sales case (see Table 4).

However, as indicated by Mil4 sales in Table 4, reduced differences in the technology adoption do not necessarily lead to a reduced difference in the imported energy. For Mil4, Δ PV changes from −12.5% to −12.4% and Δ EES from −7.1% to −6.3%, but Δ import decreases from −5.2% to −7.2% (Table 4).

These results underscore the complexity of such modelling problems and that similar technology adoption capacities can result in similar OFs, but different parts of the result can change in different directions.

Table 4. Import and export balance for the energy sale sensitivity runs as well as the original runs without sales.

	A	B	C	D	E	F
Case	Annual Export RO (MWh)	Annual Export FSO (MWh)	Δ Export Compared to FSO (%)	Annual Import RO (MWh)	Annual Import FSO (MWh)	Δ Import Compared to FSO (%)
Un2 no sales	0	0	n/a	962	1237	−22.2
Un2 sales	2720	2801	−2.9	1276	1366	−6.6
Mil4 no sales	0	0	n/a	58,694	61,920	−5.2
Mil4 sales	5088	4456	14.2	5138	5538	−7.2

4.3. The Influence of Optimal Dispatch Modeling—The Hybrid Optimization (HO)

The large OF differences for Com and Un4 and the connection to extreme relative capacity deviations will be analyzed.

In both Com and Un4, RO modeling delivers higher optimal capacities of PV and DG/CHP. In particular, the RO almost doubles the PV capacity for Un4 and invests in 182 kW of PV for Com, while FSO does not select any PV in Com. Only the optimal EES in the Un4 FSO is slightly higher than for the RO. We hypothesize that the OF differences resulting from large capacity deviations in the RO can be mitigated by dispatch optimization. To test this, optimal dispatch modeling in the FSO, based on optimal capacities from the RO, is performed. This approach constitutes a two stage Hybrid Optimization (HO) approach, in which we refer to the FSO as TSO to indicate that the full-scale time-series optimization of the second state will optimize the operational planning (i.e., dispatch), but not the capacities. The HO approach allows assessing the impact of dispatch on the OF differences. Additionally, the HO allows assessing the feasibility of a Microgrid designed by an RO when modeled using raw time-series data.

Taking the optimal investment capacity results from the RO and fixing them in the TSO model while allowing for dispatch optimization, yields OFs that are very similar to the FSO, but with better run-times as indicated in Table 5. The Com RO case shows a 5.7% OF deviation compared to the FSO OF. Using the 2-stage HO process with TSO reduces the deviation to 1.4%. Similarly, the Un4 RO deviation of 13.2% is reduced to a 0.6% difference with HO. Every other case also shows a reduction in the OF difference compared to the FSO with better run-times than the FSO.

These results are important for Microgrid planning and operation since they indicate that even though there occasionally are higher deviations between RO and FSO capacity results, the fast HO is very viable and will result in a similar OF as the slower FSO. The unit dispatch optimization absorbs OF deviations that arise from differences in the capacities between the models. Note that the TSO dispatch optimization in the HO is similar to dispatch modeling in real Microgrids through Model Predictive Controllers (MPC). Thus, actual Microgrid dispatch is economically robust (i.e., will achieve similar revenues and costs) to capacity differences introduced by sub-optimal RO modeling during Microgrid planning.

Table 5. Objective function as well as run-time differences between the HO and FSO. OF: Objective Function; R-time: Run-time; RO: Representative Optimization; HO: Hybrid Optimization; FSO: Full-Scale Time-Series Optimization.

Case	1	1a	2	2a	3	4	4a
	Δ OF RO Versus FSO (%)	Δ OF HO Versus FSO (%)	R-Time RO (mins)	R-Time HO (mins)	R-Time FSO (mins)	Δ R-Time RO Versus FSO (%)	Δ R-Time HO Versus FSO (%)
Com	5.7	1.4	0.2	0.9	1.7	−87	−48
Un4	13.2	0.6	0.2	1.0	2.4	−91	−58
Ind	−0.5	0.0	0.2	1.0	1.6	−89	−40
Res	−0.3	0.3	0.3	1.1	121.0	−100	−99
Man	0.0	0.0	0.3	1.0	7.4	−97	−87
Un1	1.3	0.2	1.0	1.8	121.1	−99	−98
Un2	−6.8	0.8	0.0	0.3	0.4	−92	−23
Un3	−0.8	0.7	0.2	1.1	88.5	−100	−99
Mil1	0.3	0.0	0.2	1.2	1.5	−86	−25
Mil2	0.4	0.3	0.3	1.1	2.2	−89	−50
Mil3	−0.1	0.0	0.2	0.9	1.4	−85	−34
Mil4	−1.2	0.3	0.2	0.9	1.2	−85	−27
Mil5	−0.2	0.0	0.2	0.9	1.1	−85	−20

5. Conclusions

This paper advances the field of Microgrid planning and operation through a comprehensive analysis of objective function and technology adoption results of peak preserving day-types Representative Optimization (RO). Not only does the analysis include a large number of Microgrids (13) with different load and renewable resource time-series, but also a great diversity of tariffs and technology assumptions. The uniqueness of the paper also stems from industry relevance in that the Microgrids are actually being considered for construction or are already being built and were analyzed in commercial applications.

The results support the widespread application of RO in Microgrid planning. The special peak-preserving day types approach represents a full time-series of 8760 h with 3 days in each month or 864 time-steps. For all but three Microgrids the objective function differences are less than 1.5%, yet run-time savings are from 85% to almost 100% compared to full-scale time-series optimization (FSO). Such run-time savings enable more detailed analysis through sensitivity studies, probabilistic parameter inputs (Monte Carlo Simulation) and decision-making, and multi-year horizon analysis.

Three analyzed Microgrids have larger OF differences at 5.7%, −6.8%, and 13.2% in the RO. All these outliers experience higher solar variability than others. However, three other Microgrids with similar or even higher solar variability experience very small OF differences of 0.0%, −0.2%, and −0.5%, indicating that there is no clear trend on how solar variability impacts the results of both models. The impact of load variability seems to be minimal, indicating that the peak-preserving day-type RO is very effective. While such larger OF differences may still be tolerable given other uncertainties in Microgrid planning, they can be mitigated through hybrid optimizations (HO) that optimize technology dispatch in a second stage, using capacity results from the first stage RO and the full time-series data in the second stage. HO still supports run-time savings of 20–99%, but reduces OF differences to less than 1.5% across the board.

The choice of RO versus HO depends on individual preferences of prioritizing run-time or OF accuracy and optimized dispatch might be one of the most important features in a Microgrid Design tool since it provides the possibility to mitigate design problems and sub-optimal capacity selections. This hypothesis will be tested in follow-on research, comparing different dispatch strategies for different DER capacities in built Microgrids.

We would also like to acknowledge a limitation of this paper: the day-type approach cannot simulate energy transfer between days and months and should not be used if seasonal storage is anticipated to be part of the solution. In such cases the presented day-type MILP needs to be modified, which will be discussed in future work.

Author Contributions: M.S. contributed to the data collection process, optimization runs, analyses, interpretations, and writing. Z.P. contributed to the data collection process, optimization runs, analyses, interpretations, and writing. P.M. contributed to the optimization runs, analyses, interpretations, and writing. K.F. contributed to the data collection process, optimization runs, and writing. J.K. contributed to analyses, interpretations, and writing. All authors have read and agreed to the published version of the manuscript.

Funding: This research received no external funding.

Acknowledgments: The authors want to thank the Microgrid team at WorleyParsons and Advisian for the numerous opportunities to experience Microgrid planning and design challenges firsthand and being able to work on exciting real Microgrid projects. We also extend our gratitude to our colleagues Adib Naslé and Scott Mitchell for their outstanding support on implementing the optimization algorithms in our optimization platform and enabling us to perform the optimization runs in such an effective manner.

Conflicts of Interest: The authors declare no conflict of interest. The funders had no role in the design of the study; in the collection, analyses, or interpretation of data; in the writing of the manuscript, or in the decision to publish the results.

Appendix A Tariff Data

For each site, the proper utility tariff was collected or provided by the client and used in the optimization. Table A1 summarizes this information.

Table A1. Summary of electric tariffs and supporting documents.

	Tariff Data				
Case	*Type*	*State*	*Utility Name*	*Tariff Name*	*Source Document*
Ind	Industrial/Pharmaceutical	Puerto Rico	Puerto Rico Electric Power Authority (PREPA)	LIS	[28]
Res	Residential/Public	Connecticut	Eversource Energy	Rate 56—Intermediate Time of Day	[29]
Man	Industrial/Materials	Puerto Rico	PREPA	GST	[28]
Com	Commercial/Public	Washington State	Seattle City Light	MDC—Medium General Service: city	[30]
Un1	University	Colorado	Black Hills Energy	CO935—LPS-PTOU	[31]
Un2	University	Hawai'i	HECO	HECO-P	[32]
Un3	University	California	SDGE	AL-TOU	[33]
Un4	University	Vermont	CBE	Rate 08—General Service	[34]
Mil1	Military	Texas	Confidential	Confidential	Confidential
Mil2	Military	New Mexico	Confidential	Confidential	Confidential
Mil3	Military	Maryland	Confidential	Large Power Schedule	Confidential
Mil4	Military	California	Confidential	Confidential	Confidential
Mil5	Military	Massachusetts	Confidential	Industrial Service	Confidential

Appendix B Technology Data

PV technology costs are presented in Table A2. Costs are based on client input or literature data. The PV costs include soft costs (e.g., labor costs) and inverter costs. The PV costs as well as Operation and Maintenance (O&M) costs are generally within the ranges reported by [35]. Per client request the O&M costs for Res and the PV costs for Un2 are outside of the range reported by [35].

Table A2. PV technology assumptions used in the Microgrid projects. "Max. space for PV" represents the maximum available onsite space for PV generation.

Case	PV Technology Assumptions						
	PV Costs ($/kW$_{DC}$)	*O&M Costs ($/kW and Month)*	*Lifetime (yrs.)*	*Electric Efficiency (%)*	*Tilt (Degrees/Confidential)*	*Orientation (South/North, West, East, Confidential)*	*Max. Space for PV (m^2)*
Ind	2150	0	30	16%	20	South	10,000
Res	2100	2.2	30	19%	Confidential	Confidential	3760
Man	2100	1.4	30	16%	17	South	31,876
Com	1470	0	30	16%	35	South	Unrestricted
Un1	1969	0.8	25	19%	Confidential	Confidential	Unrestricted
Un2	5000	0.8	25	15%	22	South east	20,000
Un3	1700	1.4	30	16%	Confidential	Confidential	40,000
Un4	2400	0	30	19%	30	South	41,806
Mil1	1470	1.5	20	15%	Confidential	Confidential	Unrestricted
Mil2	1470	1.5	20	15%	Confidential	Confidential	Unrestricted
Mil3	1700	1.4	20	15%	Confidential	Confidential	Unrestricted
Mil4	1700	1.4	20	15%	Confidential	Confidential	Unrestricted
Mil5	1700	1.4	20	15%	Confidential	Confidential	Unrestricted

Effective Electric Energy Storage (EES) costs and assumptions are shown in Table A3. Effective EES costs consider incentives and are, therefore, low compared to [36]. Lifetime numbers can also vary significantly depending on allowed operational conditions, meaning allowed max charging or discharging rates or minimum levels of the state of charge.

Table A4 summarizes the natural gas and diesel fired DG and CHP assets. For most cases, multiple options have been provided, mostly distinguished by different unit sizes, unit costs, electric efficiencies or the heat to power ratios. The heat to power ratio specifies the amount of heat generated from 1 kWh electricity. The data is based on vendor data and inputs from the project partners. DG and CHP capacity costs, electric efficiencies, and heat to power ratios broadly agree with the assumptions for the commercial demand model from the Annual Energy Outlook 2020 report [37]. The lifetime numbers seem to be more conservative (smaller) compared to EIA, with the exception of the microturbine lifetimes, which are higher than reported by EIA. The diesel genset costs are in line with [38].

The maximum annual operating hours are based on project constraints such as air regulation or technical constraints.

Table A3. Electric Energy Storage (EES) technology assumptions. The max. allowed charging and discharging rates are constraints within the MILP. Max allowed charge and discharge rates are defined as a function of the EES power capacity.

Case	EES Technology Assumptions									
	Effective EES Costs ($/kWh)	O&M Cost ($/kW Month)	Lifetime (yrs.)	Charging/Respectively Discharge Efficiency (%)	Max. Allowed Charge Rate (-)	Max. Allowed Discharge Rate (-)	Min. SOC (-)	Max. SOC (-)	Maximum Allowed Cycles Per Year (-)	Self-Discharge Per Hour (-)
Ind	250	0	5	90%	0.3	0.3	0.3	1	n/a	0.001
Res	350	0	15	90%	0.3	1	0.1	1	n/a	0.0001
Man	500	0	20	94%	0.2	0.2	0.1	1	n/a	0
Com	350	0	20	94%	0.2	0.2	0.1	1	n/a	0.001
Un1	675	0.2	25	92%	0.3	0.3	0.1	1	110	0
Un2	566	0.2	25	90%	0.3	0.3	0.1	1	n/a	0.0001
Un3	500	0	20	94%	0.2	0.2	0.1	1	n/a	0
Un4	350	0	20	90%	0.5	0.3	0.1	1	n/a	0.0001
Mil1	212	0.3	18	87%	0.3	0.3	0	1	n/a	0.01
Mil2	212	0.3	18	87%	0.3	0.3	0	1	n/a	0.01
Mil3	212	0.3	18	87%	0.3	0.3	0	1	n/a	0.01
Mil4	212	0.3	18	87%	0.3	0.3	0	1	n/a	0.01
Mil5	212	0.3	18	87%	0.3	0.3	0	1	n/a	0.01

Table A4. Summary Fuel fired Distributed Generation (DG) and Combined Heat and Power (CHP) data.

Case	DG/CHP Assumptions									
	Type (/)	Unit Capacity (kW)	Lifetime (yrs.)	Capacity Costs Installed ($/kW)	O&M Fixed Costs ($/kW/year)	O&M Variable Cost ($/kWh)	Efficiency (%)	Heat to Power Ratio (%)	Max. Annual Operating Hours (hrs.)	Backup Only (Yes/No)
Ind	n/a	n/a	n/a	n/a	n/a	n/a	n/a	n/a	n/a	n/a
Res	Microturbine	60	15	3220	0.0	0.001	25%	n/a	8760	no
Res	Microturbine	100	15	3500	0.0	0.002	40%	n/a	8760	no
Man	CHP	3304	20	3281	0.0	0.009	24%	175%	8760	no
Man	CHP	3325	20	3750	0.0	0.009	44%	94%	8760	no
Man	CHP	5670	20	3750	0.0	0.009	28%	135%	8760	no
Man	CHP	7480	20	3705	0.0	0.009	45%	33%	8760	no
Com	Microturbine CHP	61	15	3220	0.0	0.013	25%	189%	8760	no
Com	Microturbine CHP	190	15	3150	0.0	0.016	28%	133%	8760	no
Com	Microturbine CHP	242	15	2700	0.0	0.012	26%	145%	8760	no
Com	Microturbine CHP	950	15	2500	0.0	0.012	28%	130%	8760	no
Un1	Distributed Generation	250	25	2191	0.0	0.022	23%	n/a	160	no
Un1	Distributed Generation	250	25	2191	0.0	0.022	23%	n/a	200	no
Un2	n/a	n/a	n/a	n/a	n/a	n/a	n/a	n/a	n/a	n/a
Un3	Internal combustion engine	125	30	2000	0.0	0.020	26%	n/a	8760	no
Un4	Microturbine	100	15	2900	0.0	0.002	30%	n/a	8760	no
Mil1	Diesel genset	2000	20	600	10.0	0.000	32%	n/a	8760	yes
Mil2	Diesel genset	750	20	750	9.3	0.000	28%	n/a	8760	yes
Mil2	Diesel genset	750	20	750	9.3	0.000	28%	n/a	1091	no
Mil3	Diesel genset	750	20	750	9.3	0.000	28%	n/a	8760	yes
Mil4	Diesel genset	750	20	750	9.3	0.000	28%	n/a	8760	yes
Mil5	Diesel genset	750	20	750	9.3	0.000	28%	n/a	8760	yes

References

1. Navigant. *Microgrid Deployment Tracker 2Q18*; Navigant: Chicago, IL, USA, 2018.
2. Navigant. *Microgrid Deployment Tracker 2Q19*; Navigant: Chicago, IL, USA, 2019.
3. Wood, E. Whats Driving Microgrids toward a $30.9B Market? Microgrid Knowledge. 30 August 2018. Available online: https://microgridknowledge.com/microgrid-market-navigant/ (accessed on 4 March 2020).
4. Tozzi, P.; Jo, J.H. A comparative analysis of renewable energy simulation tools: Performance simulation model vs. system optimization. *Renew. Sustain. Energy Rev.* **2017**, *80*, 390–398. [CrossRef]
5. Lund, H.; Arler, F.; Alberg Østergaard, P.; Hvelplund, F.; Connolly, D.; Vad Mathiesenm, B.; Karnøe, P. Simulation versus Optimisation: Theoretical Positions in Energy System Modelling. *Energies* **2017**, *10*, 840. [CrossRef]
6. Fescioglu-Unver, N.; Barlas, A.; Yilmaz, D.; Demli, U.O.; Bulgan, A.C.; Karaoglu, E.C.; Atasoy, T.; Ercin, O. Resource management optimization for a smart microgrid. *J. Renew. Sustain. Energy* **2019**, *11*, 065501. [CrossRef]
7. Stadler, M.; Naslé, A. Planning and Implementation of Bankable Microgrids. *Electr. J.* **2019**, *32*, 24–29. [CrossRef]
8. REopt. Available online: https://reopt.nrel.gov/ (accessed on 4 March 2020).
9. DER-CAM. Available online: https://building-microgrid.lbl.gov/projects/der-cam/ (accessed on 4 March 2020).
10. Gabrielli, P.; Gazzani, M.; Martelli, E.; Mazzotti, M. Optimal design of multi-energy systems with seasonal storage. *Appl. Energy* **2018**, *219*, 408–424. [CrossRef]
11. Schütz, T.; Schraven, M.H.; Fuchs, M.; Remmen, P.; Müller, D. Comparison of clustering algorithms for the selection of typical demand days for energy system synthesis. *Renew. Energy* **2018**, *129*, 570–582. [CrossRef]
12. Marquant, J.F.; Mavromatidis, G.; Evins, R.; Carmeliet, J. Comparing different temporal dimensions representations in distributed energy system design models. Lausanne. *Energy Procedia* **2017**, *122*, 907–912. [CrossRef]
13. Fahy, K.; Stadler, M.; Pecenak, Z.K.; Kleissl, J. Input data reduction for microgrid sizing and energy cost modeling: Representative days and demand charges. *Renew. Sustain. Energy* **2019**, *11*, 065301. [CrossRef]
14. Bahl, B.; Kümpel, A.; Seele, H.; Lampe, M.; Bardow, A. Time-series aggregation for synthesis problems by bounding error in the objective function. *Energy* **2017**, *135*, 900–912. [CrossRef]
15. Pecenak, Z.K.; Stadler, M.; Mathiesen, P.; Fahy, K.; Kleissl, J. Robust Design of Microgrids Using a Hybrid Minimum Investment Optimization. *Appl. Energy* **2020**, *276*, 115400. [CrossRef]
16. Mashayekh, S.; Stadler, M.; Cardoso, G.; Heleno, M. A Mixed Integer Linear Programming Approach for Optimal DER Portfolio, Sizing, and Placement in Multi-Energy Microgrids. *Appl. Energy* **2017**, *167*, 154–168. [CrossRef]
17. Cardoso, G.; Stadler, M.; Mashayekh, S.; Hartvigsson, E. The impact of Ancillary Services in optimal DER investment decisions. *Energy* **2017**, *130*, 99–112. [CrossRef]
18. Milan, C.; Stadler, M.; Cardoso, G.; Mashayekh, S. Modelling of non-linear CHP efficiency curves in distributed energy systems. *Appl. Energy* **2015**, *148*, 334–347. [CrossRef]
19. Stadler, M.; Groissböck, M.; Cardoso, G.; Marnay, C. Optimizing Distributed Energy Resources and Building Retrofits with the Strategic DER-CAModel. *Appl. Energy* **2014**, *132*, 557–567. [CrossRef]
20. Cardoso, G.; Stadler, M.; Bozchalui, M.C.; Sharma, R.; Marnay, C.; Barbosa-Póvoa, A.; Ferrão, P. Optimal investment and scheduling of distributed energy resources with uncertainty in electric vehicle driving schedules. *Energy* **2014**, *64*, 17–30. [CrossRef]
21. Hanna, R.; Disfani, V.R.; Haghi, H.V.; Victor, D.G.; Kleissl, J. Improving estimates for reliability and cost in microgrid investment planning models. *J. Renew. Sustain. Energy* **2019**, *11*, 045302. [CrossRef]
22. Schittekatte, T.; Stadler, M.; Cardoso, G.; Mashayekh, S.; Sankar, N. The impact of short-term stochastic variability in solar irradiance on optimal microgrid design. *IEEE Trans. Smart Grid* **2018**, *9*, 1647–1656. [CrossRef]
23. Pecenak, Z.K.; Stadler, M.; Fahy, K. Efficient Multi-Year Economic Energy Planning in Microgrids. *Appl. Energy* **2019**, *255*, 113771. [CrossRef]
24. XENDEE. Available online: https://xendee.com (accessed on 4 March 2020).
25. OpenEI. Available online: https://openei.org/doe-opendata/dataset/commercial-and-residential-hourly-load-profiles-for-all-tmy3-locations-in-the-united-states/ (accessed on 12 February 2020).

26. Helioscope. Available online: https://www.helioscope.com/?gclid=EAIaIQobChMI8rXA-52E6AIVT_IRCh0aQwrNEAAYASAAEgJXY_D_BwE/ (accessed on 12 February 2020).
27. Dobos, A.P. *PVWatts Version 5 Manual*; NREL/TP-6A20-62641; National Renewable Energy Laboratory: Golden, CO, USA, September 2014.
28. PREPA. Available online: https://aeepr.com/es-pr/QuienesSomos/Ley57/Facturaci%C3%B3n/Tariff%20Book%20-%20Electric%20Service%20Rates%20and%20Riders%20Revised%20by%20Order%2005172019%20Approved%20by%20Order%2005282019.pdf (accessed on 27 January 2020).
29. Eversource. Available online: https://www.eversource.com/content/docs/default-source/rates-tariffs/ct-electric/rate-56-ct.pdf?sfvrsn=e941c762_32 (accessed on 22 January 2020).
30. Seattle City Light. Electric Rates and Provisions. Available online: https://www.seattle.gov/light/Rates/docs/2020/Schedule%20MDC%20Jan%201%202020.pdf (accessed on 23 January 2020).
31. Black Hills Energy. Schedule of Rates, Rules and Regulations for Electric Service. Available online: https://www.blackhillsenergy.com/sites/blackhillsenergy.com/files/coe-rates-tariff_0.pdf (accessed on 22 January 2020).
32. Maui Electric. Schedule "P" Large Power Service. Available online: https://www.mauielectric.com/documents/billing_and_payment/rates/hawaii_electric_light_rates/helco_rates_sch_p.pdf (accessed on 23 January 2020).
33. SDGE. Schedule AL-TOU. Available online: Sdge.com/sites/default/files/elec_elec-scheds_al-tou.pdf (accessed on 13 January 2020).
34. Burlington Electric. Available online: https://www.burlingtonelectric.com/rates-fees#large-general-service-lg (accessed on 13 January 2020).
35. Fu, R.; Feldman, D.J.; Margolis, R.M. *U.S. Solar Photovoltaic System Cost Benchmark: Q1 2018*; NREL/TP-6A20-72399; National Renewable National Laboratory (NREL): Golden, CO, USA, 2018.
36. Cole, W.J.; Frazier, A. *Cost Projections for Utility-Scale Battery Storage*; NREL/TP-6A20-73222; National Renewable National Laboratory (NREL): Golden, CO, USA, 2019.
37. EIA. US Energy Information Administration (EIA). Available online: https://www.eia.gov/outlooks/aeo/assumptions/pdf/commercial.pdf (accessed on 11 March 2020).
38. Ericson, S.J.; Olis, D.R. *A Comparison of Fuel Choice for Backup Generators*; National Renewable Energy Laboratory: Golden, CO, USA, 2019.

Article

Design of a Methodology to Evaluate the Impact of Demand-Side Management in the Planning of Isolated/Islanded Microgrids

Juan Carlos Oviedo Cepeda [1,*]**, German Osma-Pinto** [1]**, Robin Roche** [2]**, Cesar Duarte** [1]**, Javier Solano** [1]**, Daniel Hissel** [2]

[1]   Escuela de Ingenierías Eléctrica, Electrónica y de Telecomunicaciones, Universidad Industrial de Santander, Bucaramanga 680002, Colombia; gealosma@uis.edu.co (G.O.-P.); cedagua@uis.edu.co (C.D.); jesolano@uis.edu.co (J.S.)

[2]   FEMTO-ST, CNRS, University Bourgogne France Comte, UTBM, 90000 Belfort, France; robin.roche@utbm.fr (R.R.); daniel.hissel@univ-fcomte.fr (D.H.)

*   Correspondence: juan.oviedo@correo.uis.edu.co; Tel.: +57-316-246-6426

Received: 28 April 2020; Accepted: 28 June 2020; Published: 4 July 2020

Abstract: The integration of Demand-Side Management (DSM) in the planning of Isolated/Islanded Microgrids (IMGs) can potentially reduce total costs and customer payments or increase renewable energy utilization. Despite these benefits, there is a paucity in literature exploring how DSM affects the planning and operation of IMGs. The present work compares the effects of five different strategies of DSM in the planning of IMGs to fulfill the gaps found in the literature. The present work embodies a Disciplined Convex Stochastic Programming formulation that integrates the planning and operation of IMGs using three optimization levels. The first level finds the capacities of the energy sources of the IMG. The second and third levels use a rolling horizon for setting the day-ahead prices or the stimulus of the DSM and the day-ahead optimal dispatch strategy of the IMG, respectively. A case study shows that the Day-Ahead Dynamic Pricing DSM and the Incentive-Based Pricing DSM reduce the total costs and the Levelized Cost of Energy of the project more than the other DSMs. In contrast, the Time of Use DSM reduces the payments of the customers and increases the delivered energy more than the other DSMs.

Keywords: Isolated/Islanded Microgrids; planning; operation; Demand-Side Management

---

1. Introduction

Despite the efforts of governments around the world, access to electric energy in isolated regions remains a challenge [1,2]. Isolated/Islanded Microgrids (IMGs) could play a significant role in providing power to these areas where extending the utility grid is not economically feasible [3]. The implementation of Demand-Side Management (DSM) in the planning of Microgrids (MGs) reduces total costs, Levelized Cost of Energy (LCOE), and customer payments, or increases renewable energy utilization [4–9]. In this regard, it seems interesting to investigate if the application of DSMs in the planning of IMGs can bring similar benefits. Despite this, there is a paucity of literature exploring how DSMs can affect IMGs' planning and operation.

The implementation of DSM aims to affect the patterns of consumer consumption using direct or indirect strategies [10,11]. Direct strategies are composed of Direct Load Control and Interruptible/Curtailable Programs. In Direct Load Control strategies, there is a remote controller sending signals to customers' appliances, like air conditioners, heating systems, water heaters, or public lighting, on short notice. The signals can turn the appliances on/off, switch tariffs, or inform about current electricity prices. Interruptible/Curtailable Programs offer alternatives as biding programs, Emergency Demand Response (DR) programs, Capacity Market programs, and ancillary services, such as frequency support [12,13]. Indirect DSMs are composed of pricing programs, rebates/subsidies, and education programs. Pricing programs charge dynamic tariffs for energy, which can be power-based, energy-based, or a combination of both [14,15]. Energy-based tariffs incentivize energy conservation, and, therefore, are desired in IMG applications, where the energy generation is limited [16]. Instead of having a fixed flat rate, dynamic fares vary in time to reveal the actual costs of producing energy. These rates include the Time of Use (ToU) rate, Critical Peak Pricing (CPP), Extreme Day Pricing (EDP), Extreme Day Critical Peak Pricing (ED-CPP), Day-Ahead Dynamic Pricing (DADP), and Real-Time Pricing (RTP). Properly designed tariffs motivate the customers to shift their demand to off-peak periods, when the electricity price is lower and it is more convenient to produce electricity [17].

Some works in the literature explore how DSM affects the planning of MGs. Kahrobaee et al. propose a sizing approach to determine the capacity of a Wind Turbine and a Battery Energy Storage System (BESS) for a smart household considering price variations in the tariffs [18]. The authors designed a three-step process combining a rule-based controller, a Monte Carlo approach, and a Particle Swarm Optimization to perform the sizing of the components. However, the uncoordinated combination of multiple stages and the lack of an optimization formulation for energy management can lead to sub-optimal results. Erdinc et al. [19] aimed to address these drawbacks by providing a Mixed Integer Linear Programming (MILP) formulation to design an optimal energy management strategy. The work considers the seasonal and weekly variations in the load profiles in the presence of a Real-Time Pricing tariff scheme. However, it does not consider how to design the DSM itself and how different DSMs will impact the sizing of the energy sources. Kerdphol et al. propose a sizing approach for BESS using Particle Swarm Optimization to improve the frequency stability of an MG [20]. The work integrates a dynamic DSM considering load shedding of non-critical loads to rapidly restore the system frequency and reduce the BESS capacity. A rule-based controller used for the load shedding and a Particle Swarm Optimization formulation used for the sizing of the BESS prove to be adequate to regulate the frequency of the MG. However, the rule-based controller and the lack of forecast models to anticipate the critical events can lead to sub-optimal results.

Nojavan et al. [21] propose a bi-objective Mixed-Integer Non-Linear Programming (MINLP) formulation to optimally site and size a BESS in an MG considering DSM. The authors designed two optimization objectives to reduce total costs and Loss of Load Expectation. The work uses an ϵ-constraint method to draw the Pareto optimal curve and a fuzzy satisfying technique to find the best solution. Nevertheless, the authors assume that 20% of the load reacts to a Time of Use (ToU) tariff, ignoring the effects of the demand's self-elasticity. Majidi et al. use a Monte Carlo Scenario reduction technique to determine the size of a BESS in an MG [22]. The work considers the effects of uncertainties in the forecasted renewable generated power and forecasted consumption. However, similarly to [21], the authors do not consider how the customers react to the DSM; they assume that 20% of the load will react to a ToU tariff. Amir et al. [23] propose a combined algorithm to find the size and energy management strategy of a Multi-Carrier Microgrid. The work proposes a mathematical model with high sophistication that uses an MINLP formulation to obtain the optimum dispatch strategy and Genetic Algorithms to obtain the capacities of the energy sources. The work measures the changes in the patterns of consumption of the customers considering varying prices for the different forms of energy. The planning of the Multi-Carrier Microgrid considers demand and price growth over a five-year optimization horizon. Nevertheless, this work does not design the DSM. It only considers the effects of the prices of the energy providers on the Multi-Carrier Microgrid.

Planning of IMGs refers to the set of decisions that the planner must make to design an IMG project. Such decisions include: Setting the energy mix, computing the sizing of the energy sources, and defining the energy dispatch strategy, the economic incentives, and the energy tariffs, amongst others [24–26]. This set of decisions has significant consequences on the performance of IMG projects, where high penetration of renewable energy sources can reduce system inertia, thus challenging system frequency regulation, control schemes, and transient stability [13]. DSMs can partially solve some of the inherent challenges of planning IMGs.

Chauhan et al. propose to compute the sizing of the energy sources of an IMG considering a DSM that reschedules shiftable loads depending on if it is the winter or summer season [27]. The work uses an Integer Linear Programming (ILP) formulation to find the optimal rescheduling of shiftable loads and a Discrete Harmony Search algorithm to compute the sizing. A considerable drawback of the work is that the DSM only focuses on reducing the peak demand while ignoring maximizing exploitation of renewable energy. Amrollahi et al. combine an MILP formulation and the capabilities of HOMER software to compute the sizing of an IMG composed only of renewable energy sources [28]. Due to the lack of dispatchable energy sources, the authors propose the use of a DSM to reschedule shiftable loads. Rescheduling helps to balance mismatch between electric energy generation and consumption. Mehra et al. propose a work to measure the economic value of applying DSM in the sizing of a nanogrid [29,30]. The work considers the dis-aggregation of electrical demand in critical and non-critical appliances. In addition, the work takes advantage of low-cost computation intelligent devices, such as the "utility-in-a-box" solution, to implement active DSM [31]. The authors use an exhaustive search algorithm to determine the capacities of the Photo-Voltaic (PV) system and the BESS. Nevertheless, the work considers the effects of only one kind of DSM over a small-sized grid.

Prathapaneni et al. propose a multi-objective stochastic sizing algorithm that aims to minimize lifetime costs and degradation of the energy sources [32]. The work considers the effects of a DSM that uses shiftable loads, like electric vehicles or pumped hydro storage in an IMG. The work uses an Accelerated Particle Swarm Optimization (APSO) to compute the sizing of energy sources. Despite considering lifetime costs of the IMG and degradation of energy sources, the work considers a basic DSM over reduced amounts of loads that are not always present in IMG applications. Luo et al. propose a sizing methodology for an IMG using a bi-level optimization algorithm [33]. The first level computes the energy sources' capacities, considering the effects of different combinations of public subsidies for the installation of energy sources. The second level performs the dispatch strategy for the energy sources of the IMG using an MINLP formulation. In the second level of optimization, the authors implement a rescheduling mechanism of shiftable loads. A study case shows that DSM reduces installed capacities of the energy sources for the IMG.

Kiptoo et al., similarly to [28], aimed to implement a DSM to balance generation and electricity demand in an IMG only composed of renewable energy sources [34]. The DSMs consider rescheduling shiftable loads. However, the authors aim to improve the work of [28] by adding an electrical demand forecasting module using a Random Forest (RF) regression forecasting approach. The work shows that the proposed methodology reduces the total costs of the IMG project by 12.41%. Rehman et al. used HOMER software to find capacities of energy sources in an IMG [35]. The work considers a DSM capable of rescheduling shiftable loads and uses Simulink to evaluate the operation of the IMG. The use of Simulink allows the authors to design and test a model predictive control. The model predictive control controls the power during grid-connected operation and regulates load voltage in the islanding operation of the MG. Table 1 summarizes the works found in the literature that deal with the integration of DSM in the planning of IMGs, and that highlight knowledge gaps and the characteristics of the present work. It is vital to notice that Table 1 presents only the articles that consider IMGs because they are strictly related to the present work.

Table 1. Summary of the literature review.

Features	2017	2018	2019	2020	Literature Gaps	Proposed Work
Integration of sizing and Demand-Side Management (DSM)	[27,28]	[29,30]	[32,33]	[34,35]		✓
Stochastic optimization formulation			[32]			✓
Study of subsidies impacts over economic feasibility			[33]			✓
Forecasting impacts in the operation				[34]		✓
Validation of operation after sizing				[35]		✓
Tariff setting for Isolated/Islanded Microgrids (IMGs) for economic feasibility					✓	✓
Utilization of tariffs as DSMs in IMGs					✓	✓
Comparison of different DSMs using one test bench					✓	✓
Influence of public subsidies on tariff setting for IMGs					✓	✓

Despite that some of the works found in the literature evaluate the effects of DSM in the planning phase of MGs and IMGs, none of them compare the effects of different DSMs using the same test-bench. The works found by authors do not focus on design and impact evaluation of DSM over total costs and operational aspects of IMG projects. Moreover, few of the works consider the financial aspects of the cooperation between private and public capital to fund IMG projects. Additionally, none of them allow defining tariffs that guarantee the sustainability of the IMG project over time. In this regard, the present article aims to fulfill gaps found in the literature review by providing a methodology capable of:

- Obtaining the optimal sizing and the optimal energy dispatch strategy of an IMG project using a Disciplined Convex Stochastic Programming formulation.
- Obtaining the optimal energy tariffs and stimulus for the DSM to guarantee the financial viability of an IMG project.
- Evaluating the impacts of different strategies of DSMs over sizing, energy management, and costs of an IMG project in a case study.
- Implementing and evaluating different DSMs in the planning of IMGs using the same test-bench.

The formulation uses flat, ToU, CPP, DADP, and Incentive-Based Pricing (IBP) tariffs as DSM strategies. It also proposes a Direct Load Curtailment (DLC) strategy that curtails customers' electrical demand if required.

The formulation assumes that the DSMs modify the patterns of consumption of the customers, which will lead to a change in the capacities of energy sources [36–39]. The results of the application of the methodology provide the optimal size of the energy sources, the optimal energy dispatch, the optimal tariffs, the economic incentives, and the load curtailment. The rest of the article proceeds as follows: Section 2 presents the definition of the problem and the proposed solution. Section 3 presents a case study as an example of the application of the methodology, and Section 4 includes its results and analysis. Finally, Section 5 presents the conclusions of the work and future directions.

2. Definition of the Problem and Proposed Solution

The present work aims to illustrate for planners and policymakers the benefits of applying DSM for IMG planning. For that purpose, the methodology integrates sizing and IMG operation using three different optimization levels, as shown by Figure 1. The first level obtains the sizes of energy sources using a Monte Carlo analysis. The second level uses a day-ahead rolling horizon over the same optimization horizon of the first level to define incentives, tariffs, and load curtailment of each of the strategies of DSM. Finally, the third level simulates the microgrid operation by iterating in the same rolling horizon that the second level uses. The third level performs the simulation to compute the optimal dispatch strategy after weather and demand profiles are known.

The rolling horizon computes one day in advance at each time and rolls over one year. Each day, the second level computes the proper day-ahead DSM stimulus that the third level uses to compute the day-ahead response of the customers. The second level computes these stimuli using day-ahead forecasts of electrical demand. The third level applies the stimulus found in the second level to compute the customers' response. While the first and second level assess the uncertainties in day-ahead forecasts of the demand and in weather variables, the third level assumes perfect knowledge of these variables. This assumption allows the methodology to compute the impacts of errors in forecasts over the DSM performance. The formulation computes the total costs of operation on the third level. Sections 2.1–2.3 present a detailed explanation of each of the levels. Appendix A, present in Table A1 a description of all the variables used in the following sections with their respective names and units.

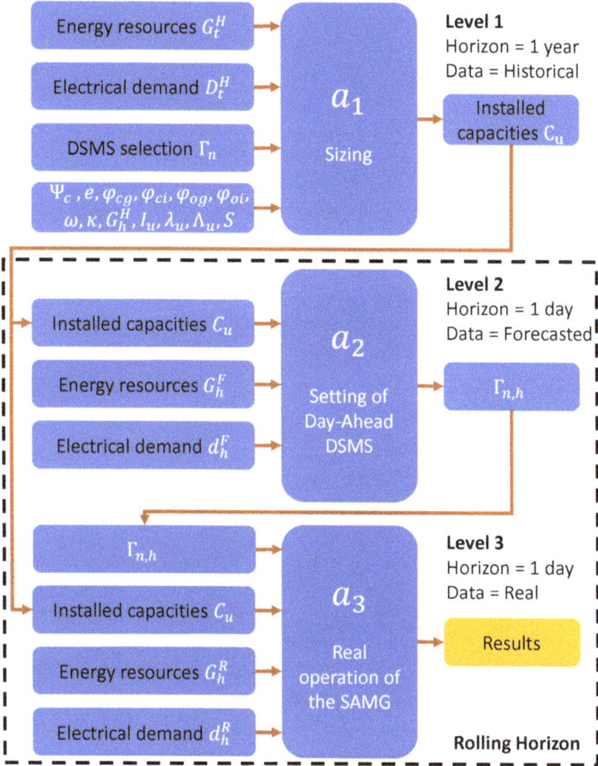

Figure 1. Graphical description of the proposed methodology.

2.1. First Level: Sizing

The formulation of the first optimization level a_1 can be stated as:

$$
\begin{aligned}
J(x^*) = \underset{x}{\text{minimize}} \quad & a_1(x,\xi) \\
\text{subject to} \quad & b_i(x,\xi) = 0 \; i = 1,\ldots,B, \\
& c_i(x,\xi) \geq 0 \; i = 1,\ldots,C
\end{aligned}
\tag{1}
$$

where x represents the decision variables, ξ represents the uncertainties of the electrical demand, $b_i, i = 1,\ldots,B$ are convex functions in x for each value of the random variable ξ, and $c_i, i = 1,\ldots,C$ are deterministic affine functions. Since a_1 , $b_i, i = 1,\ldots,B$ and $c_i, i = 1,\ldots,C$ are convex on x, the definition of Formulation (1) is a convex optimization problem [40] ([41], Chapter 7).

IMG projects can receive funding from public or private capital. To compute the effects of the funding sources over the total costs, profits, and customer payments, the formulation of a_1 introduces factors φ_{cg}, φ_{ci}, φ_{og}, and φ_{oi}, where $\varphi_{ci} + \varphi_{cg} = 1$ and $\varphi_{oi} + \varphi_{og} = 1$. The formulation of a_1 is designed to minimize the Capital Expenditures (CAPEX) and the Operational Expenditures (OPEX) of the IMG. Equations (2)–(10) describe the formulation of a_1.

$$
X_1 = \underset{C_u, E_{u,t}}{\arg\min} \; \varphi_{cg} \sum_{u=1}^{U} C_u I_u + \varphi_{og} \sum_{t=1}^{T} \sum_{u=1}^{U} (\lambda_{u,t} + \Lambda_{u,t}) E_{u,t}
\tag{2}
$$

where the CAPEX (ζ) and OPEX (ϑ) refer to:

$$
\zeta = \sum_{u=1}^{U} C_u I_u
\tag{3}
$$

$$
\vartheta = \sum_{t=1}^{T} \sum_{u=1}^{U} (\lambda_{u,t} + \Lambda_{u,t}) E_{u,t}
\tag{4}
$$

and C_u, I_u, $\lambda_{u,t}$, $\Lambda_{u,t}$, and $E_{u,t}$ represent the installed capacity, unitary investment cost, unitary dispatch costs, unitary maintenance costs, and dispatched energy by the u energy source, respectively.

It is worth noting that the formulation of Equation (2) replaces *minimize* with *argmin*, and assigns the results to the variable X_1. This replacement occurs because the second and third methodology levels require the values of the decision variables , but not the value of the achieved minimum. The rest of the optimization formulations in the document maintain the replacement.

The proposed formulation considers the energy prices as the only revenue stream for the investors. Equation (5) introduces a constraint to guarantee that the private investors recover their investments and the expected Internal Rate of Return R.

$$
-(\varphi_{ci}\zeta + \varphi_{oi}\vartheta)(1+R) + \sum_{t=1}^{T} \pi_{n,t} D_{f,t} \geq 0
\tag{5}
$$

where φ_{ci} and φ_{oi} represent the percentage of payments of the private investor for the CAPEX and OPEX costs, and $\pi_{n,t}$ represents the prices of the n tariff. Equation (6) considers the elasticity (e_t) of the customers at a time t, the initial price of the energy (π_{flat}), and the initial demand ($D_{o,t}$) to compute the final demand ($D_{f,t}$). Equation (7) introduces an energy conservation factor Ψ_c to define how the total energy consumption over the optimization horizon changes after the introduction of DSM. Values of $\Psi_c \leq 1$ decrease the total energy consumption, while values of $\Psi_c \geq 1$ increase the total energy consumption over the optimization horizon. A value of $\Psi_c = 1$ indicates that the total energy consumption over the optimization horizon remains constant after the introduction of DSMs.

$$e_t = \frac{\pi_{flat}(D_{f,t} - D_{o,t})}{D_{o,t}(\pi_{n,t} - \pi_{flat})} \tag{6}$$

$$\sum_{t=1}^{T} D_{f,t} - \Psi_c \sum_{t=1}^{T} D_{o,t} = 0 \tag{7}$$

The formulation of a_1 also includes the energy balance Equation (8), a constraint that limits energy excess (EE_t), and a constraint that limits the lack of energy (LE_t), (9) and (10), respectively. Equations (9) and (10) introduce the parameter z to control the desired level of reliability in the IMG.

$$\sum_{t=1}^{T} \sum_{u=1}^{U} E_{u,t} - EE_t + LE_t - D_{f,t} = 0 \tag{8}$$

$$\sum_{t=1}^{T} EE_t \leq (1-z) \sum_{t=1}^{T} D_{f,t} \tag{9}$$

$$\sum_{t=1}^{T} LE_t \leq (1-z) \sum_{t=1}^{T} D_{f,t} \tag{10}$$

Additionally, the a_1 formulation includes Equations (14) till (26) in order to evaluate the impact of DSM strategies on the sizing of the IMG (changing the horizon from 24 to 8760 h, respectively).

2.2. Second Level: Setting of Day-Ahead DSM Values

The formulation of the second optimization level a_2 solves the following problem:

$$X_2 = \underset{E_{u,h}^{F}, EE_{h}^{F}, LE_{h}^{F}}{\arg\min} \quad \varphi_{og} \sum_{h=1}^{24} \sum_{u=1}^{U} (\lambda_{u,h} + \Lambda_{u,h}) E_{u,h}^{F} + \omega EE_{h}^{F} + \omega LE_{h}^{F} \tag{11}$$

$$s.t. \sum_{h=1}^{24} \sum_{u=1}^{U} E_{u,h}^{F} - EE_{h}^{F} + LE_{h}^{F} - d_{f,h}^{F} = 0, \tag{12}$$

where EE_{h}^{F} and LE_{h}^{F} are the 24 day-ahead forecasted energy excess (a non-positive variable) and the forecasted lack of energy (a non-negative unrestricted variable), respectively, and ω is a penalization factor. $E_{u,h}^{F}$ and $d_{f,h}^{F}$ represent the 24 day-ahead forecasted dispatch of the u energy sources and the 24 day-ahead forecasted electrical demand, respectively.

The formulation of the second optimization level a_2 uses the capacities C_u, the day-ahead forecasts of energy resources G_{h}^{F}, and forecasts of electric demand $d_{o,h}^{F}$ as inputs in order to compute the day-ahead stimulus for the five $\Gamma_{n,h}$ DSM strategies. Four of the DSM strategies use π_n in Equation (5) as an indirect stimulus to modify the customer consumption patterns. Those four DSM strategies are: Time of Use pricing (ToU), Critical Peak Pricing (CPP), Day-Ahead Dynamic Pricing (DADP), and Incentive-Based Pricing (IBP). The last DSM uses a Direct Load Curtailment strategy that sheds a percentage of load when required. The baseline case for comparisons uses a flat tariff and no DSM. The description of the baseline case and each of the DSMs proceeds in the following subsections [42].

2.2.1. Flat Tariff (Baseline Case)

In general terms, the unitary value of a flat tariff is the sum of all the costs of producing the energy divided by the total amount of energy produced [43]. Equation (13) describes the yearly payments using a regular flat tariff.

$$\Theta_{flat} = \frac{\zeta + \sum_{t=1}^{T} \vartheta_t}{\sum_{t=1}^{T} D_{f,t}} (1+R) \sum_{t=1}^{T} D_{f,t} \tag{13}$$

However, this traditional approach does not set an optimal tariff to recover investments while minimizing energy costs. Here, we propose the introduction of a decision variable π_{flat} into the formulation to find the optimum price for the tariff.

$$\Theta_{flat} = \pi_{flat} \sum_{t=1}^{T} D_{f,t} \tag{14}$$

2.2.2. Time of Use Tariff

ToU tariffs vary daily or seasonally on a fixed schedule, using two or more constant prices [44]. One of the main benefits of this type of tariff is its stability over long periods, which gives the customer a better ability to adapt to it [45,46]. To create a ToU tariff, the planner must define the number of Y blocks and the starting and ending hours of each y block [45]. The optimization problem considers the prices π_y of the Y number of blocks as decision variables to be computed. Equation (15) presents the yearly payments using Y different block hours of prices.

$$\Theta_{tou} = \sum_{t=1}^{T} \sum_{y=1}^{Y} \pi_y D_{f,t} \tag{15}$$

The methodology computes the ToU and flat tariffs in the first optimization level a_1 and the demand response of the customers in the third level a_3. The second level is not used for the flat tariff and the ToU tariff because they do not have daily variations. The algorithm computes the flat and ToU tariffs following the same process used to find the capacities of the energy sources C_u, using adapted versions of Equations (30) and (31).

2.2.3. Critical Peak Pricing

The CPP tariff can be 3 to 5 times higher than the usual tariff, but is allowed only a few days per year [46]. In Equation (16), π_{base} is a scalar variable that is chosen to be equal to the flat tariff π_{flat}. π_{peak} is a decision variable of dimension 24 and is computed one day in advance. Equation (16) defines the day-ahead forecasted payments using a CPP tariff, and Equation (17) defines the day-ahead hourly critical peak price.

$$\Theta_{cpp}^{F} = \pi_{base} \sum_{h=1}^{\tau_{base}} d_{f,h}^{F,base} + \sum_{h=1}^{\tau_{peak}} \pi_{peak,h}^{F} d_{f,h}^{F,peak} \tag{16}$$

$$\pi_{cpp,h}^{F} = \pi_{base} + \pi_{peak,h}^{F} \tag{17}$$

A critical forecasted event, such as high demand or low generation capacity, triggers the critical peak price in a CPP tariff. In this regard, the CPP tariff must include a predictor of the critical event and a decision mechanism to set the value of the critical price. The first optimization level formulation a_1 uses historical data, which implies that the formulation has full knowledge over the optimization horizon (T = 8760 h). The perfect knowledge allows the formulation to state constraint (18), which limits the apparition of the critical price only to a few hours in a year. Equation (18) uses variable φ_{peak} to control the number of hours with critical price allowed and δ_{peak} to define how many times the base price π_{base} is scaled up. The planner defines $\varphi_{peak} \delta_{peak}$, and π_{base}, π_{peak}, τ_{base}, and τ_{peak} are decision variables that the optimization formulation computes.

$$\sum_{t=1}^{T} \pi_{peak,t} \leq \varphi_{peak} T \delta_{peak} \pi_{base} \tag{18}$$

However, in order to simulate the operation of the IMG, the rolling horizon will only know the forecasts one day in advance. The formulation must define a mechanism to determine the conditions that allow the critical peak price to take place. Thus, it defines the critical event as low daily forecasted primary energy resources (lower than a predefined threshold ϱ). The decision mechanism sets the day-ahead value of the critical price using the variable $\pi^F_{peak,h}$. Equation (19) describes the mechanism to set the CPPs in the operational phase of the IMG.

$$\pi^F_{cpp,h} = \begin{cases} \pi_{base} + \pi^F_{peak,h}, & \text{if } \sum_{h=1}^{24} G^F_h \leq \varrho \\ \pi_{base}, & \text{otherwise} \end{cases} \tag{19}$$

2.2.4. Day-Ahead Dynamic Pricing

DADP refers to a tariff that is announced one day in advance to customers and has hourly variations. This scheme offers less uncertainty to customers than "hour- ahead pricing" or "real-time pricing," thus allowing them to plan their activities [47,48]. Equation (20) introduces the day-ahead payments under a DADP tariff, using π^F_h as a decision variable vector of dimension 24.

$$\Theta^F_{dadp} = \sum_{h=1}^{24} \pi^F_h d^F_{f,h} \tag{20}$$

2.2.5. Incentive-Based Pricing

The IBP tariff provides discounts in the tariff to the customers to increase the electric energy consumption or an extra fare to penalize it. The planner can decide the IBP base price to be equal to the flat tariff π_{flat} to guarantee a constant value each day. Variable $\pi^F_{inc,h}$ computes the day-ahead hourly incentives and can take positive or negative values. Equation (21) defines the day-ahead payments using the IBP tariff.

$$\Theta^F_{inc} = \sum_{h=1}^{24} d^F_{f,h} (\pi_{base} + \pi^F_{inc,h}) \tag{21}$$

All of the N tariffs must have restrictions to avoid null or excessive pricing. Governments, policymakers, or IMG owners can guarantee fair tariffs to the customers with constraint (22).

$$\pi^{min}_n \leq \pi_n \leq \pi^{max}_n \tag{22}$$

2.2.6. Direct Load Curtailment Strategy

The DLC strategy curtails a portion ϵ^F_h out of forecasted demand if required. The planner of the IMG decides the percentage of curtailed demand κ. The final demand and day-ahead payments are defined as follows:

$$d^F_{f,h} = d^F_{o,h} - \epsilon^F_h \tag{23}$$

$$\Theta^F_{dlc} = \sum_{h=1}^{24} d^F_{f,h} \pi_{flat} \tag{24}$$

The general restrictions for the DLC strategy are defined as follows:

$$\epsilon^F_h \leq \kappa d^F_{f,h} \tag{25}$$

$$\sum_{h=1}^{24} \epsilon^F_h \leq \kappa \sum_{h=1}^{24} d^F_{f,h} \tag{26}$$

2.3. Third Level: Real Operation of the IMG

The formulation of the third optimization level a_3 solves the following problem:

$$X_3 = \underset{E^R_{u,h}, EE^R_h, LE^R_h}{\arg\min} \quad \varphi_{og} \sum_{h=1}^{24} \sum_{u=1}^{U} (\lambda_{u,h} + \Lambda_{u,h}) E^R_{u,h} + \omega EE^R_h + \omega LE^R_h \tag{27}$$

$$s.t. \sum_{h=1}^{24} \sum_{u=1}^{U} E^R_{u,h} - EE^R_h + LE^R_h - d^R_{f,h} = 0, \tag{28}$$

where the formulation computes the real dispatch of energy sources using capacities C_u, real energy resources G^R_h, real final electric demand $d^R_{f,h}$, and the energy prices π_n of each DSM in order to compute the real dispatch of the U energy sources of the IMG.

In addition to Equations (27) and (28), the formulation of a_3 must include physical restrictions for all the U energy sources used to design the IMG (maximum battery charge and discharge rates, maximum power generator output, amongst others). It is essential to highlight that EE^R_h and LE^R_h in the third level refer to energy that generators produce in excess and energy that the generators can not provide, respectively. The first level constrains the allowed quantity of excess (Equation (9)) and lack (10) of energy. The second level uses a penalization factor for these variables (Equation (11)). However, the third level is just an accumulator, a counter of these quantities.

3. Case Study

The case study aims to illustrate the capabilities and performance of the proposed methodology and considers the design of an IMG composed of a PV, a BESS, and a Diesel Generator (DG) System, as Figure 2 shows. The case study assumes that the microgrid can have two different types of load. The case study uses the load type one when the planner chooses a DSM based on price. The load type one has Smart Meters. The case study uses the second type of load when the planner decides to use the DSM based on DLC. The second type of load has a device as "GridShare" to perform the curtailment of the electrical demand [31]. The case study considers six IMG designs: Baseline case (flat tariff and no DSM) and one design for each of the proposed DSM (ToU, CPP, DADP, IBP, DLC). The results of the designs using DSM are compared with the baseline case design. All of the optimization formulation was written in Python 3.7 using the CVXPY 1.0 package [49,50]. The selected solver is MOSEK, due to its flexibility, speed, and accuracy [51,52].

The case study includes a Monte Carlo Sampling (MCS) approach to deal with the uncertainties of the stochastic formulation. The MCS approach builds different scenarios by sampling the Probability Distribution Functions (PDFs) of electrical demand. In order to build scenarios, a pre-processing step fits the historic electrical demand into monthly/hourly PDFs. For simplicity and for the sake of reduction in computational burden, the case study assumes the demand follows a Gaussian process without a covariance matrix. Afterwards, a random sampling process of the monthly/hourly PDFs builds the demand for each sample s of the MCS approach. Equation (29) describes the sampling process. Figure 3 shows monthly/hourly fitted distributions using a continuous line to represent the mean and a shaded area to represent the standard deviation.

$$D_t | m, h \sim f(\psi_{m,h}) \tag{29}$$

In Equations (2), (11) and (27), X_1, X_2, and X_3 represent the S solutions of minimizing a_1, a_2, and a_3, respectively. The C_u capacities of the energy sources selected for the IMG in the first level must supply 95% of the S electrical demands with z level of reliability (as defined by Equation (10)). A post-processing step fits the C_u results to a PDF ϕ_u, and obtains the Cumulative Distribution Function (CDF) Φ_u. The evaluation of the inverse of the CDF Φ_u at 0.95 provides the values of energy source capacities C_u. These values will supply electrical demand with the desired reliability level 95% of the time (95% of all the scenarios).

$$\Phi_u = \int_{-\infty}^{\infty} \phi_u dC_u \tag{30}$$

$$C_u = \Phi_u^{-1}(0.95) \tag{31}$$

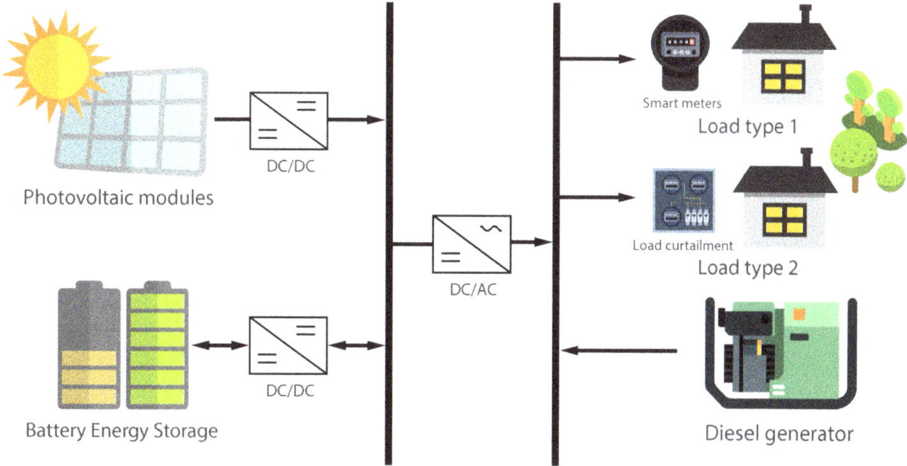

Figure 2. Architecture of the islanded/isolated microgrid of the case study.

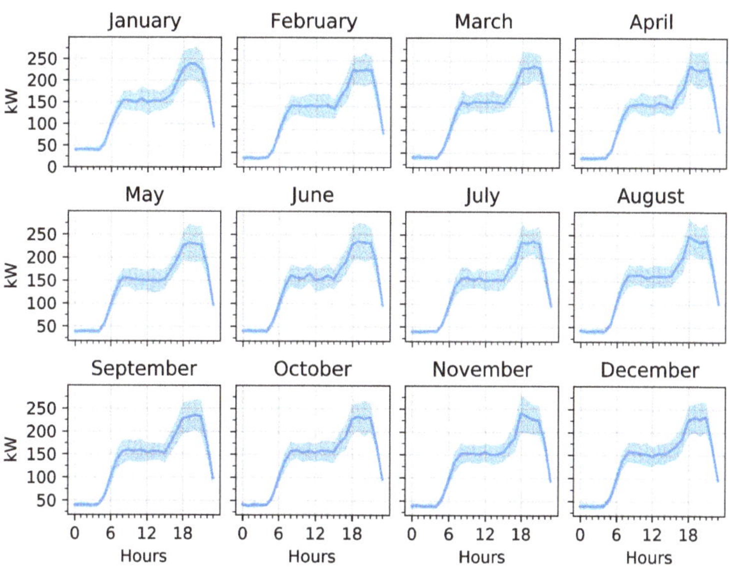

Figure 3. Fitting of the electrical demand.

Geographic and Weather Conditions of the Case Study

The case study is located at longitude $77'16'8''$ West and latitude $5'41'36''$ North (Nuquí, Colombia). The study case uses the Meteonorm database of the PvSyst software to obtain the Global Horizontal Radiation (GHI) and temperature conditions of the geographical region. Additionally, the study case uses Homer Pro software to obtain a standard community electrical demand. The standard community electrical demand that Homer Pro provides has hourly steps over a one-year horizon. Figure 4 shows the historic yearly standard profile of the electrical demand that Homer Pro provides. Figure 5 shows the yearly GHI. Figure 6 shows the yearly temperature.

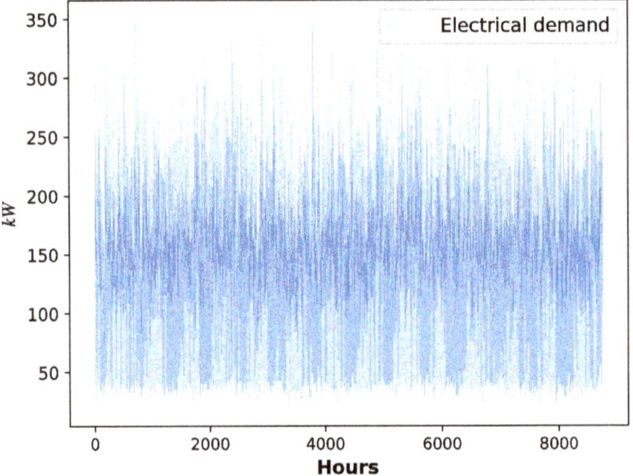

Figure 4. Yearly electrical demand.

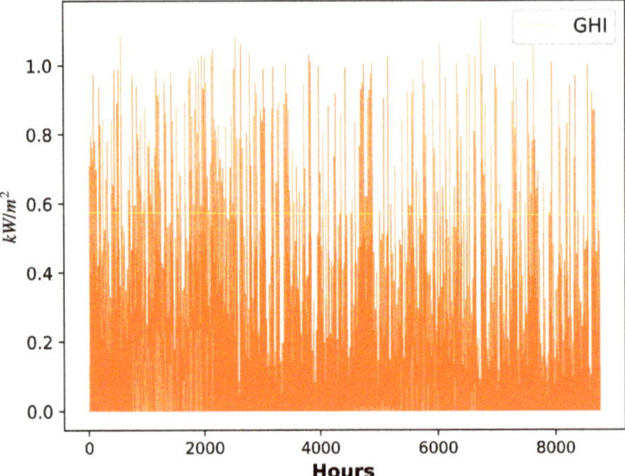

Figure 5. Yearly Global Horizontal Radiation.

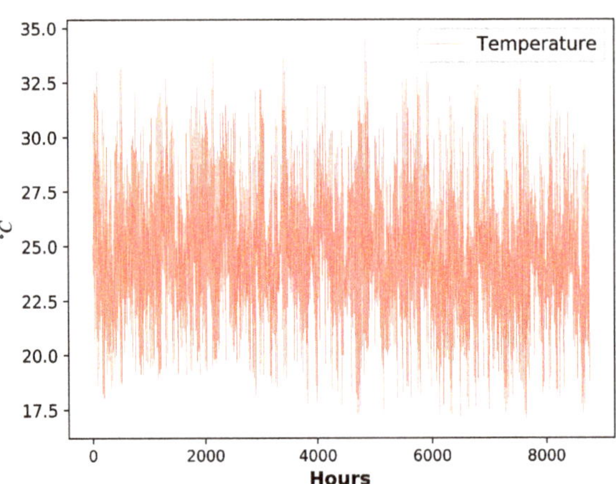

Figure 6. Yearly temperature.

The Monte Carlo Sampling analysis shown in Equation (29) builds the scenarios for the stochastic analysis using the standard community electrical demand obtained from Homer Pro (shown in Figure 4) and a scale factor of 20. The cost of diesel used for the optimization is 0.75 USD/liter. The case study takes the Diesel Generator model from [53], the PV system model from [54–56], and the BESS model from [57]. Table 2 summarizes the unitary installation and maintenance costs of the equipment obtained from regional providers. The values assigned to $\pi_{n,min}$ and $\pi_{n,max}$ in constraint (22) are 0 USD/kWh, and two times the price of the current flat tariff of urban areas in Colombia, 0.34 USD/kWh, respectively [58].

Table 2. Unitary system costs for simulations.

System	Initial Investment	Maintenance	Operation
PV	1300 USD/kW	60 USD/kW	0 USD
BESS	420 USD/kWh	23 USD/kWh	0 USD
DG	550 USD/kW	30 USD/kWh	$f(E^2_{DG,t}, \psi_L)$

Additionally, the methodology takes as inputs the values of Ψ_c, e, φ_{cg}, φ_{ci}, φ_{og}, φ_{oi}, ω, κ, G^H_h, I_u, λ_u, Λ_u, and S. Planners or policymakers can decide these values or perform sensitivity analyses over each of them. Table 3 shows the values used for simulations in this work. The following section uses the MCS approach and the inputs of Table 3 to compute the results and for the case study.

Table 3. Values of the input parameters for the simulations.

Input	Value	Input	Value
Ψ_c	1	κ	10%
e	0.3	G^H_h	Figures 5 and 6
φ_{cg}	0.9	I_u	See Table 2
φ_{ci}	0.1	λ_u	See Table 2
φ_{og}	0.9	Λ_u	See Table 2
φ_{oi}	0.1	S	100
ω	0.4		

4. Results and Analysis

The case study aims to evaluate the effects of five different DSMs over the optimization results of the proposed formulation. The five considered DSMs are ToU, CPP, DADP, IBP, and DLC. The study case evaluates different aspects of the effects of the DSMs. Section 4.1 shows the average of each of the tariffs and the curtailment of the DLC strategy. Section 4.2 shows the effects of the DSMs over the sizing of the energy sources of the IMG. Section 4.3 aims to analyze the impacts of the DSMs over the economic aspects of the microgrid. This section analyzes the impacts of DSMs over total costs, profits of private investors, customer payments, and LCOE. Additionally, the section considers the delivered energy and fuel consumption. Section 4.4 presents the effects of the forecast errors over the operation of the IMG. Section 4.5 presents percentage variations in crucial indicators as total cost of the project and LCOE between the first and the third optimization levels. Finally, Section 4.6 shows a comparison of the performance of all the DSMs.

4.1. Demand Side Management Analysis

Each of the Γ_n DSM strategies uses a different stimulus to modify customer consumption patterns. Γ_{ToU}, Γ_{CPP}, Γ_{DADP}, and Γ_{IBP} use tariffs as an indirect stimulus to modify those patterns. Figure 7 shows the average daily stimulus and the Standard Deviation (STD) of the DSMs. The lines represent daily averages of the DSM strategies, and shaded area represents STDs.

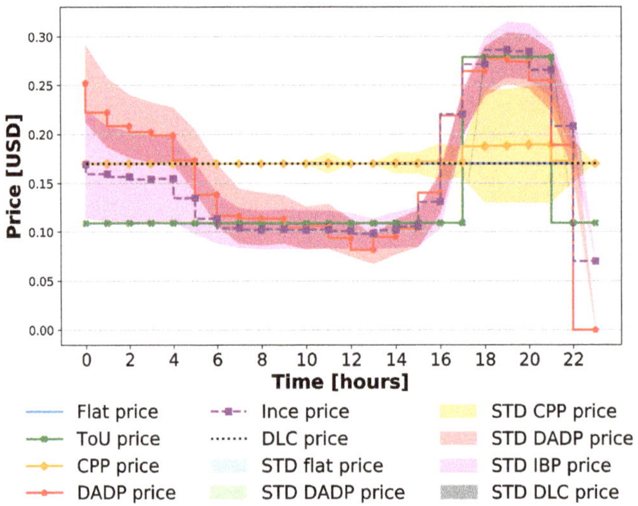

Figure 7. Daily average price of the selected tariffs.

Figure 7 presents energy prices. It is interesting to notice that IBP and DADP tariffs reduce the energy price in the middle of the day. The reduction occurs due to the presence of photovoltaic generation in the IMG. IBP and DADP DSMs incentivize customers to increase energy consumption when it is cheaper to generate electric energy.

The Γ_{DLC} DSM curtails a percentage of the demand. Figure 8 shows the daily average of the curtailed values in a continuous line and the STD of the curtailed energy in a shaded area.

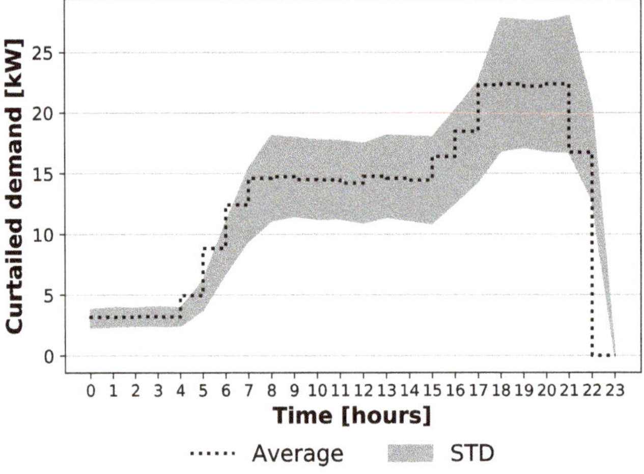

Figure 8. Daily average load curtailment for the Γ_{DLC} DSM.

The stimulus introduced by DSM strategies modifies customers' consumption patterns. Using Equation (6) and the stimulus computed using Equations (15)–(26), it is possible to compute the demand response. Figure 9 shows the demands after the application of DSM. The lines represent daily averages of the electrical demand, and shaded area represents the STDs.

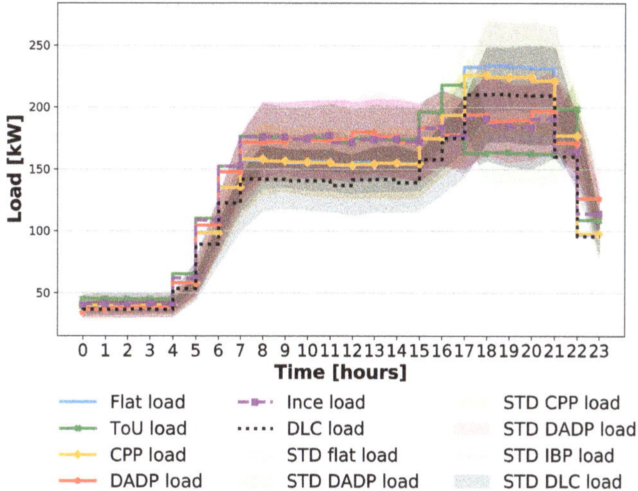

Figure 9. Daily average load for each of the DSMs and the base case.

It is interesting to note in Figure 7 that the IBP rate tends to be similar to the DADP rate. Therefore, it produces similar effects over electrical demand (see Figure 9). The lack of hourly restrictions on the appearance of the incentive of the IBP tariff causes this to occur. However, the design of hourly restrictions will rely on the experience of the IMG planner, which may ultimately lead to sub-optimal results.

4.2. Sizing Analysis

The variations in the customers' consumption patterns modify the IMG sizing. Figure 10 presents the variations in the sizing of the Diesel Generator, the photovoltaic system, and the BESS for the five DSMs.

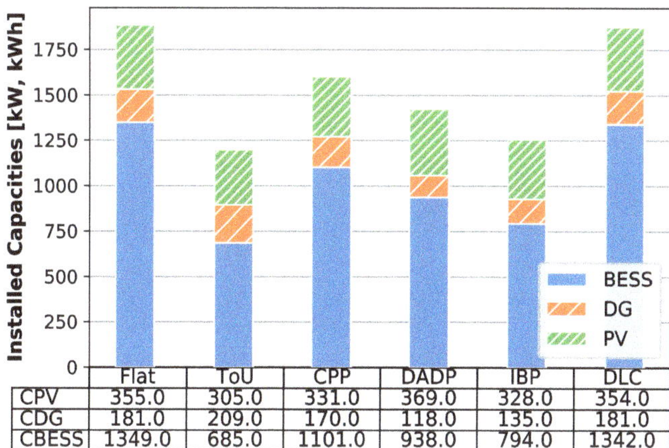

	Flat	ToU	CPP	DADP	IBP	DLC
CPV	355.0	305.0	331.0	369.0	328.0	354.0
CDG	181.0	209.0	170.0	118.0	135.0	181.0
CBESS	1349.0	685.0	1101.0	938.0	794.0	1342.0

Figure 10. Comparison of the sizing of the energy sources for the DSM against the base case. Diesel Generator and photovoltaic capacities are in kW, and the Battery Energy Storage System (BESS) capacity is in kWh.

On the one side, Figure 10 shows that ToU- and IBP-based DSMs require less installed capacity than the other alternatives. On the other side, Figure 10 shows that DLC and CPP DSMs do not considerably reduce the energy sources' installed capacities. However, reductions in installed capacities do not necessarily mean that one DSM is better than others. The following sections contribute with different analyses to determine which of the DSMs can be more suitable for IMG applications.

4.3. Economic Analysis

The DSM introduction in the IMG planning modifies total costs, investors' profits, customers' payments, total delivered energy, and LCOE, among others. Equations (32)–(36) present how to compute these values, and Figure 11 shows the results for the five DSM strategies and the base case.

$$\text{Total costs} = \zeta + \vartheta \tag{32}$$

$$\text{Profits} = \sum_{t=1}^{T} \pi_{n,t} D_{f,t} - (\varphi_{ci}\zeta + \varphi_{oi}\vartheta) \tag{33}$$

$$\text{Payments} = \sum_{t=1}^{T} \pi_{n,t} D_{f,t} \tag{34}$$

$$\text{Energy} = \sum_{t=1}^{T} D_{f,t} - |EE_{f,t}| - |LE_{f,t}| \tag{35}$$

$$\text{LCOE} = \frac{\text{Energy}}{\text{Total costs}} \tag{36}$$

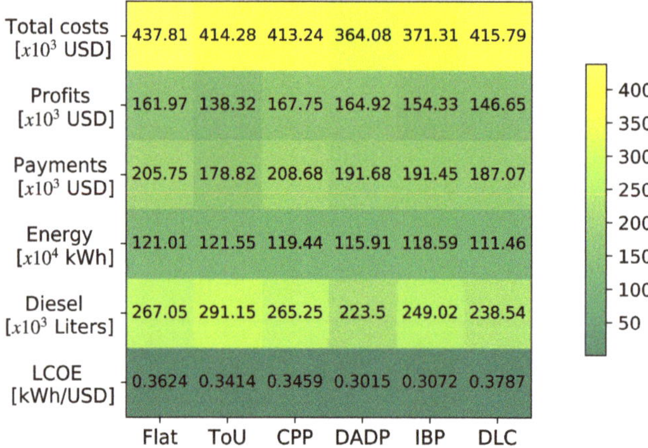

Figure 11. Comparison of the costs and the Levelized Cost of Energy (LCOE) of the five DSMs against the base case.

4.4. Assessment of the Impact of Forecast Errors

In the operational stage of the IMG, the proposed formulation computes the DSM stimulus using day-ahead load forecasts. Instead of using a particular method to perform the forecasts, the approach adds Gaussian noise to the real demand to build the forecasted demand, as is stated by Equations (37) and (38). This approach allows measurement of the impact of forecast errors over the final results in the third stage (after knowing the real values of the load).

$$\nu \sim \mathcal{N}(\mu, \sigma^2) \tag{37}$$

$$d_h^F = d_h^R \nu \tag{38}$$

Thus, this section presents a sensitivity analysis of the impact of forecast errors. By computing the simulations again, considering forecast errors of 0%, 5%, 10%, and 15%, this approach computes forecast errors using the Mean Absolute Percentage Error. Table 4 relates the percentage of error with the STD used in Equation (37).

Table 4. input parameters for simulations.

Error	σ^2	Error	σ^2
0%	N/A	5.01%	0.0628
10.01%	0.1258	15.01%	0.1881

It is significant to notice that the reported errors correspond to the average error for all the forecasts of all the simulated scenarios. In the 0% error case, the forecasted demand values are equal to the real values $d_h^F = d_h^R$. The case study found that the methodology is unable to compute the day-ahead stimulus of the DSMs when the forecast errors are near to 20% ($\sigma^2 = 0.2512$).

Figure 12 shows that the impact of the forecast errors in the total costs, the delivered energy, the investors' profits, the customers' payments, and the LCOE is not significant. The variation between the case with perfect forecasts and 15% error in the forecasts is less than 1%. The DADP tariff presents the highest variation in profits and payments of the customers, which drop 1% as compared to the case where the forecast errors are zero.

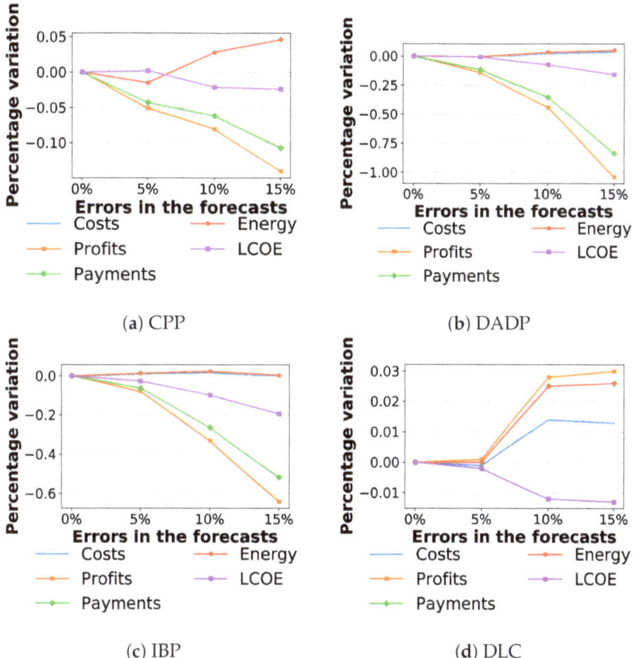

Figure 12. Effects of the forecast errors over the main results. (**a**) Forecast errors effects for a Critical Peak Pricing (CPP) DSM. (**b**) Forecast errors effects for a Day-Ahead Dynamic Pricing (DADP) DSM. (**c**) Forecast errors effects for an Incentive-Based Pricing (IBP) DSM. (**d**) Forecast error effects for a Direct Load Curtailment (DLC) DSM.

4.5. Assessment of the Relation between the First and Third Optimization Levels

This article presents the design of a methodology to compute the effects of five different DSMs over the sizing of IMGs. However, in order to calculate the energy sources' capacities, only the first optimization level of the proposed methodology is required. The second and the third optimization level formulations evaluate the performance of the IMG once it is in operation. Figure 13 reveals the percentage variations between the results from the first and third optimization levels for the five DSMs and the base case.

The first level of the proposed methodology uses a scenario approach built upon historical data and considers an optimization horizon of one year. The second and third levels use a scenario approach built upon forecasts to predict DSMs and consider a rolling horizon with an optimization horizon of one day over a year. Figure 13 presents the comparison between the average results from the first and third levels when the error in the forecasts is 10%. The extra costs, payments, and LCOE, as well as the reductions in profits and payments, are the result of the change of the optimization horizons and the use of historical instead of forecast data. Planners can also compute percentage variations between the first and third levels for different forecast errors and can utilize trends in percentage variations of each of the values to avoid executing the second and third levels of the methodology. Just by executing the first level and considering the percentages' variations in their calculations will be enough to estimate the total costs of the IMG project.

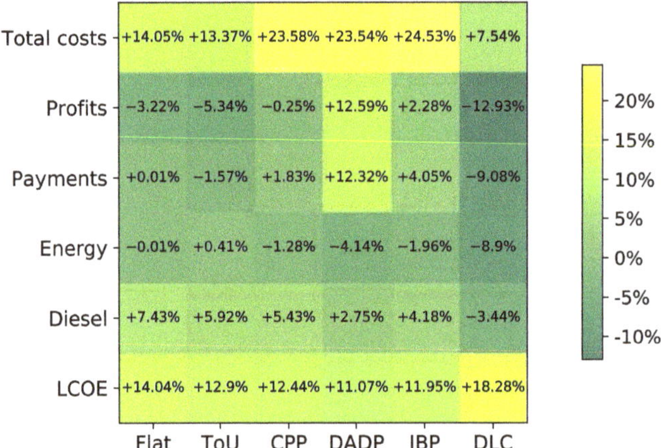

Figure 13. Percentage differences between the results of the first level and the third level for the five DSMs and the base case.

4.6. Performance Comparison of the Five DSMs

The five DSMs have different performance in different aspects. Equation (39) is adopted to measure the performance of each of the DSMs.

$$\text{Performance} = \frac{\text{worst} - \text{current}}{\text{worst} - \text{best}} \tag{39}$$

Figure 14 shows that DADP and IBP tariffs perform better than the other DSMs. However, these rates require announcing energy prices one day in advance, so customers reorganize their consumption daily. In the context of IMG, hourly variations of the tariffs might not be the best option in some scenarios. In those scenarios, a ToU tariff or CPP tariff can give a satisfying solution as well.

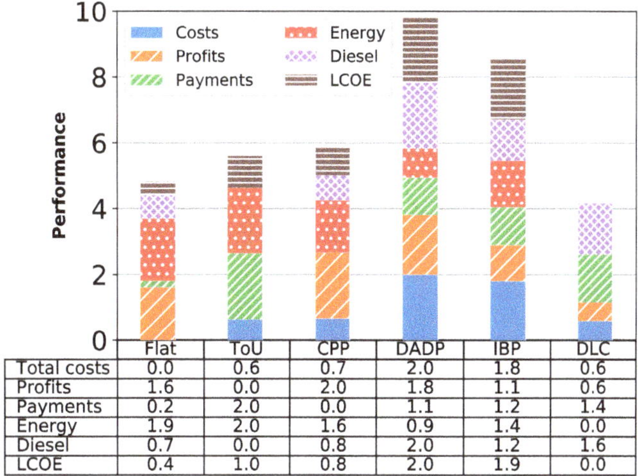

	Flat	ToU	CPP	DADP	IBP	DLC
Total costs	0.0	0.6	0.7	2.0	1.8	0.6
Profits	1.6	0.0	2.0	1.8	1.1	0.6
Payments	0.2	2.0	0.0	1.1	1.2	1.4
Energy	1.9	2.0	1.6	0.9	1.4	0.0
Diesel	0.7	0.0	0.8	2.0	1.2	1.6
LCOE	0.4	1.0	0.8	2.0	1.9	0.0

Figure 14. Performance comparison of the base case and the five DSMs

5. Conclusions

The present work proposes a methodology to design and evaluate five DSMs in the planning and operation of IMGs. The methodology allows determination of the optimal size, optimal energy dispatch strategy, and optimal stimulus for the DSMs using a Disciplined Convex Stochastic Programming approach. The work designs and evaluates the effects of the five DSMs using one case study as a test-bench, which makes this work the first attempt to do so in the literature known by the authors.

The proposed methodology can help policymakers design proper regulations for IMG projects that consider the social conditions of customers and private investors. Additionally, the methodology can be useful for IMG planners or entrepreneurs that want to build profitable business models providing energy to isolated communities. In this regard, the methodology allows policymakers to:

- Compute the effects of applying one of the five DSMs over the total costs of IMG projects in the planning phase.
- Control the revenue of private investors or entrepreneurs to prevent excessive profits.
- Minimize the total amount of subsidies paid by the government for IMG projects.
- Compute the effects over the sizing and the total costs of IMG projects for different values of customer elasticities.

Additionally, the methodology allows IMG planners or entrepreneurs to:

- Compute the expected expenses and revenues of an IMG project considering any of the five DSMs.
- Compute the sizing of the energy sources considering any of the five DSMs.
- Consider the effects of using different combinations of energy sources to supply the electrical demand.
- Obtain the optimal day-ahead energy dispatch strategy for the microgrid considering any of the five DSMs.

The methodology can provide the benefits mentioned above to its users if the assumptions that it was built upon are fulfilled. In this regard, by sharpening the assumptions, the methodology will adapt better to the conditions of IMG projects. Considering more energy sources, sophisticated models of customer elasticities, and demand response models adapted to local conditions, among others, will improve the methodology as well.

Finally, it is essential to highlight the technical characteristic of the present study, which aims to inform planners and policymakers about the benefits of applying DSMs in the planning of IMGs. However, policymakers should perform comprehensive social and behavioral studies to evaluate the potential of acceptance of price-based or direct load curtailment DSMs in the context of IMGs.

Author Contributions: Conceptualization, J.C.O.C.; Formal analysis, J.C.O.C.; Funding acquisition, G.O.-P.; Investigation, J.C.O.C.; Methodology, J.C.O.C., R.R., C.D., J.S., and D.H.; Project administration, G.O.-P.; Software, J.C.O.C.; Supervision, C.D.-G. and J.S.; Validation, J.C.O.C., R.R., J.S., and D.H.; Writing—original draft, J.C.O.C.; Writing—review and editing, J.C.O.C., R.R., Cesar Duarte, J.S., and Daniel Hissel. All authors have read and agreed to the published version of the manuscript.

Funding: Administrative Department of Science, Technology, and Innovation (Departamento Administrativo de Ciencia, Tecnología e Innovación)-COLCIENCIAS, Contracts No. 80740-191-2019 and FP44842-450-2017 (ECOS-NORD).

Acknowledgments: The authors wish to thank the Department of Electrical, Electronics, and Telecommunications Engineering (Escuela de Ingenierías Eléctrica, Electrónica y de Telecomunicaciones) of the Vice-Rectorate for Research and Extension (Vicerrectoría de Investigación y Extensión) from the Universidad Industrial de Santander (Project 8593).

Conflicts of Interest: The authors declare that they have no known competing financial interests or personal relationships that could have appeared to influence the work reported in this paper. The funders had no role in the design of the study; in the collection, analyses, or interpretation of data; in the writing of the manuscript, or in the decision to publish the results.

Abbreviations

The following abbreviations are used in this manuscript:

DSM	Demand-Side Management
MG	Microgrid
IMG	Isolated/Islanded Microgrid
LCOE	Levelized Cost of Energy
BESS	Battery Energy Storage System
PV	Photovoltaic
DG	Diesel Generator
MILP	Mixed Integer Linear Programming
MINLP	Mixed Integer Non-Linear Programming
CAPEX	Capital Expenditures
OPEX	Operational Expenditures
MCS	Monte Carlo Sampling
PDF	Probability Distribution Function
CDF	Cumulative Distribution Function
STD	Standard Deviation
ToU	Time of Use
CPP	Critical Peak Pricing
DADP	Day-Ahead Dynamic Pricing
IBP	Incentive-Based Pricing
DLC	Direct Load Curtailment

Appendix A

Table A1. Variable declaration.

First stage optimization variables		
a_1	Optimization formulation of the first stage	Unitless
φ_{ci}	Percentage of the CAPEX paid by the investor	Unitless
φ_{cg}	Percentage of the CAPEX paid by the government	Unitless
φ_{oi}	Percentage of the OPEX paid by the investor	Unitless
φ_{og}	Percentage of the OPEX paid by the government	Unitless
X_1	Results of the optimization formulations of the first stage	Unitless
t	Hour of optimization	Hours
T	Total number of hours to optimize	Hours
u	Specific generator or storage system of the microgrid	Unitless
U	Total number of generators and storage systems of the microgrid	Unitless
n	Specific DSM	Unitless
N	Total number of DSMs	Unitless
C_u	Installed capacity of the u device	kW, kWh
I_u	Unitary initial investment of the u device	USD/kW
λ_u	Unitary costs of generation of the u device	USD/kWh
Λ_u	Unitary maintenance costs of the u device	USD/kWh
$E_{u,t}$	Quantity of energy delivered with the u device	kWh
ζ	Total capital expenditures	USD
ϑ	Total operational expenditures	USD
R	Internal Rate of Return for the investors	Unitless
$\pi_{n,t}$	Price of the n tariff scheme at time t	USD/kWh
$D_{f,t}$	Final electrical demand of the community	kWh
e_t	Self-elasticity of the customers	Unitless
π_{flat}	Flat tariff	USD/kWh
$D_{o,t}$	Initial electrical demand of the community	kWh
Ψ_c	Electric energy conservation factor	Unitless
EE_t	Amount of energy in excess	kWh
LE_t	Lack of energy to fulfill the demand	kWh
z	Reliability level	Unitless
Second stage optimization variables		
a_2	Optimization formulation of the second stage	Unitless
X_2	Results of the optimization formulations of the second stage	Unitless
h	Hours of the day	Hours
$E_{u,h}^F$	Quantity of forecasted delivered energy with the u device	kWh
EE_h^F	Amount of forecasted energy in excess	kWh
LE_h^F	Lack of forecasted energy to fulfill the demand	kWh
ω	Penalization factor	Unitless
$d_{f,h}^F$	Final electrical day-ahead forecasted demand of the community	kWh
Θ_{flat}	Payments with flat tariff	USD
Θ_{tou}	Payments with ToU tariff	USD
y	Specific hourly block of the ToU tariff	Unitless
Y	Total number of hourly blocks of the ToU tariff	Unitless
π_y	Price at hour y of the ToU tariff	USD/kWh
Θ_{cpp}^F	Day-ahead forecasted payments of the customers under the CPP tariff	USD
π_{base}	Base price of the CCP tariff	USD/kWh
τ_{base}	Time under base price for the CPP tariff	Hours
$d_{f,h}^{F,base}$	Forecasted final electrical demand at base price	kWh
τ_{peak}	Time under peak price for the CPP tariff	Hours
$\pi_{peak,h}^F$	Forecasted peak price of the CCP tariff	USD/kWh
$d_{f,h}^{F,peak}$	Forecasted final electrical demand at peak price	kWh
$\pi_{cpp,h}^F$	Forecasted Critical Peak Price tariff	USD/kWh
$\pi_{peak,t}$	Peak price of the CCP tariff	USD/kWh
φ_{peak}	Percentage of the horizon T allowed to have a peak price	Unitless
δ_{peak}	Times that π_{base} is scaled in the CPP tariff	Unitless
G_h^F	Global horizontal solar radiation	W/m^2
ϱ	Threshold to trigger the CPP price	kW/m^2
Θ_{dadp}^F	Day-ahead forecasted payments of the customers under the DADP tariff	USD
π_h^F	Forecasted hourly price of the DADP tariff scheme	USD/kWh
Θ_{ince}^F	Day-ahead forecasted payments of the customers under the incentive-based tariff	USD
$\pi_{ince,h}^F$	Forecasted incentive price of the IBP tariff	USD/kWh
$\pi_{n,min}$	Minimum value of the n tariff	USD/kWh
π_n	Price of the n tariff scheme	USD/kWh
$\pi_{n,max}$	Maximum value of the n tariff	USD/kWh
$d_{o,h}^F$	Forecasted initial electrical demand	kWh
ϵ_h^F	Forecasted curtailed demand	kWh
Θ_{dlc}^F	Day-ahead forecasted payments of the customers under the DLC DSM	USD
κ	Percentage of the electrical demand to curtail	kWh

Table A1. *Cont.*

Third stage optimization variables		
a_3	Optimization formulation of the third stage	Unitless
X_3	Results of the optimization formulations of the third stage	Unitless
$E_{u,h}^R$	Real quantity of delivered energy with the u device	kWh
EE_h^R	Real amount of energy in excess	kWh
LE_h^R	Real lack of energy to fulfill the demand	kWh
$d_{f,h}^R$	Real final electrical demand	kWh
Case study		
D_t	Electrical demand at time t	kW
m	Months of the year	Unitless
h	Hours of the day	Hours
$\psi_{m,h}$	PDF of the month m and hour h	kW
Φ_u	CDF of the capacity results	kW
ϕ_u	PDF of the capacity results	kW
s	Specific scenario	Unitless
S	Total number of scenarios	Unitless
Y_L	Diesel price per liter	USD/liter
L_u	Lifetime of the u technology	Years
L_p	Lifetime of the IMG project	Years

References

1. Almeshqab, F.; Ustun, T.S. Lessons learned from rural electrification initiatives in developing countries: Insights for technical, social, financial and public policy aspects. *Renew. Sustain. Energy Rev.* **2019**, *102*, 35–53. [CrossRef]

2. Ciller, P.; Lumbreras, S. Electricity for all: The contribution of large-scale planning tools to the energy-access problem. *Renew. Sustain. Energy Rev.* **2020**, *120*, 109624. [CrossRef]

3. Edwin, M.; Nair, M.S.; Joseph Sekhar, S. A comprehensive review for power production and economic feasibility on hybrid energy systems for remote communities. *Int. J. Ambient Energy* **2020**, 1–39. [CrossRef]

4. Taebnia, M.; Heikkilä, M.; Mäkinen, J.; Kiukkonen-Kivioja, J.; Pakanen, J.; Kurnitski, J. A qualitative control approach to reduce energy costs of hybrid energy systems: Utilizing energy price and weather data. *Energies* **2020**, *16*, 1401. [CrossRef]

5. Zhao, H.; Lu, H.; Li, B.; Wang, X.; Zhang, S.; Wang, Y. Stochastic optimization of microgrid participating day-ahead market operation strategy with consideration of energy storage system and demand response. *Energies* **2020**, *13*, 1255. [CrossRef]

6. Wang, Y.; Yang, Y.; Tang, L.; Sun, W.; Zhao, H. A stochastic-CVaR optimization model for CCHP micro-grid operation with consideration of electricity market, wind power accommodation and multiple demand response programs. *Energies* **2019**, *12*, 3983. [CrossRef]

7. Wang, Y.; Huang, Y.; Wang, Y.; Yu, H.; Li, R.; Song, S. Energy management for smart multi-energy complementary micro-grid in the presence of demand response. *Energies* **2018**, *11*, 974. [CrossRef]

8. Nguyen, A.D.; Bui, V.H.; Hussain, A.; Nguyen, D.H.; Kim, H.M. Impact of demand response programs on optimal operation of multi-microgrid system. *Energies* **2018**, *11*, 1452. [CrossRef]

9. Ahmad, S.; Ahmad, A.; Naeem, M.; Ejaz, W.; Kim, H.S. A compendium of performance metrics, pricing schemes, optimization objectives, and solution methodologies of demand side management for the smart grid. *Energies* **2018**, *11*, 2801. [CrossRef]

10. Zunnurain, I.; Maruf, M.; Islam, N.; Rahman, M.; Shafiullah, G. Implementation of advanced demand side management for microgrid incorporating demand response and home energy management system. *Infrastructures* **2018**, *3*, 50. [CrossRef]

11. Hussain, H.M.; Javaid, N.; Iqbal, S.; Hasan, Q.U.; Aurangzeb, K.; Alhussein, M. An efficient demand side management system with a new optimized home energy management controller in smart grid. *Energies* **2018**, *11*, 190. [CrossRef]

12. Wang, Y.; Tang, Y.; Xu, Y.; Xu, Y. A Distributed Control Scheme of Thermostatically Controlled Loads for the Building-Microgrid Community. *IEEE Trans. Sustain. Energy* **2020**, *11*, 350–360. [CrossRef]

13. Wang, Y.; Xu, Y.; Tang, Y. Distributed aggregation control of grid-interactive smart buildings for power system frequency support. *Appl. Energy* **2019**, *251*, 113371. [CrossRef]

14. Franz, M.; Peterschmidt, N.; Rohrer, M.; Kondev, B. *Mini-Grid Policy Toolkit*; Technical Report; Alliance for Rural Electrification: Eschborn, Germany, 2014.

15. Reber, T.; Booth, S.; Cutler, D.; Li, X.; Salasovich, J.; Ratterman, W. *Tariff Considerations for Micro-Grids in Sub-Saharan Africa*; Technical Report February; NREL: Golden, CO, USA, 2018.

16. Casillas, C.E.; Kammen, D.M. The delivery of low-cost, low-carbon rural energy services. *Energy Policy* **2011**, *39*, 4520–4528. [CrossRef]

17. Jin, M.; Feng, W.; Liu, P.; Marnay, C.; Spanos, C. MOD-DR: Microgrid optimal dispatch with demand response. *Appl. Energy* **2017**, *187*, 758–776. [CrossRef]

18. Kahrobaee, S.; Asgarpoor, S.; Qiao, W. Optimum sizing of distributed generation and storage capacity in smart households. *IEEE Trans. Smart Grid* **2013**, *4*, 1791–1801. [CrossRef]

19. Erdinc, O.; Paterakis, N.G.; Pappi, I.N.; Bakirtzis, A.G.; Catalão, J.P. A new perspective for sizing of distributed generation and energy storage for smart households under demand response. *Appl. Energy* **2015**, *143*, 26–37. [CrossRef]

20. Kerdphol, T.; Qudaih, Y.; Mitani, Y. Optimum battery energy storage system using PSO considering dynamic demand response for microgrids. *Int. J. Electr. Power Energy Syst.* **2016**, *83*, 58–66. [CrossRef]

21. Nojavan, S.; Majidi, M.; Esfetanaj, N.N. An efficient cost-reliability optimization model for optimal siting and sizing of energy storage system in a microgrid in the presence of responsible load management. *Energy* **2017**, *139*, 89–97. [CrossRef]

22. Majidi, M.; Nojavan, S.; Zare, K. Optimal Sizing of Energy Storage System in a Renewable-Based Microgrid Under Flexible Demand Side Management Considering Reliability and Uncertainties. *J. Oper. Autom. Power Eng.* **2017**, *5*, 205–214.

23. Amir, V.; Jadid, S.; Ehsan, M. Optimal Planning of a Multi-Carrier Microgrid (MCMG) Considering Demand-Side Management. *Int. J. Renew. Energy Res.* **2018**, *8*, 238–249.

24. Clairand, J.M.; Arriaga, M.; Canizares, C.A.; Alvarez-Bel, C. Power Generation Planning of Galapagos' Microgrid Considering Electric Vehicles and Induction Stoves. *IEEE Trans. Sustain. Energy* **2019**, *10*, 1916–1926. [CrossRef]

25. Gamarra, C.; Guerrero, J.M. Computational optimization techniques applied to microgrids planning: A review. *Renew. Sustain. Energy Rev.* **2015**, *48*, 413–424. [CrossRef]

26. Khodaei, A.; Bahramirad, S.; Shahidehpour, M. Microgrid Planning Under Uncertainty. *IEEE Trans. Power Syst.* **2015**, *30*, 2417–2425. [CrossRef]

27. Chauhan, A.; Saini, R.P. Size optimization and demand response of a stand-alone integrated renewable energy system. *Energy* **2017**, *124*, 59–73. [CrossRef]

28. Amrollahi, M.H.; Bathaee, S.M.T. Techno-economic optimization of hybrid photovoltaic/wind generation together with energy storage system in a stand-alone micro-grid subjected to demand response. *Appl. Energy* **2017**, *202*, 66–77. [CrossRef]

29. Mehra, V.; Amatya, R.; Ram, R.J. Estimating the value of demand-side management in low-cost, solar micro-grids. *Energy* **2018**, *163*, 74–87. [CrossRef]

30. Mehra, V. Optimal Sizing of Solar and Battery Assets in Decentralized Micro-Grids with Demand-Side Management. Ph.D. Thesis, Massachusetts Institute of Technology: Cambridge, MA, USA, 2017.

31. Harper, M. *Review of Strategies and Technologies for Demand-Side Management on Isolated Mini-Grids*; Technical Report; Lawrence Berkeley National Laboratory, Schatz Energy Research Center: Berkeley, CA, USA, 2013.

32. Prathapaneni, D.R.; Detroja, K.P. An integrated framework for optimal planning and operation schedule of microgrid under uncertainty. *Sustain. Energy Grids Netw.* **2019**, *19*, 100232. [CrossRef]

33. Luo, X.; Liu, J.; Liu, Y.; Liu, X. Bi-level optimization of design, operation, and subsidies for standalone solar/diesel multi-generation energy systems. *Sustain. Cities Soc.* **2019**, *48*, 101592. [CrossRef]

34. Kiptoo, M.K.; Adewuyi, O.B.; Lotfy, M.E.; Ibrahimi, A.M.; Senjyu, T. Harnessing demand-side management benefit towards achieving a 100% renewable energy microgrid. *Energy Rep.* **2020**, *6*, 680–685. [CrossRef]

35. Rehman, S.; Habib, H.U.R.; Wang, S.; Buker, M.S.; Alhems, L.M.; Al Garni, H.Z. Optimal Design and Model Predictive Control of Standalone HRES: A Real Case Study for Residential Demand Side Management. *IEEE Access* **2020**, *8*, 29767–29814. [CrossRef]

36. Choynowski, P. *Measuring Willingness to Pay for Electricity*; Technical Report 3; Asian Development Bank: Manilla, Philippines, 2002.

37. Oerlemans, L.A.; Chan, K.Y.; Volschenk, J. Willingness to pay for green electricity: A review of the contingent valuation literature and its sources of error. *Renew. Sustain. Energy Rev.* **2016**, *66*, 875–885. [CrossRef]

38. Kim, J.H.; Lim, K.K.; Yoo, S.H. Evaluating residential consumers' willingness to pay to avoid power outages in South Korea. *Sustainability* **2019**, *11*, 1258. [CrossRef]
39. Yevdokimov, Y.; Getalo, V.; Shukla, D.; Sahin, T. Measuring willingness to pay for electricity: The case of New Brunswick in Atlantic Canada. *Energy Environ.* **2019**, *30*, 292–303. [CrossRef]
40. Ali, A.; Kolter, J.Z.; Diamond, S.; Boyd, S. Disciplined convex stochastic programming: A new framework for stochastic optimization. In Proceedings of the Thirty-First Conference on Uncertainty in Artificial Intelligence, Amsterdam, The Netherlands, 12–16 July 2015; Number 3 in 31, pp. 62–71.
41. Liberti, L.; Maculan, N. Disciplined Convex Programming. In *Global Optimization, from Theory to Implementation*; Springer: Berlin/Heidelberg, Germany, 2008; Volume 105, pp. 9455–9456. [CrossRef]
42. Celik, B.; Roche, R.; Suryanarayanan, S.; Bouquain, D.; Miraoui, A. Electric energy management in residential areas through coordination of multiple smart homes. *Renew. Sustain. Energy Rev.* **2017**, *80*, 260–275. [CrossRef]
43. Inversin, A.R. *Mini-Grid Design Manual (English)*; Technical Report; World Bank: Washington, DC, USA, 2000.
44. Baatz, B. *Rate Design Matters: The Intersection of Residential Rate Design and Energy Efficiency*; Technical Report March; American Council for an Energy-Efficient Economy: Washington, DC, USA, 2017.
45. Glick, D.; Lehrman, M.; Smith, O. *Rate Design for the Distribution Edge*; Technical Report August; Rocky Mountain Institute: Boulder, CO, USA, 2014.
46. Kostková, K.; Omelina.; Kyčina, P.; Jamrich, P. An introduction to load management. *Electr. Power Syst. Res.* **2013**, *95*, 184–191. [CrossRef]
47. Joe-Wong, C.; Sen, S.; Ha, S.; Chiang, M. Optimized day-ahead pricing for smart grids with device-specific scheduling flexibility. *IEEE J. Sel. Areas Commun.* **2012**, *30*, 1075–1085. [CrossRef]
48. Borenstein, S.; Jaske, M.; Rosenfeld, A. *Dynamic Pricing , Advanced Metering and Demand Response in Electricity Markets*; Technical Report October; University of California Energy Institute: Berkeley, CA, USA, 2002.
49. Liberti, L.; Maculan, N. *Global Optimization: From Theory to Implementation*; Springer: Berlin/Heidelberg, Germany, 2006; pp. 155–210.
50. Diamond, S.; Boyd, S. CVXPY: A Python-embedded modeling language for convex optimization. *J. Mach. Learn. Res.* **2016**, *17*, 1–5.
51. Andersen, E.D.; Roos, C.; Terlaky, T. On implementing a primal-dual interior-point method for conic quadratic optimization. *Math. Program. Ser. B* **2003**, *95*, 249–277. [CrossRef]
52. Andersen, E.D.; Andersen, K.D. The Mosek Interior Point Optimizer for Linear Programming: An Implementation of the Homogeneous Algorithm. In *High Performance Optimization*; Frenk, H., Roos, K., Terlaky, T., Zhang, S., Eds.; Springer: Boston, MA, USA, 2000; Volume 33, doi:10.1007/978-1-4757-3216-0. [CrossRef]
53. Oviedo Cepeda, J.C.; Khalatbarisoltani, A.; Boulon, L.; Osma-pinto, A.; Antonio, C.; Gualdron, D.; Solano, J.E. Design of an Incentive-based Demand Side Management Strategy for Stand-Alone Microgrids Planning. *Int. J. Sustain. Energy Plan. Manag.* **2020**, *28*, 1–21. [CrossRef]
54. Li, B.; Roche, R.; Paire, D.; Miraoui, A. Sizing of a stand-alone microgrid considering electric power, cooling/heating, hydrogen loads and hydrogen storage degradation. *Appl. Energy* **2017**, *205*, 1244–1259. [CrossRef]
55. Zhang, J.; Li, K.J.; Wang, M.; Lee, W.J.; Gao, H.; Zhang, C.; Li, K. A Bi-Level Program for the Planning of an Islanded Microgrid Including CAES. *IEEE Trans. Ind. Appl.* **2016**, *52*, 2768–2777. [CrossRef]
56. Skoplaki, E.; Palyvos, J.A. Operating temperature of photovoltaic modules: A survey of pertinent correlations. *Renew. Energy* **2009**, *34*, 23–29. [CrossRef]
57. Bukar, A.L.; Tan, C.W.; Lau, K.Y. Optimal sizing of an autonomous photovoltaic/wind/battery/diesel generator microgrid using grasshopper optimization algorithm. *Sol. Energy* **2019**, *188*, 685–696. [CrossRef]
58. Grupo EPM. *Tarifas de Energía Mercado Regulado*; Grupo EPM: Medellín, Colombia, 2019.

Article

A Virtual Tool for Load Flow Analysis in a Micro-Grid

Giovanni Artale [1], Giuseppe Caravello [1], Antonio Cataliotti [1], Valentina Cosentino [1], Dario Di Cara [2,*], Salvatore Guaiana [1], Ninh Nguyen Quang [3], Marco Palmeri [1], Nicola Panzavecchia [2] and Giovanni Tinè [2]

[1] Department of Engineering, Università degli Studi di Palermo, 90128 Palermo, Italy; giovanni.artale@unipa.it (G.A.); giuseppe.caravello02@unipa.it (G.C.); antonio.cataliotti@unipa.it (A.C.); valentina.cosentino@unipa.it (V.C.); salvatore.guaiana@unipa.it (S.G.); marcopalmeri94@gmail.com (M.P.)
[2] Institute of Marine Engineering (INM), National Research Council (CNR), 90146 Palermo, Italy; nicola.panzavecchia@cnr.it (N.P.); giovanni.tine@cnr.it (G.T.)
[3] Institute of Energy Science, Vietnam Academy of Science and Technology, Hanoi 100000, Vietnam; nqninh@ies.vast.vn
* Correspondence: dario.dicara@cnr.it

Received: 29 May 2020; Accepted: 16 June 2020; Published: 18 June 2020

Abstract: This paper proposes a virtual tool for load flow analysis in energy distribution systems of micro-grids. The solution is based on a low-cost measurement architecture, which entails low-voltage power measurements in each secondary substation and a voltage measurement at the beginning of the medium voltage (MV) feeder. The proposed virtual tool periodically queries these instruments to acquire the measurements. Then, it implements a backward–forward load flow algorithm, to evaluate the power flow in each branch and the voltage at each node. The virtual tool performances are validated using power measurements acquired at the beginning of each MV feeder. The uncertainties on each calculated quantity are also evaluated starting from the uncertainties due to the used measurement instruments. Moreover, the influence of the line parameter uncertainties on the evaluated quantities is also considered. The validated tool is useful for the online analysis of power flows and also for planning purposes, as it allows verifying the influence of future distributed generator power injection. In fact, the tool is able to off-line perform the load flow calculation in differently distributed generation scenarios. The micro-grid of Favignana Island was used as a case study to test the developed virtual tool.

Keywords: smart grid; power system; distributed generation; micro-grid; load flow; measurement uncertainties

---

1. Introduction

The growing demand to increase the percentage of renewable energy and reduce diesel consumption led the distributor system operator (DSO) of micro-grids, as those of small islands, to carry out an analysis in order to verify the safety management of the power grid and to redesign it in case it does not satisfy the grid constraints. In such islands, several issues arise in terms of power system safe operation and planning [1–7], considering the specific characteristics of these kinds of distribution networks, which are relatively small, with a high variability between the winter and summer loads and not always sufficiently covered by public communication infrastructures. Apart from actual energy policies and regulatory frameworks, or technical capabilities enabled by advanced modeling and analysis tools, an important issue for effectively increasing the distributed generation and storage systems presence in distribution networks is the possibility for DSOs to achieve new simple and versatile tools for power system monitoring and management purposes. These tools have to be based on proper communication and measurement infrastructures, which should be feasible for DSOs themselves, in terms of low cost, flexibility and expandability features, in order to allow their development starting from the existing instrumentation and equipment typically employed in such networks.

Several papers can be found in the literature concerning the measurement and communication technologies in distribution systems. For example, in [8–12], a wide overview is given of measurement technologies and architectures for the smart distribution grids, including metering and communication infrastructures. For distribution networks, especially those of isolated islands, supervisory control and data acquisition (SCADA) systems are typically employed for monitoring, protection and control purposes. As regards measurement instrumentation, several kinds of equipment are considered, such as smart meters and sensors, power quality analyzers, phasor measurement units (PMUs) and micro-PMUs (µPMUs), and so on, which can be more or less suitable, depending on the considered distribution system management applications and the particular characteristics of the considered network. In particular, many recent researches have been focused on PMUs and µPMUs for distribution network monitoring, control and diagnostic applications [13–19]. However, such solutions can be unsuitable for small island micro-grids, because power lines are short and/or the intrinsic costs of such instrumentation are high. To reduce the installation costs, some authors propose to use a few measurement points and to integrate them with load estimations [20–27]; however, when dealing with load estimations (or pseudo-measurements), higher uncertainty levels are generally expected and more sophisticated algorithms can be needed for the distribution system state's estimation, which also may entail higher computational costs. The integration of differently distributed measurement solutions have also been investigated, for example, considering the possibility of smart meter and power quality meter exploitation or SCADA- and PMU-enhanced integration, for a number of applications (load forecasting, optimization, demand side management, fault detection and so on) [28–40]. If the application of such solutions is envisaged for small distribution networks, such as those addressed in this paper, the main problems are related to the processing of algorithms' accuracy and complexity, considering the reasonable computational capabilities of the DSOs control centers. Another fundamental element to enable network observability is the communication between these measurement devices and the control room of the micro-grid. Different solutions can be used for this purpose: optic fiber, power line communications, GSM, wireless and so on. The different communication solutions must be compared in terms of cost, reliability, security, environmental impact and power quality effects [41–47], considering also their availability and suitability in islanded micro-grids.

In summary, the main issues to apply these technologies in the case of small islanded grids are:

- the cost of the measurement infrastructure;
- the availability and the cost of communication systems from public or private providers or eventually the installation cost of a dedicated infrastructure;
- the high-load variability connected with the seasonal tourist influx;
- the high sensitivity to distributed generation uncertainty, especially in low load seasons, which can cause a high variation of the required power and consequently high voltage and frequency variation in the network.

In this paper, these aspects are considered, for proposing solutions especially tailored for the case of an island's distribution network. The basic aim of this work was the development of a feasible tool for load flow analysis and power system planning, based on a low-cost distributed measurement system. The proposed solution has been implemented on-field, on the real distribution network of the Island of Favignana (Italy, Mediterranean Sea). To fully investigate the suitability of the proposed solution, it has to be characterized by the means of an uncertainty analysis, considering both the real measurement data and the network parameters uncertainty propagation on the power flow estimations. Experimental results are presented to validate the algorithm and verify its capability to evaluate in real time the power flows with good accuracy. Furthermore, off-line simulations based on real measurement data have been carried out, showing how the proposed virtual tool is also useful to study the impact of distributed generation in an isolated network. The proposed solution allows addressing all the aforementioned issues in terms of reduced costs, thanks to the use of low-cost instrumentation communication infrastructures, without affecting the capability for the real-time evaluation of the power flows and to plan the network improvement, avoiding critical situations due to the distributed generation and loads variability.

In detail, an ad hoc monitoring system is used, which is particularly suitable for a small island, because of its reduced cost and simplicity of installation. The solution is based on the use of power quality analyzers (PQAs) in distribution network substations, which are less expensive than PMUs. Moreover, they are installed at the low-voltage (LV) side of power transformers, thus reducing installation costs; these last could be even null, if smart meters are already installed by DSO for energy theft detection purposes. Starting from LV active and reactive power measurements acquired by these instruments in each secondary substation, the authors developed a backward–forward load flow (BF-LF) algorithm, which allows determining all the other network state variables [48–50]. The algorithm requires only one additional voltage measurement at medium voltage (MV) bus-bars of the central generating station; typically, this measurement can already be available in a real distribution network, thus no extra costs are needed for further instrumentation. Thanks to its low computation cost and simplicity of implementation, the BF-LF algorithm is a good solution for this kind of small distribution network. As regards the communication between the substation meters and DSO control center, a wireless HiperLAN-based architecture is used which is particularly suitable in the case of small islands, thanks to their orography. Starting from the aforementioned distributed measurement infrastructure, a virtual instrument (VI) has been developed and implemented, which is able to query all the installed PQAs to acquire the measurement data and to perform the load flow calculation of the whole network. The algorithm was implemented in a LabVIEW environment, because it allows assuring a highly readable and simple use. As regards the metrological characterization of the developed system, a Monte Carlo procedure is also implemented to perform an uncertainty analysis of the calculated power flows, considering the input uncertainties on both the real measurement data and the network parameter knowledge. The presented results show how the proposed architecture allows monitoring the power system in real-time and with good accuracy. Furthermore, the measurement data are acquired and stored in a database; this allows running the offline simulation; in this way, the validated software tool can also be used to perform the simulation of photovoltaic penetration scenarios and to observe its impact on power flows.

This paper is organized as follows. In Section 2, the monitoring architecture and the developed virtual tool for load flow analysis are described, including the implementation of the uncertainty evaluation algorithm, used for the characterization of the load flow analysis accuracy starting from on-field measurements. In Section 3, the implementation of the proposed solution is presented for the real case study of Favignana Island micro-grid. In Section 4, the algorithm is validated with real measurement data, evaluating also all the uncertainty contributions. Finally, in Section 5 the virtual tool is used to perform a simulation of different scenarios of photovoltaic generation.

2. Monitoring Architecture and Virtual Tool for Load Flow Analysis

The observability of distribution networks in real time is the base element for a proper grid management, maintaining its stability and correct operation. Currently, most MV distribution networks around the world are scarcely monitored. Few measurement instruments are usually installed and DSOs are not able to perform a real-time load flow of the whole network. As already mentioned, this is mainly due to the intrinsic costs of a monitoring solution, which entails the costs of the measuring equipment, MV transducers and communication network. A solution to reduce these costs is shown in Figure 1. In this solution, the authors proposed to use PQAs at the LV side of the power transformer in each secondary substation and a further PQA at the MV bus-bars in the generating substation [48]. The developed virtual tool is installed in the DSO control center. It queries all PQAs and performs the load flow analysis.

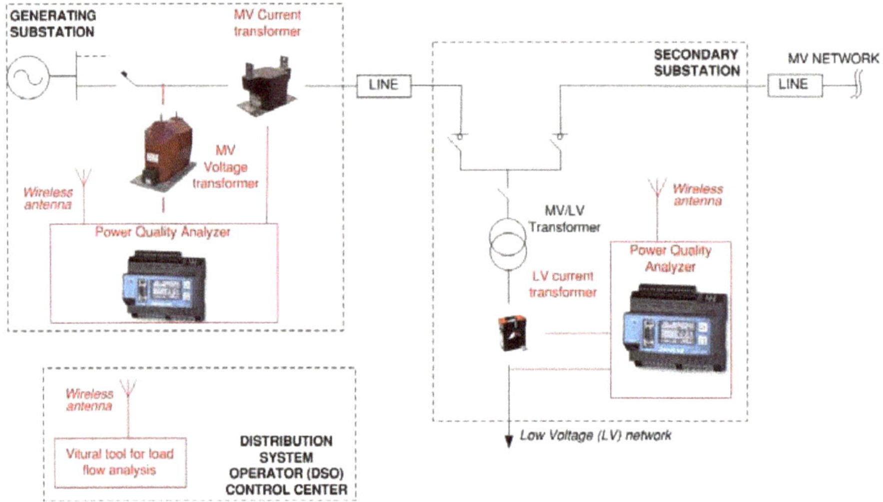

Figure 1. Proposed architecture for the medium voltage (MV) distribution network monitoring.

To this aim, a proper BF-LF algorithm was implemented. The algorithm uses as input the measured active and reactive powers of each load and the voltage at the MV bus-bars of the generating substation. Thanks to these measurements, the algorithm is able to univocally determine all the unknown state variables, i.e., node voltages and branch power flows.

The PQA at the MV bus-bars also allows measuring active and reactive powers at the beginning of the feeder; these data are not used by the BF algorithm; they will be used in Section 4 to validate the algorithm performances instead. The block diagram for the BF algorithm implementation in LabVIEW is shown in Figure 2, where:

- *V* is the array of node voltages;
- *FP* and *FQ* are the arrays of the active and reactive power flows, respectively;
- *PL measured* and *QL measured* are the arrays of the measured active and reactive powers, respectively;
- *DeltaP* is the array of the calculated power losses in the network;
- *VMT* is the voltage used as reference for the slack bus: its module is equal to the value measured at MV bus-bars, i.e., *V measured*, and its phase is assumed as 0;
- *Tol module* and *Tol phase* are the thresholds used as tolerance in the load flow algorithm;

Two different sequence frames are used for backward and forward sweeps on the whole network: the backward sweep calculates the power flows in each branch, and the forward sweep calculates the node voltages. These two frames are included in a while loop, thus they are repeated until a convergence condition is met on both the amplitude and the phase of the voltage at each node (at each iteration, this condition is verified in the third sequence frame).

To analyze the algorithm formulation in more detail, the single-phase network model used is shown in Figure 3 [50–52]. In the model, the voltages are the medium values of the three phases while the active and reactive powers are the total powers of the three phases. Network parameters are shown in Figure 3 and listed in Table 1; they have to be known for each branch and node of the network. As regards this, in practical cases this is a source of uncertainty (as these data are affected by uncertainty); in this viewpoint, in Section 4 the impact of such uncertainty on power flow results is analyzed.

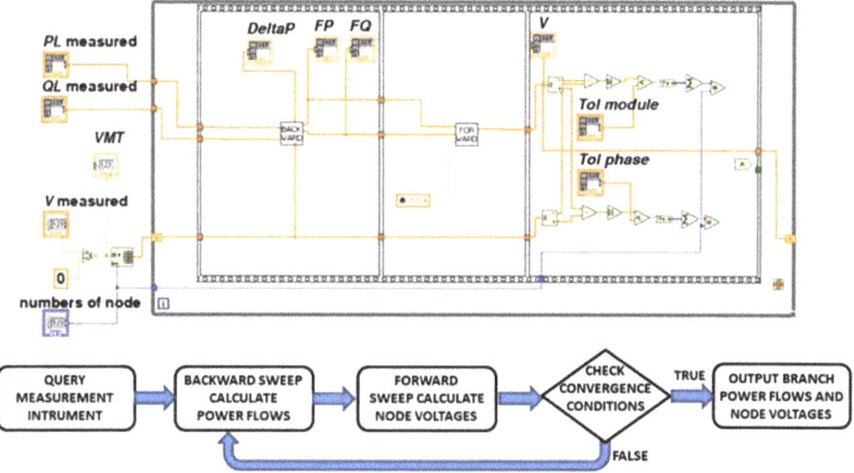

Figure 2. Block diagram for the power flow calculation.

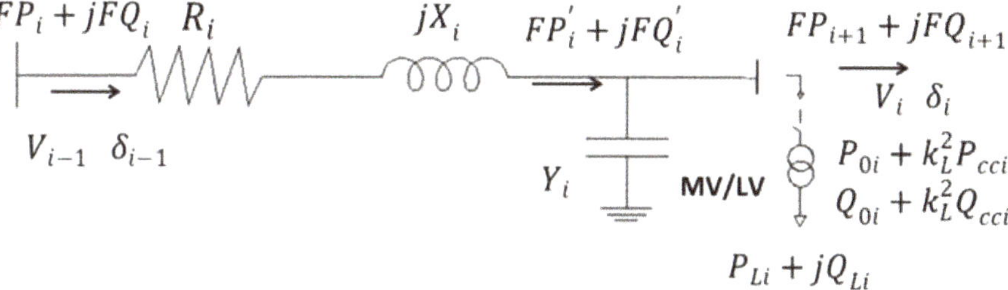

Figure 3. Scheme of the *i*-th brunch of the network model.

Table 1. Required network parameters.

Branch Line Parameters	R_i	Longitudinal resistance
	X_i	Longitudinal reactance
	Y_i	Shunt admittance
Transformer Parameters	$A_{n,i}$	Rated power
	$P_{0,i}$	No load active power losses
	$Q_{0,i}$	No load reactive power losses
	$P_{cc,i}$	Short circuit active power losses
	$Q_{cc,i}$	Short circuit reactive power losses

2.1. Backward Sweep

As already mentioned, firstly the backward sweep is performed from the last node to the beginning of the MV feeder. Active and reactive power flows in the longitudinal impedance of each branch are calculated as:

$$FP'_i = FP_{i+1} + \left(P_{Li} + P_{0_i} + k_L^2 P_{cc_i}\right) \tag{1}$$

$$FQ'_i = FQ_{i+1} + \left(Q_{Li} + Q_{0_i} + k_L^2 Q_{cc_i}\right) - V_i^2 Y_i \tag{2}$$

whcrc:

- V_i is the voltage amplitude at the *i*-node;
- FP_{i+1} e FQ_{i+1} are the power flows downstream from the *i*-node; these terms are null in the case of a terminal node;
- P_{Li} and Q_{Li} are active and reactive powers measured at LV side of power transformers;
- to obtain the equivalent MV load, the power transformer losses are added as $P_{0_i} + k_L^2\, P_{cc_i}$ and $Q_{0_i} + k_L^2\, Q_{cc_i}$ for the active and reactive power, respectively (these terms are not added in the case of MV users, because they are included in P_{Li} and Q_{Li} measured at the MV side of the transformer);
- k_L is the load factor, i.e., the ratio between the actually drained apparent power and its rated value:

$$k_L^2 = \frac{P_{Li}^2 + Q_{Li}^2}{A_{n_i}^2} \tag{3}$$

When the virtual tool is used to simulate the distributed generators' connection to the LV network, the generated powers will be summed to P_{Li} and Q_{Li}.

The active and reactive power flows in each branch are finally obtained by summing the line losses as:

$$FP_i = FP_i' + \Delta P = FP_i' + R_i \frac{FP_i'^2 + FQ_i'^2}{V_i^2} \tag{4}$$

$$FQ_i = FQ_i' + \Delta Q = FQ_i' + X_i \frac{FP_i'^2 + FQ_i'^2}{V_i^2} \tag{5}$$

2.2. Forward Sweep

In the forward sweep, node voltages are calculated starting from the measured voltage at MV bus-bars and the calculated power flows in each branch. The voltage phasor at node *i* is calculated as:

$$\overline{V}_i = \overline{V}_{i-1} - \sqrt{3}\,\dot{Z}_i\,\overline{I}_i \tag{6}$$

where $\overline{I}_i$ is the phasor of the current flowing in the longitudinal impedance.

It can be obtained from the following expression:

$$\overline{I}_i = \frac{FP_i - jFQ_i}{\sqrt{3}\cdot \overline{V}_{i-1}^*} \tag{7}$$

Combining these last two expressions, the voltage phasor can be finally obtained as:

$$\overline{V}_i = \frac{V_{i-1}^2 - (P_i \cdot R_i + Q_i \cdot X_i) - j(P_i \cdot X_i - Q_i \cdot R_i)}{\overline{V}_{i-1}^*} \tag{8}$$

2.3. Convergence Condition

The convergence condition is verified on both the voltage amplitude and phase. In further detail, the difference is calculated between the amplitudes and the phases of two subsequent cycles. If these differences were below a tolerance threshold for all the nodes, the while cycle is stopped, otherwise a further iteration is performed.

2.4. Uncertainty Analysis

To evaluate the uncertainty on power flows' calculated values, the propagation of uncertainties was studied starting from the measurement uncertainties of the input quantities [53], i.e., the load powers of secondary substations and the voltage of MV bus-bars of a generating station.

The uncertainties on the power measurements acquired at the LV side of power transformers are calculated taking into account the following contributions:

- The PQAs uncertainty of the power measurements, $u_{P\%}$ and $u_{Q\%}$;
- The uncertainty introduced by the current transformers (CTs), due to the ratio and phase angle errors, $\eta_{CT\%}$ and ε_{CT}, respectively [54].

The uncertainty of power measurements acquired at the MV level in MV user substations is determined taking into account the following contributions:

- The PQAs uncertainty of the power measurements, $u_{P\%}$ and $u_{Q\%}$;
- The uncertainty introduced by the MV CTs [54];
- The uncertainty introduced by MV voltage transformers (VTs), due to the ratio and phase angle errors, $\eta_{VT\%}$ and ε_{VT}, respectively [55].

The uncertainty on the voltage measurement at the MV bus-bars of generating stations is determined taking into account the following contributions:

- The PQAs uncertainty of the voltage measurements;
- The uncertainty introduced by the MV VTs.

More in detail, the uncertainties on the active and reactive power measurements for the MV users are calculated, considering a type B evaluation and a rectangular distribution, through the following formulas [50]:

$$u_{P_{MV\%}} = \frac{\sqrt{\eta_{CT\%}^2 + (\tan\theta\,100\sin\varepsilon_{CT})^2 + \eta_{VT\%}^2 + (\tan\theta\,100\sin\varepsilon_{VT})^2 + u_{P\%}^2}}{\sqrt{3}} \tag{9}$$

$$u_{Q_{MV\%}} = \frac{\sqrt{\eta_{CT\%}^2 + (\cot\theta\,100\sin\varepsilon_{CT})^2 + \eta_{VT\%}^2 + (\cot\theta 100\sin\varepsilon_{VT})^2 + u_{Q\%}^2}}{\sqrt{3}} \tag{10}$$

where θ is the phase shift between the current and the voltage. For the uncertainties of active and reactive power measurements at the LV level, $u_{P_{LV\%}}$ and $u_{Q_{LV\%}}$, a similar expression is used (where the terms related to the VTs are omitted).

To assess the uncertainty of the load flow output, the law of propagation of uncertainties should be applied to determine the partial derivatives of the measurement model. An alternative solution proposed in the standard [56] performs an iterative analysis with a Monte Carlo method. In more detail, the Monte Carlo procedure suggests repeating the calculation and iteratively varying the input quantities in their uncertainty range, thus obtaining the uncertainty distribution of the output quantities. Following this approach, a second VI was designed to be used offline to validate the load flow algorithm and evaluate its performances in terms of accuracy in the calculated power flows. The VI performs 10^5 times the load flow aforementioned algorithm; at each iteration the input quantities are randomly varied within the related uncertainty intervals through the following expressions:

$$P'_{Li} = P_{Li}\cdot\left(1 + u_{P_{LV\%}}\cdot 100\cdot R_p\right) \tag{11}$$

$$Q'_{Li} = Q_{Li}\cdot\left(1 + u_{Q_{LV\%}}\cdot 100\cdot R_q\right) \tag{12}$$

$$V'_{MT} = V_{MT}\cdot\left(1 + u_{V_{MT}}\cdot R_v\right) \tag{13}$$

where:

- P_i^L e Q_i^L are the estimated values of the power measurements;
- $u_{V_{MT}}$ is the relative uncertainty of the voltage measurements;

- R_p R_q and R_v are the random numbers chosen within a standard normal distribution.

The high number of iterations guarantees that random numbers do not affect the results. Active and reactive powers were considered as uncorrelated quantities. At each run, active and reactive power flows on each branch and node voltages were calculated. At the end of the 10^5 iterations, the frequency distributions of the calculated power flows were evaluated. The average value and the expanded uncertainty were then calculated (confidence level of 95.45%, coverage factor $k = 2$).

Figure 4 shows the implementation of the Monte Carlo analysis in LabVIEW. The sub-VI implementing the load flow algorithm is inside a "for" cycle, which is used to iteratively repeat the calculations. A second "for" cycle is used to extract the frequency distribution for each node and then calculate the mean values and the standard deviations. The expended uncertainties are then obtained as twice ($k = 2$) the standard deviations.

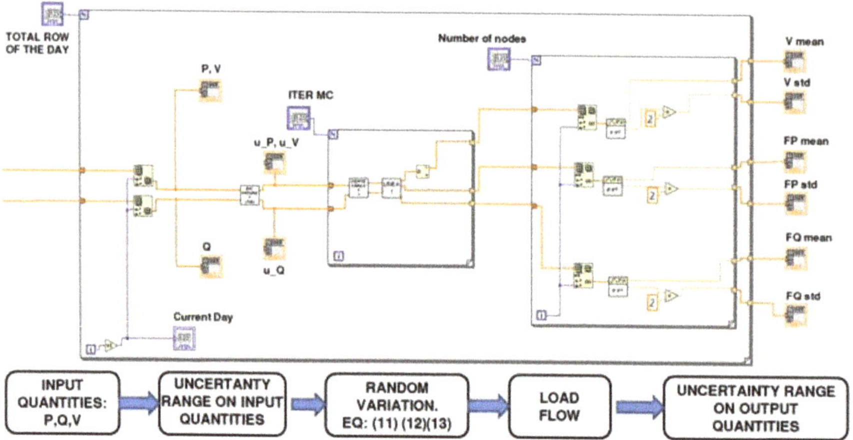

Figure 4. Implementation of the Monte Carlo analysis in LabVIEW.

3. Case Study: Microgrid of Favignana ISLAND

The aforementioned virtual instrument and the related measurement architecture were implemented in the real case study of the microgrid of Favignana island.

The production, distribution and energy sale on Favignana island are managed by the SEA S.P.A. (Società Elettrica di Favignana). The electricity network of the island of Favignana is currently composed of three medium voltage lines (in the following, named MTL1, MTL2 and MTL3), which depart from a central generating station. The three lines feed both MV/LV secondary substations (which LV lines depart from, to supply LV users), and MV user substations. The central station has seven generation units for a total installed power of 16,120 kVA. The MV lines are mostly equipped with MV cables. Only a few feeder sections are equipped with overhead lines.

The electrical scheme of the MV line named "MT L1" is shown in Figure 5. It is the longest MV line of the island (25,640 m). It powers 28 secondary substations (21 MV/LV secondary substations and seven MV users) mostly placed outside the city center. In Figure 5, each node of the "MTL1" line is indicated with a number, which will be used to show the load flow results. The branches will be indicated with the numbers of the nodes in which they end instead. The black numbers indicate the principal nodal substations. Line parameters and power transformer-rated data were already reported in a first study focused on line "MTL1" [49].

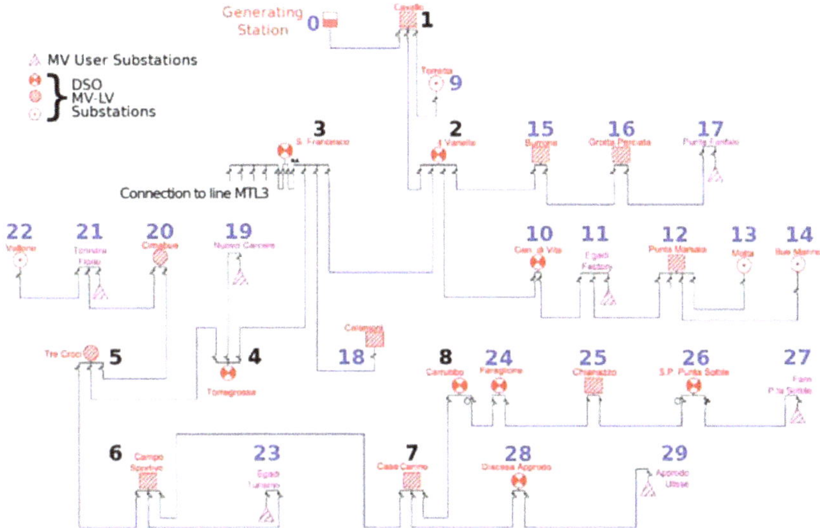

Figure 5. Electrical scheme of the feeder "MT L1" of the Favignana MV distribution network.

The second line of the Favignana MV distribution network was named "MTL2" and it is shown in Figure 6. It is 2281 m long, and this line powers the city center with four MV/LV secondary substations. Its line parameters and related power transformer-rated data are reported in Tables 2 and 3, respectively.

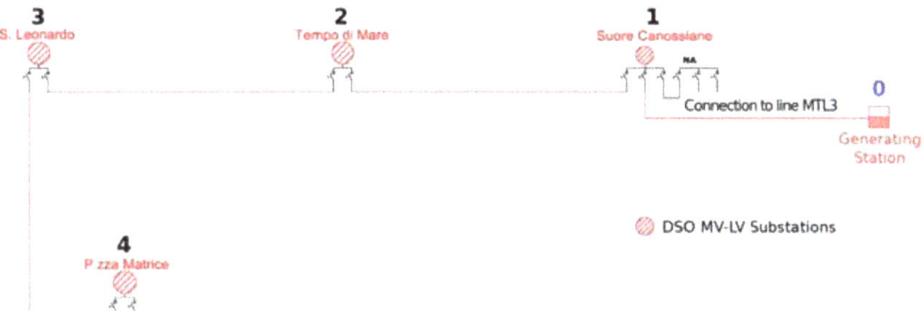

Figure 6. Electrical scheme of the feeder "MT L2" of the Favignana MV distribution network.

Table 2. Line parameters of the "MTL2" MV feeder.

Branch	From Node To Node	R (Ω)	X (Ω)	Y (μS)
1	0–1	0.401	0.105	48.35
2	1–2	0.330	0.053	17.84
3	2–3	0.381	0.062	20.61
4	3–4	0.272	0.072	32.83

Table 3. Rated data of the MV/LV transformers powered by MTL2.

Node	A_n (kVA)	P_0 (W)	Q_0 (VAR)	P_{cc} (W)	Q_{cc} (VAR)
1	160	460	3651	2350	5953
2	800	1900	8592	9000	47,148
3	800	1500	12,717	8500	47,241
4	630	1650	7378	7800	36,986

The third MV line is shown in Figure 7 and it is named "MTL3". It is 3785 m long, and it powers five MV/LV secondary substations and one MV user substation. Its line parameters and related power transformer-rated data are reported in Tables 4 and 5, respectively.

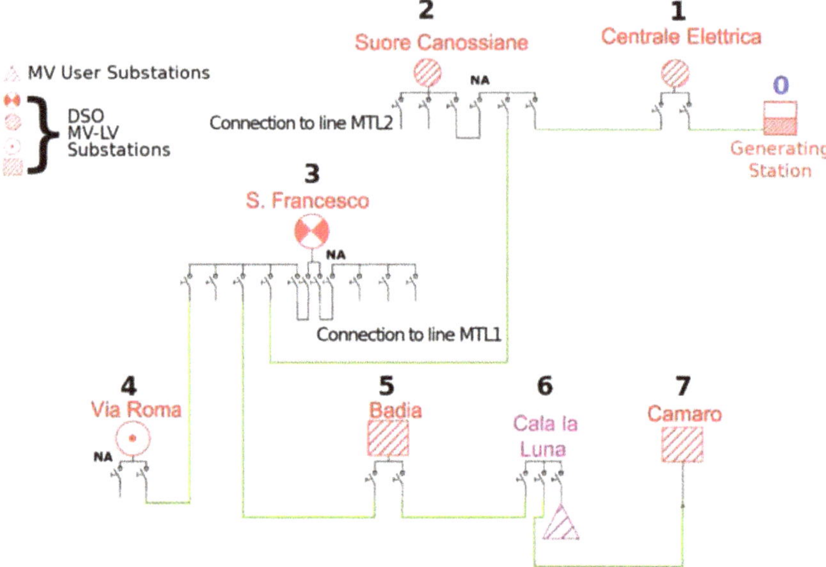

Figure 7. Electrical scheme of the feeder "MT L3" of the Favignana MV distribution network.

Table 4. Line parameters of the "MTL3" MV feeder.

Branch	From Node To Node	R (Ω)	X (Ω)	Y (μS)
1	0–1	0.01	0.003	1.2
2	1–2	0.401	0.105	48.35
3	2–3	0.574	0.151	69.24
4	3–4	0.195	0.032	10.56
5	3–5	0.771	0.125	42
6	5–6	0.135	0.022	7.29
7	6–7	0.353	0.057	19.1

Table 5. Rated data of the MV/LV transformers powered by MTL3.

Node	A_n (kVA)	P_0 (W)	Q_0 (VAR)	P_{cc} (W)	Q_{cc} (VAR)
1, 2, 7	160	460	3651	2350	5953
5	630	1650	7378	7800	36,986
4	800	1900	8592	9000	47,148
3	1250	950	17,474	11,000	74,189

A PQA Janitza UMG 604 is installed in each secondary substation of the network. All the PQAs are linked to the DSO-monitoring control center via a HiperLAN network. This solution was chosen as it was the best economical solution due to the orography of the island. The developed VI can query each instrument via Modbus over TCP/IP. The VI periodically queries each PQA at time intervals of 2 s, to acquire the measured active and reactive powers. The collected measurements are used to run the power flow calculation; then they are stored along with the results. The considered application requires a time accuracy in the millisecond range, thus a network time protocol (NTP) over the Ethernet network is used to synchronize all PQAs. Moreover, a further PQA is installed at the beginning of each MV feeder. As already mentioned, the voltage measurements of these last PQAs are used in the load flow algorithm, while their active and reactive power measurements are used for the algorithm power flow output validation.

4. Experimental Validation and Uncertainty Analysis of the Proposed Algorithm in the Case Study

4.1. Algorithm Validation

The experimental validation of the BF-LF algorithm results was carried out by comparing calculated active and reactive power flows with active and reactive power values measured by the PQAs installed at the beginning of each MV line (for the assessment of the uncertainty on the powers measured in MV, see Section 4.2). Figure 8 shows the comparison between the measured and the calculated values, for the first branch of the "MTL1" line. The comparison is performed every 2 s for the 24 h of the 31 May 2018. As can be seen, the measured and estimated values are superimposed. To highlight their differences, they are reported in Figure 9. It can be observed that the difference between the measured and estimated values is always very small, in comparison with the measurement uncertainty, thus confirming the correctness of the power flow calculations. Similar graphs are reported for reactive power flows (see Figures 10 and 11). The results obtained for the MTL2 and MTL3 lines are very similar to those by MTL1, thus they are omitted.

The same analysis was carried out for different days. For each day, Figure 12 shows the maximum and average values of the difference between the measured and the calculated values in the percentage of the measured active power flow. Differences of less than 0.1% and 0.2% were observed. These results demonstrate how the values calculated by the virtual instrument were very close to the measured values. Similar results were obtained for the reactive power flows and for the other lines.

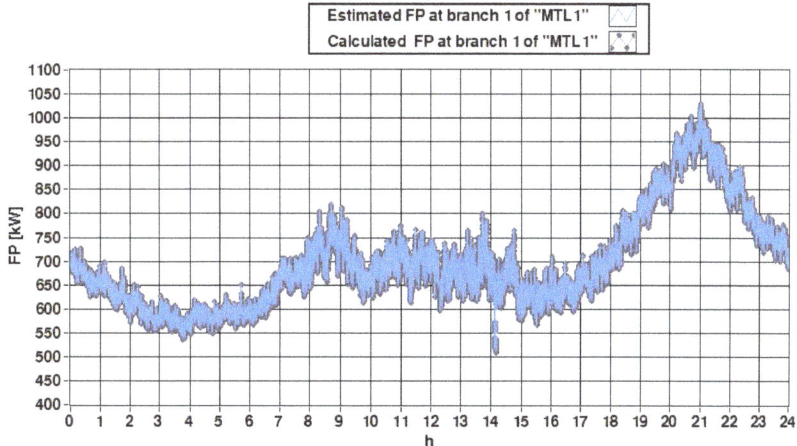

Figure 8. Comparison between the measured and the calculated values of the active power flow of first branch 1 of the "MTL1" line during the day 31 May 2018.

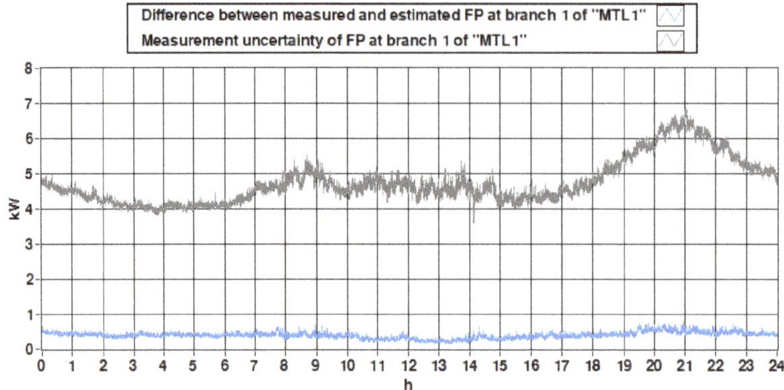

Figure 9. Difference between the measured and the calculated active power flow of branch 1 of the "MTL1" line and the uncertainty in its measurement (31 May 2018).

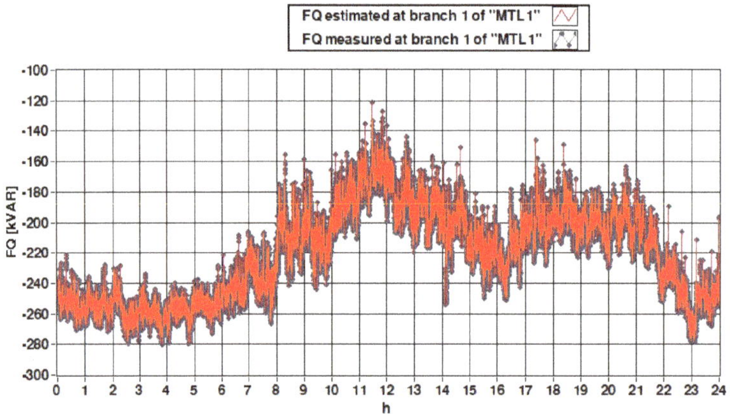

Figure 10. Comparison between the measured and the calculated values of the reactive power flow of branch 1 of the "MTL1" line during the 31 May 2018.

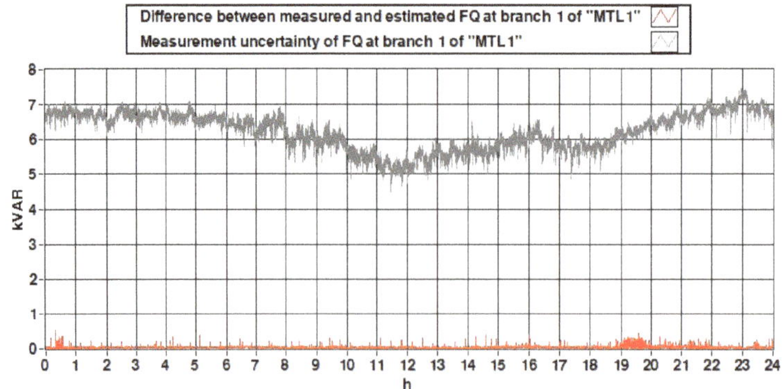

Figure 11. Difference between the measured and the calculated reactive power flow of branch 1 of the "MTL1" line and the uncertainty in its measurement (31 May 2018).

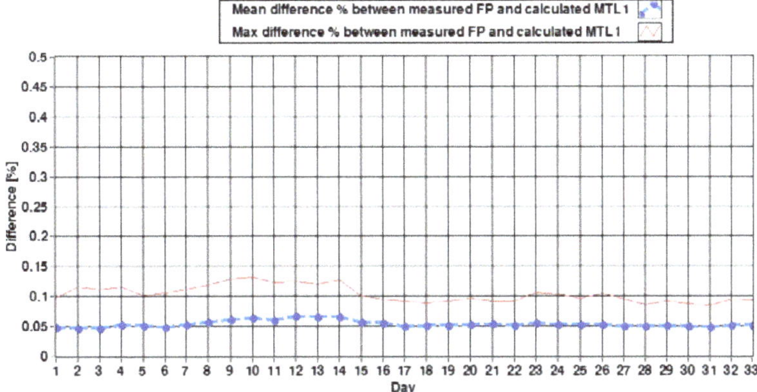

Figure 12. Average and maximum value of the relative difference between the measured and the calculated values of the active power flow of branch 1 of the "MT L1" line for the 33 analyzed days.

4.2. Uncertainty Analysis

To evaluate the uncertainty in the output quantities, a Monte Carlo analysis was performed using the VI described in Section 2.4. The Monte Carlo procedure was performed for a whole day, the 31 May 2018; the power flows and related uncertainties were evaluated every 2 s. For each set of measured input quantities, the load flow algorithm was performed 10^5 times. At the end of the 10^5 iteration, an average value and an extended uncertainty are obtained. Then, the process is repeated for the following set of input data, related to the subsequent 2 s.

Figure 13 shows the comparison between the calculated uncertainty with the Monte Carlo procedure and the measurement uncertainty obtained with (9) for the active power flows measured by the PQA installed at the MV level of branch 1 of the "MTL1" line. Figure 14 shows a similar graph for the reactive power flows. It can be seen that calculated uncertainties are comparable with those of measurement power flows.

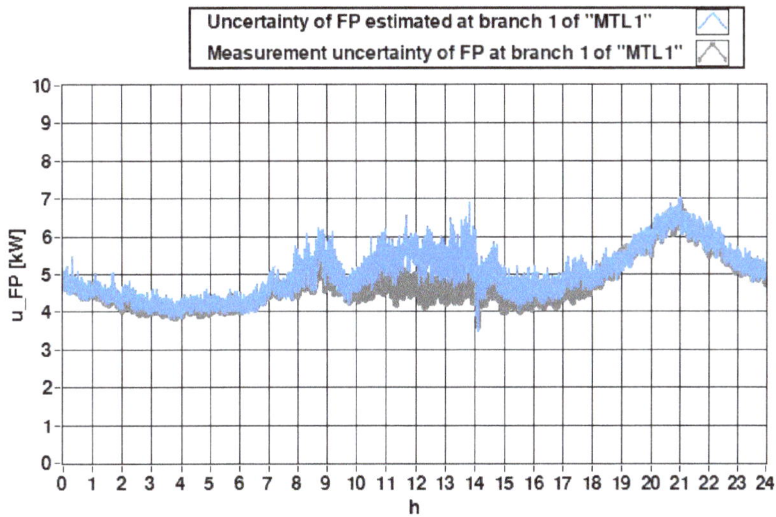

Figure 13. Comparison between the measurement and the calculated uncertainties in the active power flow of branch 1 of the "MTL1" line (31 May 2018).

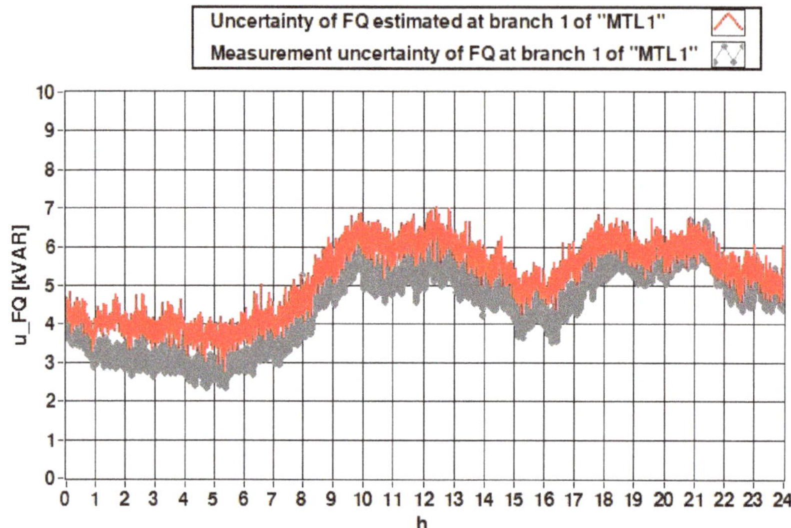

Figure 14. Comparison between the measurement and the calculated uncertainties in the reactive power flow of branch 1 of the "MTL1" line (31 May 2018).

Maximum, minimum and average uncertainties on the power flows of each branch of "MTL1" line are reported in Figures 15 and 16. The same graphs are shown for the "MTL2" line (Figures 17 and 18) and for the "MTL3" line (Figures 19 and 20). In all cases, the results of the measured and calculated power flows were compatible and the calculated uncertainties are comparable with the measurement ones.

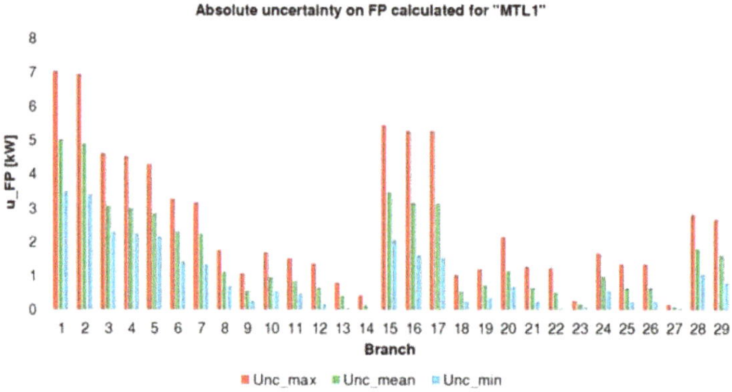

Figure 15. Maximum, average and minimum values of the absolute uncertainties in the calculated active power flows of the "MTL1" line (31 May 2018).

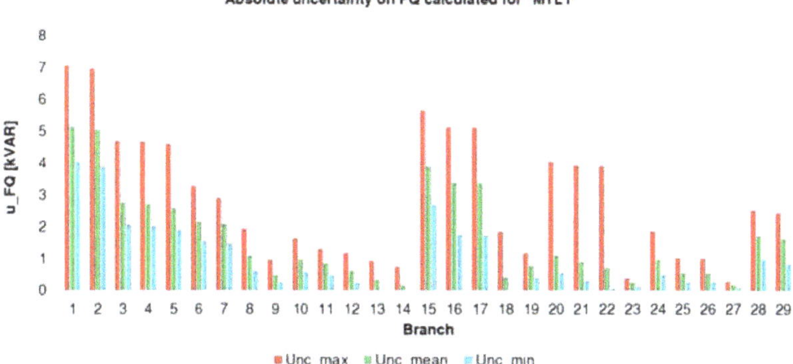

Figure 16. Maximum, average and minimum values of the absolute uncertainties in the reactive power flows calculated of the "MTL1" line (31 May 2018).

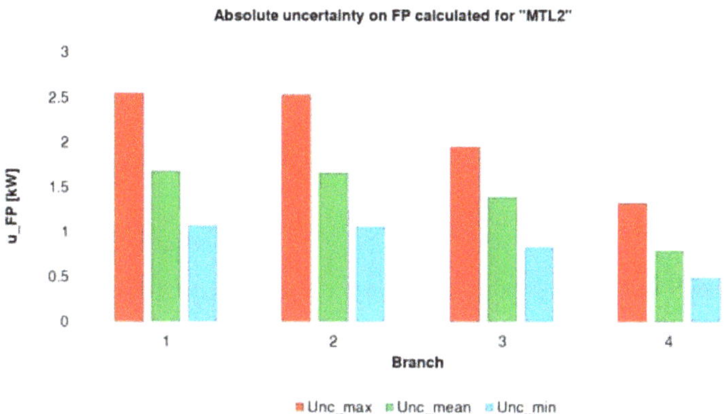

Figure 17. Maximum, average and minimum values of the absolute uncertainties in the calculated active power flows of the "MTL2" line (31 May 2018).

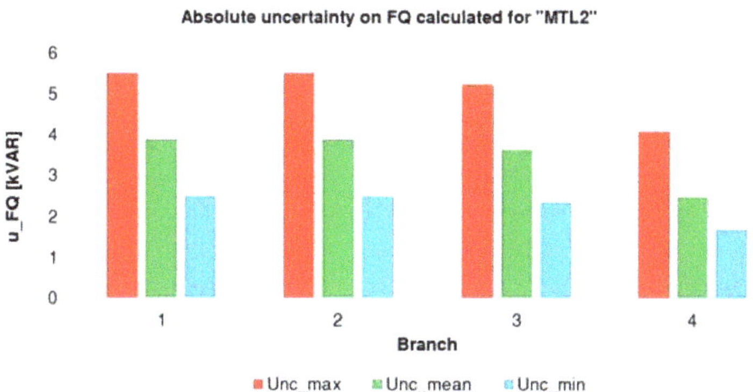

Figure 18. Maximum, average and minimum values of the relative uncertainties in the reactive power flows calculated of the "MTL2" line (31 May 2018).

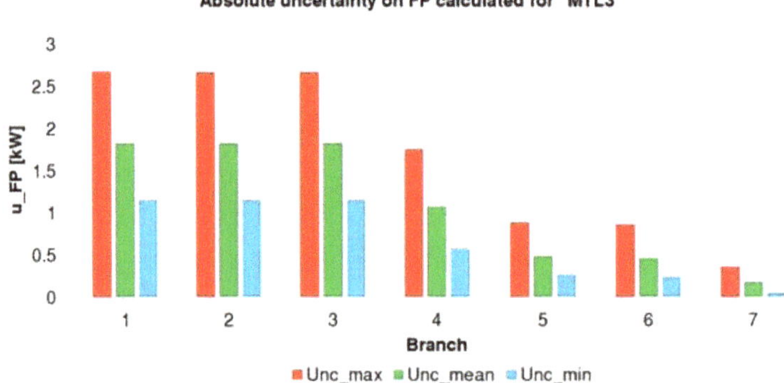

Figure 19. Maximum, average and minimum values of the absolute uncertainties in the calculated active power flows of the "MTL3" line (31 May 2018).

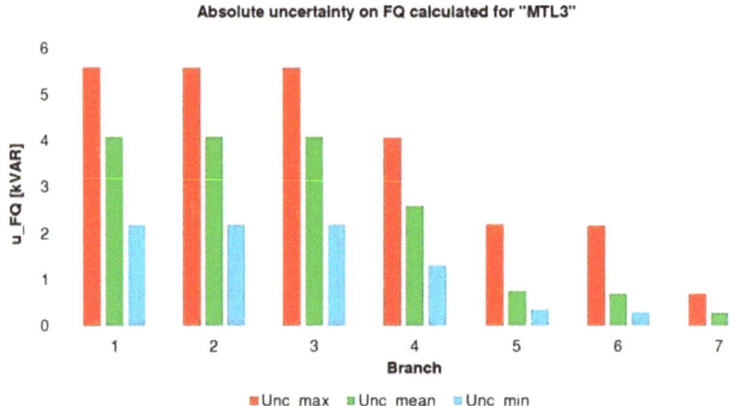

Figure 20. Maximum, average and minimum values of the absolute uncertainties in the reactive power flows calculated of the "MTL3" line (31 May 2018).

4.3. Impact of Line Parameters Uncertainties

Since network parameters are also input quantities of the load flow algorithm, their uncertainty influence on the load flow calculation results was also studied. In fact, the network parameter-rated data provided by DSO cannot be exactly equal to the real ones. Thus, an analysis was conducted to assess the influence of the possible deviation of actual line parameter values from the rated ones. In addition, in this case, the evaluation was carried out by means of a Monte Carlo procedure.

In more detail, an uncertainty value was assumed equal for all the network parameters (R, X and Y). Then, for a given condition of all input quantities, the load flow calculation was repeated 10^5 times. At each execution, in addition to the random variation of all the measured quantity inputs, network parameters were also varied inside the assumed uncertainty range. At the end of the 10^5 iterations, an average value and an uncertainty value were obtained for each power flow. The procedure was repeated 16 times, varying the uncertainty value on the network parameters from 0 to 15% with a step equal to 1. The value 15% was chosen as upper limit value. It was determined considering a 10% uncertainty on line length. Moreover, it is lower than the parameter variation correspondent to two subsequent cable subsections (i.e., 25 and 50 mm^2).

Figure 21 shows the trend of the relative uncertainties on the active power flow of branch 1 of the "MTL3" line, due to network parameter uncertainties. Figure 22 shows the same trend for reactive power. The input data are those measured at 13:00:00 on 31 May 2018. It can be seen that the relative uncertainty in the active power flow remains constant. This result can be justified by analyzing the dependence of the active power flow from the network parameters; in fact, it basically depends on the line losses on longitudinal resistance R_i (see Figure 3); these losses represent an extremely small percentage of the active power flows. This explains the low dependence of the active power flow from the network parameter uncertainties. A different behavior was observed for the reactive power flow in Figure 22. It is more sensitive to network parameter variability. This is due to the reactive power flow dependence from the line transversal capacitive admittance, i.e., from the term $(-V_i^2/Y_i)$ of Equation (2). It can be seen that network parameter uncertainty variation from 0 to 15% corresponds to a reactive power flow variation from 2% to 9%. Thus, reactive power flow calculations are more sensitive than active power flow ones to grid parameter uncertainty. However, the reactive power flow remains lower than 10% even with a maximum network parameter variation of 15%.

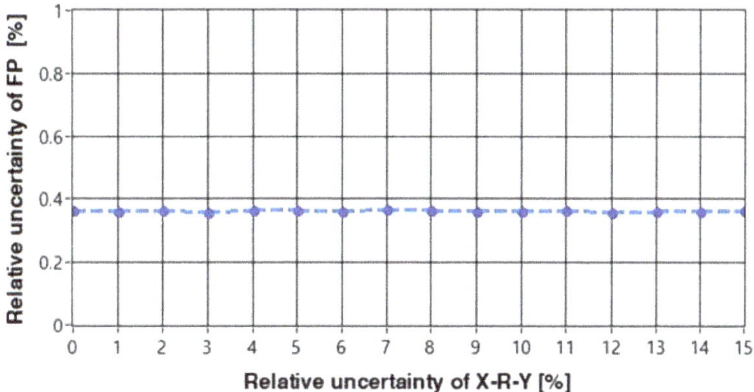

Figure 21. Relative uncertainty in the calculated active power flow of branch 1 of the "MTL3" line in dependence of the line parameter uncertainties.

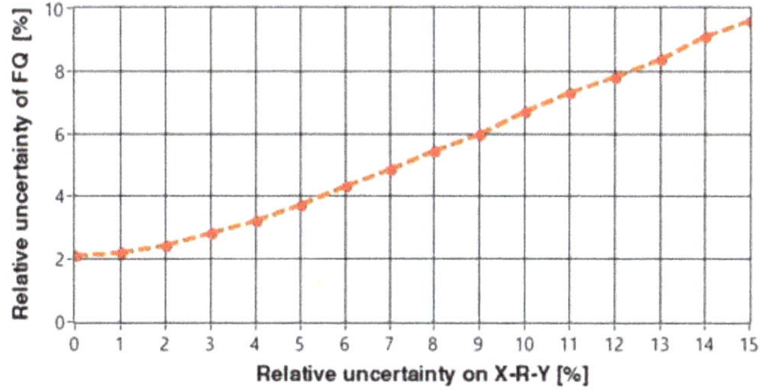

Figure 22. Relative uncertainty in the calculated reactive power flow of branch 1 of the "MTL3" line in dependence of the line parameters uncertainties.

5. Simulation with Distributed Generation

In previous sections, the virtual tool measurement performances were validated. Thus, it can be used offline for planning purposes, i.e., to evaluate the network quantities (voltages, power flows and losses) in the case of an increase in the power injected into the network from distributed generation (DG). In this way, the behavior of the network quantities can be analyzed in different operating conditions whilst varying the number, size and position of the distributed generators connected to the grid.

In order to show a possible use of the VI, in this paragraph the results of some simulation on the "MTL2" line were reported. For each scenario, a fixed power value of photovoltaic production was chosen. In the hypothesis of installing the distributed generators in the LV network, their rated powers were expressed as a percentage of the rated power of the MV/LV transformer powering the related LV line. For a given rated power, the distributed generator's actual power depended on the period of the year under test and on the time during each day. Since the aim of this paper was to study the influence of an increase in the distributed generation injected into the network, the ideal conditions were considered as a worst case scenario, neglecting the possible influence of weather disturbances. Thus, the trend of daily solar radiation was modeled with a parabola curve, whose parameters depend on the analyzed season.

The photovoltaic power plants were simulated as PQ generators. In the case study herein presented, the generators' reactive powers were set to zero, in order to investigate a worst case scenario condition. Different simulations were performed on the "MTL2" line, varying the power injected by the distributed generators in each node. In more detail, five scenarios were here reported based on the following installed powers expressed in a percentage of the nominal power of the corresponding power transformer:

- Scenario 1: no injected power from DG;
- Scenario 2: installed DG power equal to 10% of the transformer-rated power for each secondary substation;
- Scenario 3: installed DG power equal to 20% of the transformer-rated power;
- Scenario 4: installed DG power equal to 30% of the transformer-rated power for substations 3 and 4 and equal to 10% for substations 1 and 2;
- Scenario 5: installed DG power equal to 35% of the nominal transformer power for substations 3 and 4, 10% for substation 2 and 5% for substation 1.

Table 6 shows the absolute values of the injected power for the aforementioned scenarios.

Table 6. Distributed generation active powers in the different scenarios under test.

	Distributed Generation Power (kW)			
Scenario	Node 1	Node 2	Node 3	Node 4
Scenario 1	0	0	0	0
Scenario 2	16	80	80	63
Scenario 3	32	160	160	126
Scenario 4	16	80	240	189
Scenario 5	8	80	280	220.5

Figures 23 and 24 show the active power flows and nodal voltages obtained with the virtual tool. Figure 23 shows how the increase in the overall distributed generation causes a progressive reduction in the power flows of the line. Scenario 4 causes a reverse power flow in branch 4, while scenario 5 causes a reverse power flow up to branch 2. It should be considered that these power flow inversions have to be monitored and correctly managed, because they can cause frequency instability and network malfunctioning, especially in an islanded micro-grid as that of Favignana.

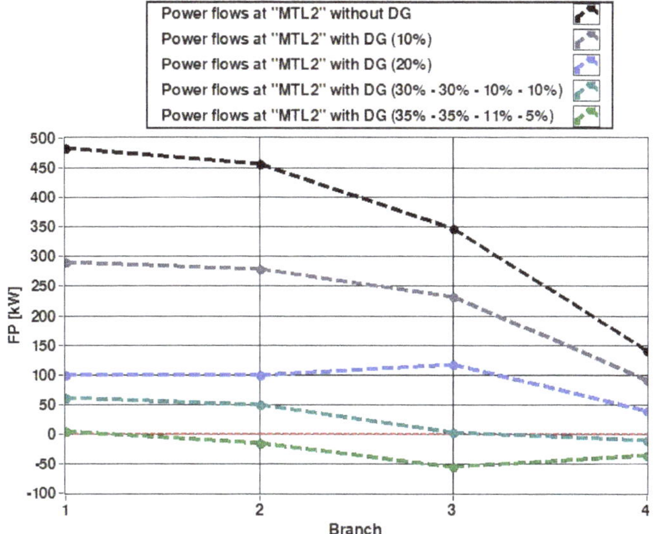

Figure 23. Active power flows of the "MTL2" line as the power fed by the distributed generation (DG) changes (31 May 2018 at 1:00 pm).

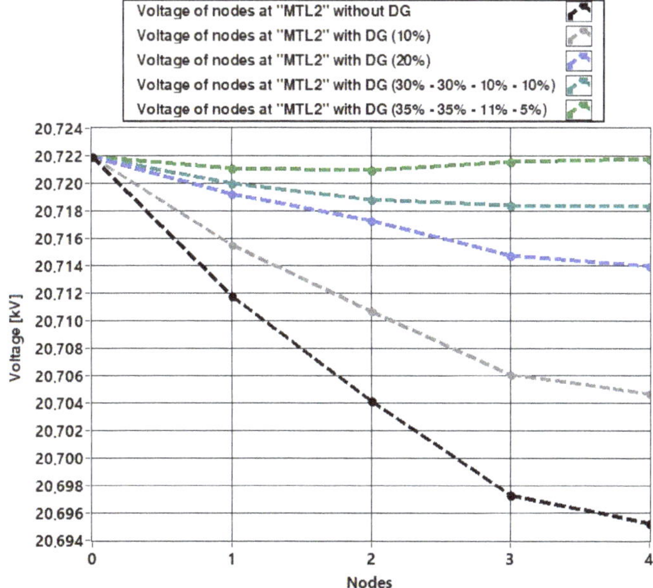

Figure 24. Nodal voltages of the "MTL2" line as the power fed by the DG changes (31 May 2018 at 1:00 pm).

Figure 24 shows the relationship between the power flow reductions and inversions and the increase in all nodal voltages, in particular in the voltages at the nodes 3 and 4 which are characterized by the higher power injection from DG. These increases cause the variations on the voltage trends which became no longer monotonous. On the other hand, differently from power flow variations due to DGs, these values of voltage variations were not necessarily considered a problem for the analyzed network.

Figures 25–28 show the comparison between the active power flow during the day in the absence of DG and those of scenario 4 for the "MTL2" line branches from 1 to 4, respectively. It can be noted that the effect of the DG injection in the morning hours causes power flow inversions in branches 3 and 4 (Figures 27 and 28).

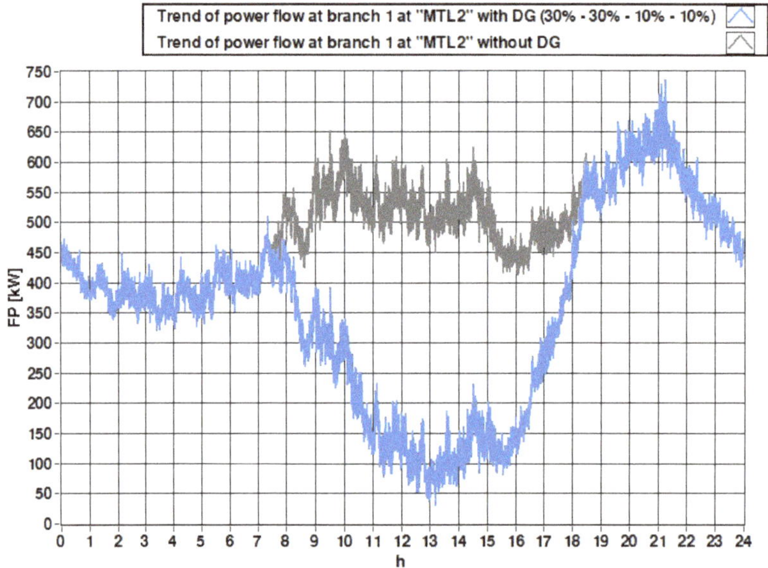

Figure 25. Active power flow during the day 31 May 2018 in branch 1 of the "MTL2" line, both in the absence and with the DG power injection corresponding to Scenario 4.

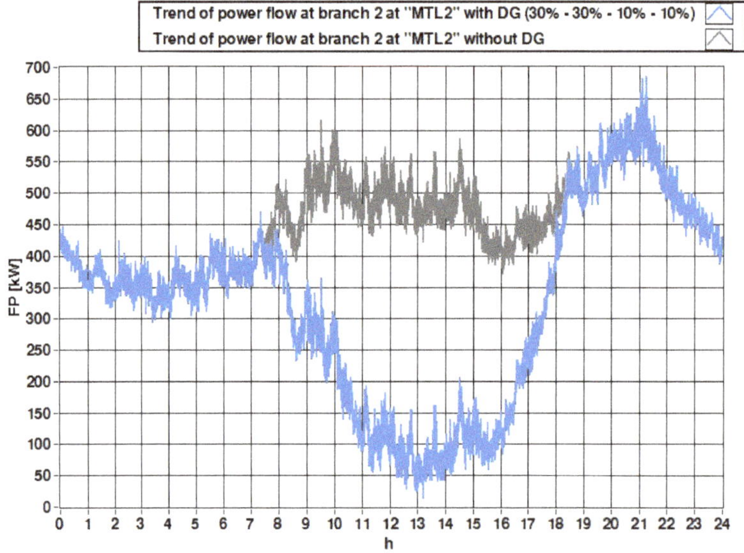

Figure 26. Active power flow during the day 31 May 2018 in branch 2 of the "MTL2" line, both in the absence and with the DG power injection corresponding to Scenario 4.

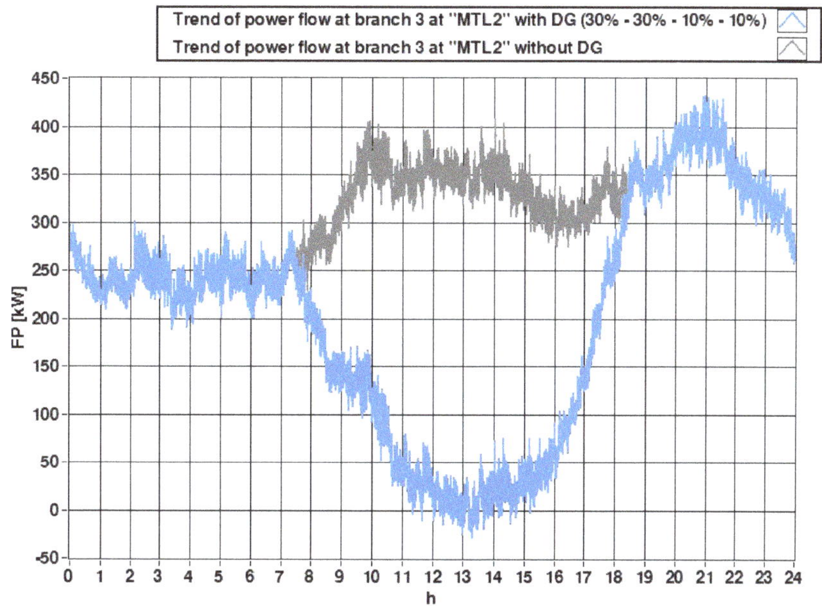

Figure 27. Active power flow during the day 31 May 2018 in branch 3 of the "MTL2" line, both in the absence and with the DG power injection corresponding to Scenario 4.

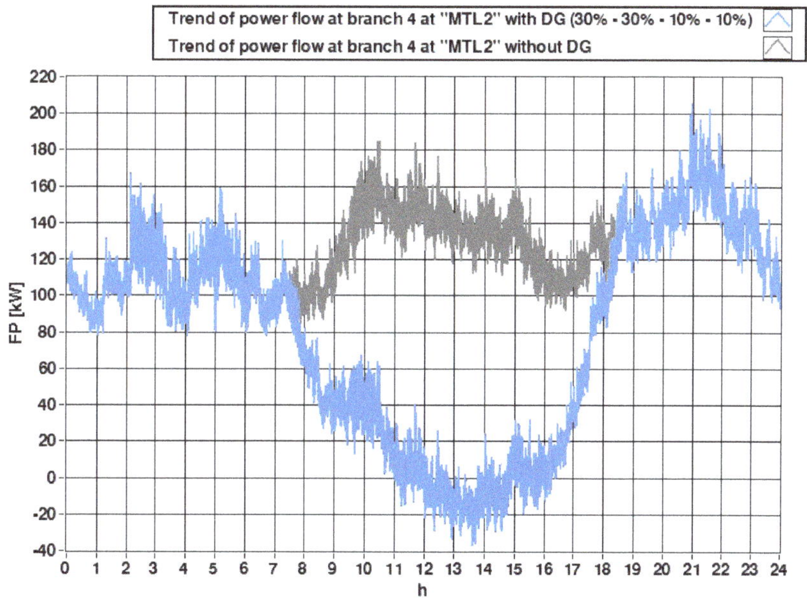

Figure 28. Active power flow during the day 31 May 2018 in branch 4 of the "MTL2" line, both in the absence and with the DG power injection corresponding to Scenario 4.

These results show how the tool can be used to know the possible consequences of installing photovoltaic DG on the network. Starting from the size and position of the generators, it is possible to study the effects of a new power plant installation on the MV network by simulating their presence and verifying the consequent variations on the voltage profiles and power flows. High-power flow inversions, especially close to the generating station, can in fact cause network instability. Thus, the virtual tool can be useful to plan for improvement in the network, such as the installation of storage systems and to define their power and energy in order to avoid or reduce power flow inversions in the network.

6. Conclusions

This paper investigated the possibility of using a feasible virtual tool for load flow analysis and power system planning, based on a low-cost distributed measurement system, specifically tailored for the case of an islanded distribution network. The proposed solution has been on-field implemented and characterized, on the real distribution network of the Island of Favignana (Italy, Mediterranean Sea).

The proposed virtual tool has been developed in a LabVIEW environment for monitoring the MV distribution network of the micro-grid; the tool allows determining the power flows and losses in all the branches and voltages in all the nodes of the MV network, even in the presence of distributed generation. The monitoring of the MV network is carried out through a simplified approach to load flow, implementing a backward/forward algorithm belonging to the class of power summation methods; the input quantities of the algorithm are active and reactive power measurements, which are taken at the LV side of the power transformers (with the exception ion the case of MV users), allowing a reduction in the costs of the measuring points.

The accuracy of the load flow calculations was experimentally verified for all three MV lines of the Favignana network, by comparing the power flow values calculated at the beginning of each line with the related values measured by the PQAs installed in the generating substation, also taking into account the related uncertainty range. This comparison, carried out using the measurement data acquired every 2 s in a period of 33 days, showed the compatibility between the calculated and measured power flows. The accuracy of the quantities calculated by the VI was also assessed by performing an uncertainty analysis. In further detail, starting from the accuracy specifications of the used instruments and measuring transducers, uncertainties in the power measurements were calculated both for the LV and MV secondary substations and for the voltage measurements at the MV bus-bars of the central generating station. The propagation of these uncertainties in the estimated power flows was investigated by implementing a Monte Carlo analysis. The uncertainty results obtained for the estimated values were comparable to that of the installed instruments and the results of the estimated and measured power flows were compatible. The same approach was also used to evaluate the dependence of the results on the variability of the network parameters. In this case, no high influence was observed in the active power flows, while the reactive powers were found sensible to the cable parameter knowledge.

The validated VI was also used as a useful tool for the network planning. It allows simulating different conditions of DG penetration (in particular from photovoltaics) starting from real load measurement data and evaluating power flows, voltage profiles and losses in the network, by varying the number, size and position of the distributed generators. A first study on one of the three lines showed how the distributed generators in an islanded micro-grid mostly affect the trend of power flows, causing power flow inversions which have to be correctly managed to avoid network instability or malfunctioning.

The obtained results confirm the feasibility of the developed solution, which allows the DSO to effectively monitor networks the power flows, by the means of a simple analysis tool and a low-cost distributed measurement system. It has been also shown that the proposed virtual tool can be used for planning purposes of distributed generation and storage increase. For example, in future works it can allow for investigating different distributed generation scenarios, where the installation of storage systems can be also investigated, in terms of both storage location and size, in order to avoid or reduce power flow inversions in the network. Furthermore, the use of power quality analyzers, as those used in the real case study implementation, can allow the expandability of the proposed system, with the implementation of power-quality additional virtual tools on the same platform, without the need of further instrumentation costs.

Author Contributions: Conceptualization, A.C., V.C., D.D.C. and G.T.; data curation, G.A., S.G., M.P., N.P. and G.C.; investigation, A.C., V.C., D.D.C. and G.T.; methodology, A.C., V.C., D.D.C., N.P., G.C., and G.T.; software, G.A., S.G., M.P., N.P. and G.C.; supervision, A.C., V.C., D.D.C., N.N.Q. and G.T.; validation, G.A., A.C., V.C., D.D.C., S.G., N.P., G.C., N.N.Q. and G.T.; writing—original draft, G.A., S.G., N.P. and G.C.; writing—review and editing, A.C., V.C., D.D.C., N.N.Q. and G.T. All authors have read and agreed to the published version of the manuscript.

Funding: This research was funded by the following grants: PO FESR Sicilia 2014–2020, Action 1.1.5, Project n. 08000PA90246, Project acronym: I-Sole, Project title: "Smart grids per le isole minori (Smart grids for small islands)", CUP: G99J18000540007 and supported by the Bilateral Agreement CNR/VAST, Joint Research Projects 2020–2021, CUP: B54I20000220001, QTIT01.03/20-21.

Acknowledgments: The authors wish to thank the local DSO of Favignana (Società Elettrica di Favignana, SEA S.p.a.) for the support in the measurement data collection and processing.

Conflicts of Interest: The authors declare no conflict of interest.

References

1. Alves, M.; Segurado, R.; Costa, M. On the road to 100% renewable energy systems in isolated islands. *Energy* **2020**, *198*, 117321. [CrossRef]
2. Kotzebue, J.R.; Weissenbacher, M. The EU's Clean Energy strategy for islands: A policy perspective on Malta's spatial governance in energy transition. *Energy Policy* **2020**, *139*, 111361. [CrossRef]
3. Silva, A.R.; Estanqueiro, A. Optimal Planning of Isolated Power Systems with near 100% of Renewable Energy. *IEEE Trans. Power Syst.* **2020**, *35*, 1274–1283. [CrossRef]
4. Eras-Almeida, A.A.; Egido-Aguilera, M.A. Hybrid renewable mini-grids on non-interconnected small islands: Review of case studies. *Renew. Sustain. Energy Rev.* **2019**, *116*, 109417. [CrossRef]
5. Mendoza-Vizcaino, J.; Sumper, A.; Galceran-Arellano, S. PV, wind and storage integration on small islands for the fulfilment of the 50-50 renewable electricity generation target. *Sustain. Switz.* **2017**, *9*, 905. [CrossRef]
6. Corsini, A.; Cedola, L.; Lucchetta, F.; Tortora, E. Gen-set control in stand-alone/RES integrated power systems. *Energies* **2019**, *12*, 3353. [CrossRef]
7. Santos, A.Q.; Ma, Z.; Olsen, C.G.; Jorgensen, B.N. Framework for microgrid design using social, economic, and technical analysis. *Energies* **2018**, *11*, 2832. [CrossRef]
8. Hernández-Callejo, L.A. Comprehensive Review of Operation and Control, Maintenance and Lifespan Management, Grid Planning and Design, and Metering in Smart Grids. *Energies* **2019**, *12*, 1630. [CrossRef]
9. Ibhaze, A.E.; Akpabio, M.U.; Akinbulire, T.O. A review on smart metering infrastructure. *Int. J. Energy Technol. Policy* **2020**, *16*, 277–301. [CrossRef]
10. Dileep, G. A survey on smart grid technologies and applications. *Renew. Energy* **2020**, *146*, 2589–2625. [CrossRef]
11. Kabalci, Y. A survey on smart metering and smart grid communication. *Renew. Sustain. Energy Rev.* **2016**, *57*, 302–318. [CrossRef]
12. Andreadou, N.; Olariaga Guardiola, M.; Fulli, G. Telecommunication Technologies for Smart Grid Projects with Focus on Smart Metering Applications. *Energies* **2016**, *9*, 375. [CrossRef]
13. Prasad, S.; Kumar, D.M.V. Trade-offs in PMU and IED deployment for active distribution state estimation using multi-objective evolutionary algorithm. *IEEE Trans. Instrum. Meas.* **2018**, *67*, 1298–1307. [CrossRef]
14. Delle Femine, A.; Gallo, D.; Landi, C.; Luiso, M. The Design of a Low Cost Phasor Measurement Unit. *Energies* **2019**, *12*, 2648. [CrossRef]
15. Bertocco, M.; Frigo, G.; Narduzzi, C.; Muscas, C.; Pegoraro, P.A. Compressive Sensing of a Taylor-Fourier Multifrequency Model for Synchrophasor Estimation. *IEEE Trans. Instrum. Meas.* **2015**, *64*, 3274–3328. [CrossRef]

16. Von Meier, A.; Stewart, E.; McEachern, A.; Andersen, M.; Mehrmanesh, L. Precision micro-synchrophasors for distribution systems: A summary of applications. *IEEE Trans. Smart Grid* **2017**, *8*, 2926–2936. [CrossRef]

17. Dusabimana, E.; Yoon, S.G. A Survey on the Micro-Phasor Measurement Unit in Distribution Networks. *Electronics* **2020**, *9*, 305. [CrossRef]

18. Liu, Y.; Wu, L.; Li, J. D-PMU based applications for emerging active distribution systems: A review. *Electr. Power Syst. Res.* **2020**, *179*, 106063. [CrossRef]

19. Hojabri, M.; Dersch, U.; Papaemmanouil, A.; Bosshart, P. A Comprehensive Survey on Phasor Measurement Unit Applications in Distribution Systems. *Energies* **2019**, *12*, 4552. [CrossRef]

20. Pokhrel, B.R.; Bak-Jensen, B.; R Pillai, J. Integrated Approach for Network Observability and State Estimation in Active Distribution Grid. *Energies* **2019**, *12*, 2230. [CrossRef]

21. Artale, G.; Cataliotti, A.; Cosentino, V.; Di Cara, D.; Guaiana, S.; Telaretti, E.; Panzavecchia, N.; Tinè, G. Incremental Heuristic Approach for Meter Placement in Radial Distribution Systems. *Energies* **2019**, *12*, 3917. [CrossRef]

22. De Din, E.; Pau, M.; Ponci, F.; Monti, A. A Coordinated Voltage Control for Overvoltage Mitigation in LV Distribution Grids. *Energies* **2020**, *13*, 2007. [CrossRef]

23. Ginocchi, M.; Ahmadifar, A.; Ponci, F.; Monti, A. Application of a Smart Grid Interoperability Testing Methodology in a Real-Time Hardware-In-The-Loop Testing Environment. *Energies* **2020**, *13*, 1648. [CrossRef]

24. Soares, T.M.; Bezerra, U.H.; Tostes, M.E.L. Full-Observable Three-Phase State Estimation Algorithm Applied to Electric Distribution Grids. *Energies* **2019**, *12*, 1327. [CrossRef]

25. Sanseverino, E.R.; Di Silvestre, M.L.; Zizzo, G.; Gallea, R.; Quang, N.N. A Self-Adapting Approach for Forecast-Less Scheduling of Electrical Energy Storage Systems in a Liberalized Energy Market. *Energies* **2013**, *6*, 5738–5759. [CrossRef]

26. Bucci, G.; Ciancetta, F.; D'Innocenzo, F.; Fiorucci, E.; Ometto, A. Development of a Low Cost Power Meter Based on A Digital Signal Controller. *Int. J. Emerg. Electr. Power Syst.* **2018**, *19*. [CrossRef]

27. Sanduleac, M.; Lipari, G.; Monti, A.; Voulkidis, A.; Zanetto, G.; Corsi, A.; Toma, L.; Fiorentino, G.; Federenciuc, D. Next Generation Real-Time Smart Meters for ICT Based Assessment of Grid Data Inconsistencies. *Energies* **2017**, *10*, 857. [CrossRef]

28. Alahakoon, D.; Yu, X. Smart electricity meter data intelligence for future energy systems: A survey. *IEEE Trans. Ind. Inform.* **2015**, *12*, 425–436. [CrossRef]

29. Wena, L.; Zhoua, K.; Yanga, S.; Lia, L. Compression of smart meter big data: A survey. *Renew. Sustain. Energy Rev.* **2018**, *91*, 59–69. [CrossRef]

30. Avancini, D.B.; Rodrigues, J.J.; Martins, S.G.; Rabêlo, R.A.; Al-Muhtadi, J.; Solic, P. Energy meters evolution in smart grids: A review. *J. Clean. Prod.* **2019**, *217*, 702–715. [CrossRef]

31. Kamyabi, L.; Esmaeili, S.; Koochi, M.H.R. Power quality monitor placement in power systems considering channel limits and estimation error at unobservable buses using a bi-level approach. *Int. J. Electr. Power Energy Syst.* **2018**, *102*, 302–311. [CrossRef]

32. Kong, X.; Chen, Y.; Xu, T.; Wang, C.; Yong, C.; Li, P.; Yu, L. A Hybrid State Estimator Based on SCADA and PMU Measurements for Medium Voltage Distribution System. *Appl. Sci.* **2018**, *8*, 1527. [CrossRef]

33. Lin, C.; Wu, W.; Guo, Y. Decentralized Robust State Estimation of Active Distribution Grids Incorporating Microgrids Based on PMU Measurements. *IEEE Trans. Smart Grid* **2019**, *11*, 810–820. [CrossRef]

34. Liu, Y.; Li, J.; Wu, L. State estimation of three-phase four-conductor distribution systems with real-time data from selective smart meters. *IEEE Trans. Power Syst.* **2019**, *34*, 2632–2643. [CrossRef]

35. Kumar, P.; Lin, Y.; Bai, G.; Paverd, A.; Dong, J.S.; Martin, A. Smart grid metering networks: A survey on security, privacy and open research issues. *IEEE Commun. Surv. Tutor.* **2019**, *21*, 2886–2927. [CrossRef]

36. Dehghanpour, K.; Wang, Z.; Wang, J.; Yuan, Y.; Bu, F. A survey on state estimation techniques and challenges in smart distribution systems. *IEEE Trans. Smart Grid* **2018**, *10*, 2312–2322. [CrossRef]

37. Ahmada, F.; Rasoola, A.; Ozsoyb, E.; Rajasekarc, S.; Sabanovica, A.; Elitaşa, M. Distribution system state estimation-A step towards smart grid. *Renew. Sustain. Energy Rev.* **2018**, *81*, 2659–2671. [CrossRef]

38. Branco, H.; Oleskovicz, M.; Coury, D.V.; Delbem, A.C.B. Multiobjective optimization for power quality monitoring allocation considering voltage sags in distribution systems. *Int. J. Electr. Power Energy Syst.* **2018**, *97*, 1–10. [CrossRef]

39. Sheibani, M.; Ketabi, A.; Nosratabadi, M. Optimal power quality meters placement with consideration of single line and meter loss contingencies. *Int. J. Ind. Electron. Control. Optim.* **2018**, *1*, 81–89.

40. Elphick, S.; Gosbell, V.; Smith, V.; Perera, S.; Ciufo, P.; Drury, G. Methods for harmonic analysis and reporting in future grid applications. *IEEE Trans. Power Deliv.* **2016**, *32*, 989–995. [CrossRef]

41. Sharma, K.; Saini, L.M. Power-line communications for smart grid: Progress, challenges, opportunities and status. *Renew. Sustain. Energy Rev.* **2017**, *67*, 704–751. [CrossRef]

42. Sendin, A.; Pena, I.; Angueira, P. Strategies for Power Line Communications Smart Metering Network Deployment. *Energies* **2014**, *7*, 2377–2420. [CrossRef]

43. Bali, M.C.; Rebai, C. Improved maximum likelihood S-FSK receiver for PLC modem in AMR. *J. Electr. Comput. Eng.* **2012**, *2012*, 452402. [CrossRef]

44. Rinaldi, S.; Pasetti, M.; Sisinni, E.; Bonafini, F.; Ferrari, P.; Rizzi, M.; Flammini, A. On the Mobile Communication Requirements for the Demand-Side Management of Electric Vehicles. *Energies* **2018**, *11*, 1220. [CrossRef]

45. Artale, G.; Cataliotti, A.; Cosentino, V.; Di Cara, D.; Fiorelli, R.; Guaiana, S.; Panzavecchia, N.; Tinè, G. A New Coupling Solution for G3-PLC Employment in MV Smart Grids. *Energies* **2019**, *12*, 2474. [CrossRef]

46. Elgenedy, M.; Papadopoulos, T.A.; Galli, S.; Chrysochos, A.I.; Papagiannis, G.K.; Al-Dhahir, N. MIMO-OFDM NB-PLC Designs in Underground Medium-Voltage Networks. *IEEE Syst. J.* **2019**, *13*, 3759–3769. [CrossRef]

47. Ouissi, S.; Ben Rhouma, O.; Rebai, C. Statistical modeling of mains zero crossing variation in powerline communication. *Meas. J. Int. Meas. Confed.* **2016**, *90*, 158–167.

48. Cataliotti, A.; Cosentino, V.; Di Cara, D.; Russotto, P.; Telaretti, E.; Tinè, G. An Innovative Measurement Approach for Load Flow Analysis in MV Smart Grids. *IEEE Trans. Smart Grid* **2016**, *7*, 889–896. [CrossRef]

49. Cataliotti, A.; Cosentino, V.; Di Cara, D.; Nuccio, S.; Panzavecchia, N.; Tinè, G. A simplified approach for load flow analysis in MV smart grids based on LV power measurements. In Proceedings of the 2017 IEEE International Instrumentation and Measurement Technology Conference (I2MTC), Torino, Italy, 22–25 May 2017; pp. 1–6.

50. Cataliotti, A.; Cosentino, V.; Di Cara, D.; Guaiana, S.; Nuccio, S.; Panzavecchia, N.; Tinè, G. Measurement uncertainty impact on simplified load flow analysis in MV smart grids. In Proceedings of the 2018 IEEE International Instrumentation and Measurement Technology Conference (I2MTC), Houston, TX, USA, 14–17 May 2018; pp. 1354–1359.

51. Haque, M.H. Efficient load flow method for distribution systems with radial or mesh configuration. *IEEE Proc. Gener. Transm. Distrib.* **1996**, *143*, 33–38. [CrossRef]

52. Haque, M.H. A general load flow method for distribution systems. *Electr. Power Syst. Res.* **2000**, *54*, 47–54. [CrossRef]

53. ISO\IEC. *Uncertainty of Measurement—Part 3: Guide to the Expression of Uncertainty in Measurement (GUM:1995)*; ISO\IEC: Geneva, Switzerland, 2008; Volume 98-3, Available online: https://www.iso.org/standard/50461.html (accessed on 16 June 2020).

54. IEC Standard. *Instrument Transformers—Part 2: Additional Requirements for Current Transformers IEC Standard 61869-2*; IEC Standard: Geneva, Switzerland, 2012.

55. IEC Standard. *Instrument Transformers—Part 3: Additional Requirements for Inductive Voltage Transformers IEC Standard 61869-3*; IEC Standard: Geneva, Switzerland, 2012.

56. Joint Committee for Guides in Metrology (JCGM). *Evaluation of Measurement Data—Supplement 1 to the 'Guide to the Expression of Uncertainty in Measurement'—Propagation of Distributions Using a Monte Carlo Method, 101:2008*; JCGM: Sevres, France, 2008.

Article

A Chronological Literature Review of Electric Vehicle Interactions with Power Distribution Systems

Andrés Arias-Londoño [1,*], Oscar Danilo Montoya [2,3] and Luis Fernando Grisales-Noreña [1]

[1]   Facultad de Ingeniería, Institución Universitaria Pascual Bravo, Calle. 73 # 73a-226, Medellín 050034, Colombia; luis.grisales@pascualbravo.edu.co

[2]   Facultad de Ingeniería, Universidad Distrital Francisco José de Caldas, Carrera 7 No. 40B-53, Bogotá D.C 11021, Colombia; o.d.montoyagiraldo@ieee.org or omontoya@utb.edu.co

[3]   Laboratorio Inteligente de Energía, Universidad Tecnológica de Bolívar, km 1 vía Turbaco, Cartagena 131001, Colombia

*   Correspondence: andres.arias366@pascualbravo.edu.co

Received: 31 March 2020; Accepted: 2 June 2020; Published: 11 June 2020

Abstract: In the last decade, the deployment of electric vehicles (EVs) has been largely promoted. This development has increased challenges in the power systems in the context of planning and operation due to the massive amount of recharge needed for EVs. Furthermore, EVs may also offer new opportunities and can be used to support the grid to provide auxiliary services. In this regard, and considering the research around EVs and power grids, this paper presents a chronological background review of EVs and their interactions with power systems, particularly electric distribution networks, considering publications from the IEEE Xplore database. The review is extended from 1973 to 2019 and is developed via systematic classification using key categories that describe the types of interactions between EVs and power grids. These interactions are in the framework of the power quality, study of scenarios, electricity markets, demand response, demand management, power system stability, Vehicle-to-Grid (V2G) concept, and optimal location of battery swap and charging stations.

Keywords: battery swap station; charging station; demand management; demand response; electric vehicle; electricity markets; power quality; Vehicle-to-Grid

---

1. Introduction

Over the last few years, electric transport has been largely promoted by governments as an effort to reduce dependence on fossil fuels and air pollution from vehicles propelled by internal combustion engines (ICEs). Electric vehicles (EVs) have become an important component in this subject due to the advantages presented compared to ICE vehicles, i.e., reduction of noise and notable decrease in greenhouse gases release. Furthermore, a massive introduction of EVs in the power distribution networks leads to adverse effects in terms of the following:

- voltage drops,
- non-desired load peaks,
- increment in energy losses,
- overload on grid components,
- load factor reduction,
- reliability indices deterioration, and
- power quality issues.

A large quantity of research around the interaction of EVs with power distribution systems is found in specialized literature. Particularly, in the IEEE Xplore database (one of the most important worldwide scientific research databases), this subject has been strongly addressed for a bit more than ten years.

Despite the extensive knowledge published in this database, few literature reviews can be found in this regard. Some of the reviews are reported in [1,2]. In [1], recent literature focusing on distribution system services provided by EVs is presented based on three categories: active power support, reactive power support, and renewable energy source integration support. Weaknesses in the control strategies are identified to encourage the exploration of new areas aligned with the current requirements of smart grids. On the other side, Dubay and Santoso in [2] perform a detailed review to evaluate and mitigate the impacts of charging EVs on residential distribution systems. Other relevant reviews are shown by [3,4]. In [3], certain operating features such as voltage stability, peak load, power quality, and transformer performance are considered as key classifiers; conversely, Jia in [4] introduces a more general classification focused on objective functions, optimization methods, and market design. Furthermore, works such as [5] develop a comprehensive review using a systematic classification considering a wide approach to the EV–power grid interaction.

Compared with the literature reviews mentioned above, the primary contributions of this paper are listed below:

- provide a chronological literature review through the end of 2019 of research on the interaction between EVs and power distribution systems, found in IEEE Xplore database;
- perform a detailed and systematic classification of the papers that address EVs and distribution networks, considering relevant categories, e.g., power quality, demand management, power system stability, Vehicle-to-Grid services, and demand response, among others;
- identify the topics that need further exploration, mindful of the upcoming increase of EVs recharging on power grids.

The present article is an updated and improved version of the conference paper in [5]. The rest of the article is divided in the following sections: A general overview of the time window under research for the development of this review is presented in Section 2. Sections 3 and 4 present a time-sequential revision subperiod, for the periods 1973–2015 and 2016–2019, respectively, considering a debugged list of papers from the IEEE Xplore database. The review is developed by using key categories in regard to the type of interactions between EVs and power distribution systems. Section 5 presents some observations and comparisons between the mentioned periods, followed by Section 6 which provides conclusions of the work and avenues of research. Finally, some final reflections are included in Section 7.

2. General Overview

The literature review relevant to the interaction of EVs and power distribution networks is based on an exhaustive search of the works published in the IEEE Xplore database from 1973 to 2019. Firstly, the term "electric vehicle" was used as a key parameter in the database browser, obtaining around 64,400 papers associated with this criterion. Secondly, it was decided whether the paper was assigned to the study according to its content, since several works are not related to EV and power grid interactions, despite the key term filter "electric vehicle" in the browser. Those papers that were not within the theme of this literature review were discarded since their subjects were generally associated with other topics in terms of vehicle operation, i.e., propulsion system, architectures, brake recuperation, power train control, velocity profile optimization, and motion planning, among others. Subsequently, the remaining papers were listed in a debugged database, according to a specific subject in relation to EVs and power grids, as shown in Table 1. The complete list of papers investigated to develop this review can be found in [6].

Table 1. Ranking by number of publications.

Identification	Topic	Number of Publications
ID1	Power quality	104
ID2	Scenarios study	498
ID3	Electricity markets	125
ID4	Demand response	83
ID5	Demand management	514
ID6	Power system stability	164
ID7	Vehicle-to-Grid (V2G)	343
ID8	BSS and/or EVCSs	239
Total		**2070**

According to Table 1, the first column identifies the work category. The works identified as ID1 assess the reliability and harmonics level, caused by the recharge of the EVs in the distribution network, providing results in terms of indices such as total harmonic distortion (THD) and current and voltage signal spectra. This category is relevant since the internal components of the EV are considered as harmonic signal sources. ID2 identifies the evaluation of network load factor, energy loss lines, and transformer overload, among other aspects, under different insertion levels of EVs. Other aspects addressed in this category are the stochastic analysis, usage politics, and EV growth trends in the automotive industry. The works belonging to ID3 consider studies framed within the EV participation in electricity markets, energy price, and cost-to-benefit ratio. The publications identified as ID4 correspond to those works in which the demand response provides an opportunity for EV owners to play a significant role in the operation of the electric grid, by reducing or shifting their EV recharge during peak periods in response to financial incentives. In ID5, the works include mathematical programming, focused on minimizing the operation and investment costs and/or maximizing the quantity of EVs that can be plugged into the network, considering operative constraints (load factor, voltage limits, and maximum current flows) and EV owners' driving patterns. ID6 is a category for the studies in which the EVs provide signals to support power system stability, including ancillary services and voltage, frequency, and small signal stability. The Vehicle-to-Grid (V2G) concept, and the interaction of EVs with distributed generation sources and power storage systems, is developed in publications with ID7. Last but not least, category ID8 presents the works that address the EVCSs (Electric Vehicle Charging Stations) planning and BSS (Battery Swap Stations) in distribution systems, supported by one or some of the following aspects: path planning, transportation network, queuing analysis, traffic flows, routing, and charging station configuration.

In accordance with Figure 1, the impact of EVs on distribution networks was seldom studied during the 1970s, with only one publication in the IEEE Xplore database reported in the year 1973. In the following decade, the scope was not notably changed, with only three publications. Furthermore, the research pipelines were expanded to study the quality of power (ID1). Later, between 1990 and 2006, another category (ID5) showed up in the list mentioned before, which signifies a starting point for mathematical modeling and optimization, focused on the timely demand management of consumers and EVs. In general, between 1973 and 2006, the efforts around this discipline involved almost twenty publications considering power quality, scenario studies, and demand management.

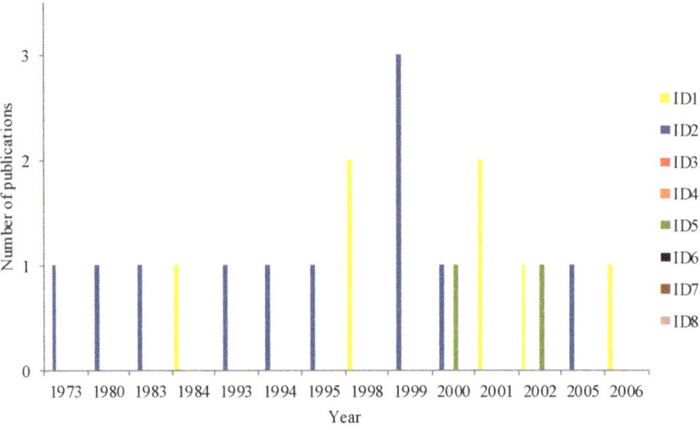

Figure 1. Research trends of EVs and distribution networks, 1973–2006.

Figure 2 provides details on the publications from 2007 to 2019. In 2007, research trends showed the same behavior as in Figure 1; however, in 2008 the range of study choices expanded to EVs participating in electricity markets, power systems stability, and grid support under the V2G concept. In 2009 another trend arose, featured by the works framed within the role played by EVs in the context of demand response. The year 2010 represents a point in which there was a large increase in the number of publications on EVs and their interactions with power networks. In the same year, the optimal locations of EV charging stations and battery swap stations were introduced in the list under study.

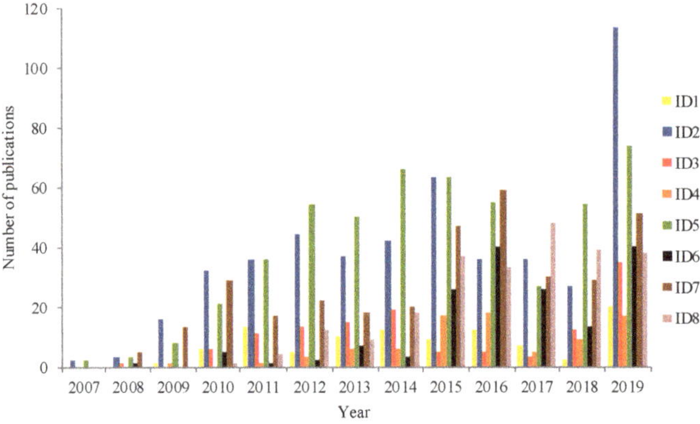

Figure 2. Research trends of EVs and distribution networks, 2007–2019.

According to Figure 3, the period of time 2010–2019 covers the largest number of publications, as the need for network operators and the academic community to manage and confront the increase of EVs plugged into the distribution network increased. In the time-lapse considered for the development of this literature review (1973–2019), the number of works included up to 2070 publications, encompassing journals and conferences. As presented in Table 1, the current state of research was classified in compliance with the research stream and number of publications. This does not imply a low importance for the category with the lowest number of publications.

In the following sections a survey of the main works is addressed, taking into account the categories mentioned above and the 1973–2019 period divided into key subperiods. Since the database is extensive in addressing the total of papers in the review, the choice of the papers addressed throughout the development of this review corresponded to a rigorous examination of each paper listed in the filtered database posted in [6]. In the examination, the contributions and the novelties applied to solve the problem of a certain topic/category were considered. From our point of view, the article was not cited if its proposal to solve a certain problem was framed within conventional strategies without any novelty.

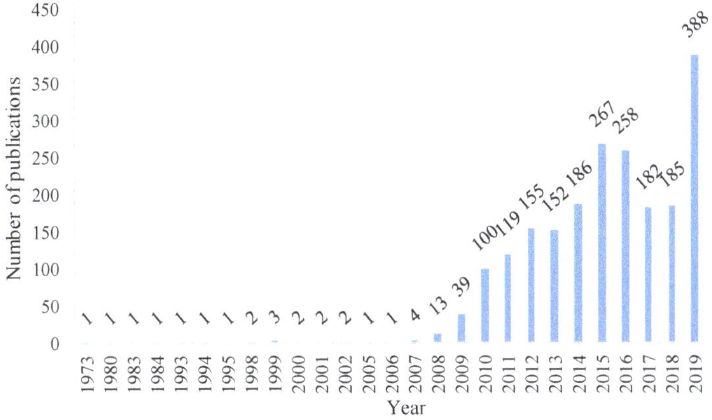

Figure 3. Growth of research from 1973 to 2019.

3. Chronological Review: Part I

This section of the chronological review is focused on the first half of the total papers listed in [6], i.e., 1055 papers for the period 1973–2015. In Figure 4, the participation percentage of the categories is depicted over this first set of works, with a major presence of works focused on demand management and study of scenarios, followed by the V2G concept with a 16% of participation.

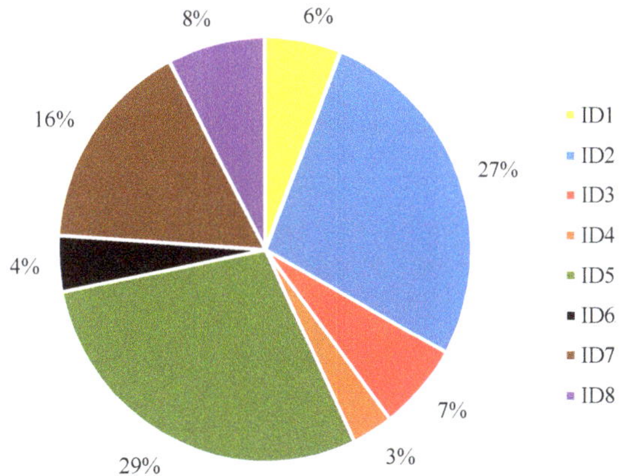

Figure 4. Participation percentage of categories: first half.

3.1. Subperiod 1973 to 1999

As earlier mentioned, the theme of EVs and electric grids interacting started to be researched in the 1970s, according to the IEEE Xplore database. Specifically, in 1973, in [7] a start point of this research was presented, expressing the relevance to conduct a study around the imminent deployment of EVs in the coming years and their impact on generation plants and distribution systems. However, it was not until 1980 [8] that electric companies become more receptive to topics related to the recharge of EVs, stimulating recharge at night time to increase the load factor of the system, and decrease the cost per kWh. Hence, supported in mathematical definitions and technical arguments, [9] presents a study about improving the load factor, opening the possibility for demand management strategies to focus on cushioning the collateral effects of EVs. Not only was load factor improvement one of the study topics in the 1980s, particularly in 1984 the impact of power quality started to be researched, due to the non-linear nature of battery chargers of EVs [10]. These devices can distort the voltage signal and generate harmonic currents, which create problems for power systems, e.g., increase in neutral current, hot spots in transformers, and inaccuracy on measure instruments. To solve these problems, in [10] a current smoothing within the EV charger circuit is proposed.

Since the 1990s, the study of scenarios of EVs in electric networks had become more popular than the power quality studies. This was due to the growing interest to determine strategies allowing EVs to be recharged at hours of low conventional demand and, thus, to flatten the load curve. In this context, Rahman and Shrestha in [11] stated that not only does having sufficient capacity of generation during valley hours need to be considered, and to provide energy to EVs without adverse effects on the electric grid, but it is also necessary to study the strength of distribution systems in order to support these load additions. Since a large quantity of EVs should be recharged during the low-demand period, a considerable EV load can generate non-desirable demand peaks at the beginning of this period. Thus, the need arises to develop fast-charging batteries to recharge some EVs at the end of the valley period and make the demand curve more uniform.

In [12], the necessary elements are successfully established to boost the EV market, among them, the development of batteries with enhanced characteristics and the willingness of distribution companies to improve their electric infrastructure to ensure reliable and timely delivery of energy to a great quantity of EVs. By the year 1995, according to [13], the EV technology remained in prototype status in terms of production and development of batteries; therefore, the number of EVs on the streets for the next twenty years aimed to speculate data. This represented an obstacle to define the load model of the EVs. Three years later, in 1998, the efforts were concentrated again on the problems of power quality, where [14] presented a statistical method to determine the maximum penetration threshold of EVs in a distribution system, so that the THD index did not exceed over 5%. The same year, with the intention to model appropriately the EV load in the system, Ref. [15] shows the modeling of EV chargers from a procedure in which a Montecarlo simulation was used, which obtained the THD in terms of an expected value and a standard deviation. Nevertheless, in contrast with the aspects mentioned in [11], from the point of view of the distribution company, a fast charge is not desired because of the resulting big demand peaks. Despite the motivation for this kind of recharge in low-demand periods, its usage is suggested in emergency cases. Under these supposed disadvantages, quick chargers receive a lot of interest when the likely effect of this type of charger in the distribution system is known [16]. In that work, assessment of the size and influence of EV charging harmonics versus the penetration level is presented. Other works such as [17] discuss the behavior of the demand curves, considering random aspects like the initial time of recharge of EV and the state of charge (SOC). At the end of this decade, the report presented in [18] explains the development of EVs in the last thirty years, the effect at the environmental level, and electricity generation, confirming the terms established in [7] when the scale is tipped again to improve the performance of these vehicles.

3.2. Subperiod 2000 to 2009

At the beginning of the 21st century, the field extended to the role played by EVs as distributed resources to supply, partially or totally, the domestic demand in periods of time where the energy price is relatively high [19]. Under this scheme, some benefits include reduction of energy cost paid by the customer, mitigation of stress perceived by the transformers and distribution lines, privileges related with tax decrease and ease to build recharge infrastructures at their homes, among others. Taking into account the aspects above, it is important to point out the work done by Ceraolo and Pede in [20], where the distance traveled by an EV is estimated with the remaining SOC and the ability of the battery to provide energetic capacity as a function of the discharge rate. In 2002, power quality issues caused by EV recharge were addressed again, showing a quadratic relation between the transformer useful life and current THD index of the battery charger, establishing a limit between 25% and 30% for the THD to provide a reasonable life expectancy for the transformer [21]. Taking up the usage of EVs as distributed resources, it is mentioned in [22] other advantages in this framework, such as mobile AC power, backup energy for homes and offices, stability ancillary services, spinning reserves, and regulation. Due to the incrase in battery activities, not only in EV mobility but also in the area of ancillary services, it is necessary to take into account the economic viability of this framework because the useful life of the battery is reduced when the charge/discharge rate is increased. In 2006 and as a consequence of doing deeper research about the potential of EVs in electric networks, Breucker et al in [23] highlight the services presented by EVs fleets:

- elimination of harmonics, as the non-linear elements of the battery chargers act as active filters,
- power factor improvement by injecting reactive power and peak shaving,
- primary and secondary control for the power balance between the generation and the demand,
- frequency regulation in low-stability grids, inclusive with a lower quantity of EVs, and
- Ancillary generation for outages and construction projects.

During the year 2007, the focus was on conventional topics (scenario studies and demand management) without great contributions. The year 2008 represents a start point for the participation of EVs in power system stability and electricity markets. A clear example is the work published in [24] where the small signal stability of a power system with EVs is analyzed, which can act as constant current or impedance load. The results show that, when the EVs are charged in constant current mode, the electric network is prone to instability. Hence, in constant impedance mode, high penetration levels of EVs can be reached before the instability point. Related with electric markets, in the same year, [25] worked with the effect of EVs on the locational marginal price (LMP). This framework of wholesale electricity prices is determined from the incremental cost of re-dispatching a system to supply an additional demand unit in a specific location, subject to generation and transmission constraints. The EVs are loads that can be recharged at different geographic points and can influence greatly on LMP. Apparently it was not until 2008 that the term V2G (Vehicle-to-Grid) was made official to characterize the ancillary services of EVs to the grid, although in previous years this topic had already been addressed. In this context, some works such as [26], study the requirements for the V2G concept. The existing information flow between the network operator and the EV encompasses the following: the ID of EV, the preferences and parking status of EV, battery storage capacity, SOC, and the power flow from the battery to grid. However, the most important aspects consider the communication range of the system and security in the information transmission, besides the fulfillment of IEEE Standard 1547 where the minimum requirements are established to introduce energy to electric grids [27]. Other publications like [28] consider the relevance of V2G concept for electricity generation balance in environments highly penetrated by distributed generation, as in the Denmark case where around 20% of its energetic capacity comes from wind generation. Conversely, a study done by [29] determined that the viability of increasing EVs in the Azores islands was related with the energy usage coming from renewable sources for recharging EVs.

Efforts focused on demand management, scenario studies, and the V2G concept increased in 2009. Mathematical modeling appears as a new alternative to study the effects of EVs in distribution systems. The topics related with demand response and power quality arise again, although in low proportions. With respect to the V2G concept, the study done in [30] proposed a mathematical programming model for optimal dispatch of generation units, which include the small thermal units and the energy stored in EVs, considering technical, spatial, and temporary constraints. Particle Swarm

Optimisation, or PSO, was used to solve the problem, obtaining an increase in benefits and reliability in the distribution system. It is necessary to point out, through the stability stream, the work performed by El Chehaly et al. in [31], which proposed a Short-Term Voltage Stability Index (SVSI) for wind generation with ancillary services provided by EVs. This index is based on the difference between pre-fault voltage and the minimum voltage reached at fault status; in this manner, with a high number of EVs present, SVSI can be reduced, and the voltage profile is improved. Similar to [30], in [32] a mathematical model was designed in order to maximize the number of EVs plugged into distribution systems, subject to voltage limits and battery energetic requirements. This same philosophy was applied in [33] where minimization of the losses in the system is sought through coordinated charging of EVs. In each iteration of the optimization problem, a conventional load flow is executed to determine the actual network status. A necessary work to highlight for its connection between electric grids and the distribution network gas is the one presented in [34], where losses of both grids are minimized through the transformer's tap control and the compressor's output pressure, in order to cushion the impact of EV load.

3.3. Subperiod 2010 to 2011

In 2010, one of the most studied topics was the V2G concept, whose proposal consists of providing power at peak demand hours and absorbing power at minimum demand hours, taking advantage of storing energy of EVs. This is established by [35], where the need to synchronize charging and discharging of EVs with the smart grid is presented in order to avoid overloading in the distribution system. A specific study of this topic is done in [36], where a known network is considered with several scenarios of EVs inclusion, from 10% to 30%. In [37], the V2G interaction is used to decrease the percentage of distribution transformers losses. There, the authors make the analysis by using a Time-Coordinated Optimized Power Flow (TCOPF), where EVs are considered as distributed generation, making an optimal dispatch of energy according to their requirements in an interval of time. Efficiency improvements of the system are achieved because the EVs consume power from the grid while the demand is low, leveling the valley of the demand profile and reducing the peaks in hours of maximum demand. Under these circumstances, events of charging/discharging the EV battery when connected to the power grid greatly influence battery aging. In this sense, P. Venet et al. presents in [38] a data-driven approach to analyze real-world usage of batteries in EVs, considering different road conditions (urban, extra-urban, and highway), which can result in a useful tool to characterize EV architectures and estimate their functionalities with the V2G concept.

Other contributions, those shown in [39], explore the economic incentives that EVs users can receive by contributing to soften the load profile curve. Incentives are made by offering refunds to EVs buyers, taking as reference the project carried out in California where each customer with photovoltaic energy capacity installed is eligible to obtain a discount of 2.5 USD/Wp. In [40], the inclusion of plug-in hybrid electric vehicles (PHEVs) in distribution systems is considered as a factor to recuperate the voltage stability; therefore, a method based on neural networks to determine a voltage stability index given a specific condition is shown. Additionally, in the same year some studies were developed with stochastic processes [41], demonstrating the importance of an intelligent strategy to charge and discharge EVs. Following this research stream, in [42] EVs are studied in different statuses—the first status presents a car in motion, the second status suggests a car parked in an industrial area, and the third status supposes a car parked in a residential zone. The status of each vehicle at a given time is assigned according to a Montecarlo simulation. Two levels of EV insertion are considered: 25% and 50%.

The next year, in 2011, some works like [43] developed a model for the market and infrastructure of EV recharge stations. As the year before, V2G interaction gained prominence, with the study of frequency control on grids with a high degree of generation by renewable energy sources [44]. The works in [45,46] present the possibility to use EVs and PHEVs as dynamic containers of electric power, which can be set up at any time; while in [47], an optimization algorithm combined with

Voronoi polygons is implemented, which locates equitably recharge stations, obtaining load balances according to the distribution of vehicles and the network topology. In [48], Falahati et al. evaluate reliability indices in an existing system with different EV insertion levels, concluding in particular that the test system used is not ready enough to supply the necessary demand for these elements in the system. Therefore, as formerly mentioned, the relevance of coordinating EVs and the electric network is confirmed. The impact of over-sizing the capacity of the network is analyzed in [49], where a general methodology using structural data for this proposal is presented.

Several works treat the interaction between EVs and power grids from the economic and technical perspectives, posing optimal charge and discharge schedules for the EVs; however, there is an issue on behalf of the EV owner related to the acceptance level of this person to use the network to charge the vehicle battery when permitted, and deliver the energy stored on it when needed. This topic is studied in [50], where the synchronization is not with the EV and its charge and discharge schedules, but with the owners of these vehicles and their needs because these EVs owners can dispose of the energy from the network at any time; therefore, regulations have to be presented to restrict the schedules and load capacity of each vehicle.

3.4. Subperiod 2012 to 2015

A large variety of studies done in 2012 used advanced optimization techniques, as the case of [51] where dynamic programming is used in order to determine the minimum current needed to achieve a desired SOC in the batteries, reducing the grid losses and the chance of wires overloading. In [52], in order to avoid an electric system saturation, a tariff plan is proposed to decrease the quantity of EVs running daily. This is done based on the day-ahead market, using a dynamic tariff that varies according to the energetic scheduling of the day. In [53], the concept of battery swap station is used; this is an idea that achieves to increase the dynamism around vehicular traffic. This scheme does not affect daily tasks of users when the batteries removed are charged at valley hours.

As it advances, some of the topics slightly missed were taken up, as the case of the power quality due to harmonic distortions, which is studied again in [54,55], demonstrating that the most important harmonics (3rd and 5th harmonics) injected into the grid cancel each other out when a large quantity of EVs are connected in the same grid. Later, in [56], the chance to use the EVs as a backup source at homes is studied, incorporating the V2H (Vehicle-to-Home) scheme to supply the individual demand during interruptions of power delivery during short periods.

In 2014, additional works represent the EVs smart charge, used to flatten the load curve [57] with diverse methods and test systems. In [58], the technical impact over the distribution networks is not the unique topic of interest to study, but also the environmental impact carried out by the EVs usage, through CO_2 reduction, which is demonstrated in the results obtained. In [59], an optimization work is done where the benefits of battery swap stations and the recharge stations are compared. It is demonstrated that the battery swap system is more suitable to apply in public transportation because the times for recharging batteries can be larger than the times taken to replace a depleted battery for a fully charged one.

Some works in 2015, such as [60], demonstrate efforts in the improvement of the distribution system under the V2H and V2G concepts, considering non-served energy indices. Thereby, in two test cases an improvement is achieved; the first case is composed of a centralized technology for EV recharge (V2G mode), and the second case is formed by disperse EV charging stations (V2H mode). Within this framework, in the V2G paradigm the reliability of the EV battery must be guaranteed, and proper state of health techniques should be applied accordingly, as presented by P. Venet et al. in [61]. In the context of energy markets, prominent works were published in 2015. As mentioned in [62], the revenues are the decisive factor in terms of integrating EVs into the energy market. In the United States and some European countries, EVs participate in several business cases framed in primary, secondary, and tertiary reserve power, peak load reduction, and day-ahead energy markets. A more detailed focus is depicted by [63], where a centralized real-time EV charging management from an EV aggregator that participates in the energy and regulation markets is developed. The EV aggregator optimizes the market bidding strategy using a two-stage stochastic optimization model, which produces optimal first-stage decisions for submission in the day-ahead market and second-stage scenario-dependent decisions for submission in the real-time market. The model can account for all uncertain day-ahead and real-time conditions as well as energy deviations between day-ahead and real-time energy markets. The storage technology implemented in EVs offers an attractive alternative for EVs to support the Short-Term Operating Reserve (STOR). According to [64], storage can help manage imbalances between electric power generation and consumption that could result in undesirable impacts across the entire network. Among the reasons for which this technology is a good option for STOR are

- storage has superior part-load efficiency,
- efficient storage can use twice its rated capacity (i.e., it can stop discharging and start charging at the same time), and
- storage output can be varied very rapidly (e.g., output can change from 0% to 100% and from 100% to 0%).

From the EV perspective, STOR implementation is highly dependent on several critical factors, among them are state of charge, connection availability at times of grid requirement, fast response, and capability of providing twice the rated capacity.

Ancillary services, such as active power control and voltage support, are expected to be provided by EVs [65]. The first service is associated with the balance between production and demand to guarantee a secure operation of the electric grid at a constant frequency. Voltage support has to be performed locally because voltage fluctuations in power systems are usually due to the variation of reactive power demand and its transmission along the power lines. Since reactive power cannot be transmitted over long distances, voltage control has to be carried out by using special devices dispersed throughout the system to produce the necessary reactive power to match demand and keep the voltage within appropriate limits. According to the tasks developed by frequency and voltage control, EVs must comply with the following four criteria: supply duration, directional shifts, response rate, and service duty. Supply duration refers to the time over which the device, in this case the EV, has to be available to provide the ancillary service. Directional shifts are associated with sudden changes in charge and discharge of the batteries, which is suitable for short and volatile services. Long directional shifts are not convenient for EV batteries due to the degradation effects on the assets. Response rate is the time within which the resource providing the ancillary service needs to initiate service, which can be from less than one minute up to one hour. Service duty refers to the intermittent or continuous nature of consumption of ancillary services. The first one enables the EV to be charged while it is not providing the service.

In the framework of frequency control, Izadkhast and Garcia in [66] propose a new model to assign a participation factor to each EV, which facilitates the incorporation of several EV fleet characteristics, i.e., minimum desired state of charge, drive train power limitations, and charging modes (constant current and constant voltage). Participation factor defines the EV's availability to provide the primary frequency control. A wider range of responsive devices, e.g., inverter-based photovoltaic systems, EVs, and domestic controllable loads, are considered in [67] for frequency and voltage control based on power sensitivity analysis. These devices are classified according to the controllability degree. Once a voltage or frequency violation is detected in the system, the most effective buses are identified and receive the most effective control signals to perform appropriate changes in their reactive or active powers. In [68], a control technique is proposed to mitigate the charging current ripple when the current shifts the reference. A different approach is presented by Poornazaryon et al. in [69], where a method for primary and secondary frequency control is proposed based on artificial neural networks to train and validate the advanced droop control. Other works are framed within the computational decrease by using modified constraints in the mathematical model, such as the approach performed in [70]. The MTZ (Miller, Tucker, and Zemlin) formulation is used as a tool to avoid sub-tours in the vehicle routing problem, in order to obtain BSS locations and improve algorithm performance.

4. Chronological Review: Part II

The second half of works correspond to the papers listed in [6] in the range of time 2016–2019, with a total of 1015 papers. According to Figure 5, demand management and study of scenarios were the most prominent and had the same percentage, followed by the V2G concept category. In comparison with proportions shown in Figure 4, the positions for the first four categories remain equal, and a variation in the last four positions can be noted. Particularly, the categories of electricity markets and power quality held one and two positions respectively. On the other hand, a rise in two positions and one position, respectively, for the categories of power system stability and demand response were observed.

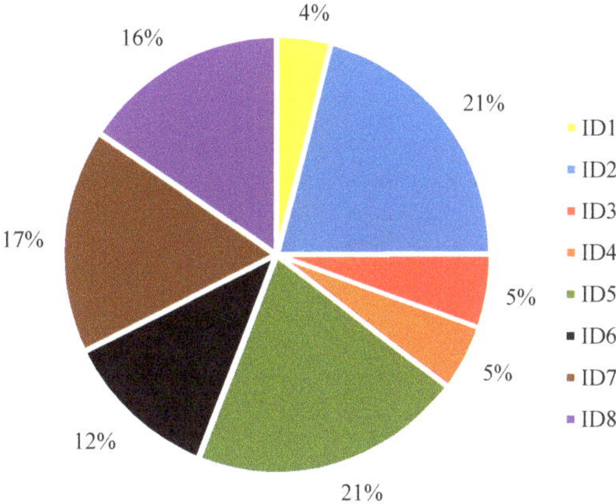

Figure 5. Participation percentage of categories: second half.

4.1. Subperiod 2016

Returning to EVs and their interaction with electricity markets via the aggregator concept, in 2016 Zhang and Kezunovic in [71] present contributions in the analytical estimation of aggregated EV charging/discharging power capacity, taking into account EV stochastic mobility and driver behavior, to improve the ramp rate of conventional generators through cooperation, and to participate in the ramp market on the system's reliability and flexibility as well as on EVs themselves. EV fleets can be aggregated in a mobile energy storage, which has the potential to compensate the un-contracted power if the contracts between the market players are breached. In this way, [72] performs an optimal strategy for both energy and reserve markets considering trade off and effect on EV battery degradation, in order to assess the expected profit that an aggregator can collect by participating in the energy and regulation market.

Continuing with the research approaches in 2016, some contributions are addressed in the context of demand response. This concept can avoid building new large-scale power generation and transmission infrastructures by improving the electric utility load factor. In [73], a demand response strategy is proposed for shaping a load profile to tackle the problem of overloading in distribution transformer when the EVs are used along with other loads. Overloading is first analyzed, and then the demand response is used to mitigate it once the total load exceeds the rated power of the distribution transformer. A more structured work is implemented in [74] from the mathematical model perspective, representing the total load at a charging station, considering a queuing model followed by a neural network. The queuing model considers arrival of EVs as a non-homogeneous Poisson process, and the service time is represented by using detailed characteristics of the battery. The charging station load (which is a function of the number and type of EVs charging at station, total charging current, arrival rate, and time) is integrated within a distribution operations framework to determine the optimal operation and smart charging schedules. Some works classified in the demand response can also be enrolled in the demand management approach. This situation is presented in [75], with a proposal to achieve a grid-friendly charging load profile based on the transactive control paradigm. In this way, EV owners can participate in real-time pricing electricity markets to reduce their charging costs. Similar efforts are presented in [76], developing a model for optimal behavior of EV parking lots in the energy and reserve markets, within the framework of price-based and incentive-based demand response programs. Concluding with this year, a practical case study is carried out in [77], evaluating the impact of EV uptake on Britain's power distribution networks by monitoring 200 customers during 1.5 years. At current projections for EV insertion, upgrading low-voltage infrastructure will cost consumers approximately USD 3 billion by 2050. This cost can be largely avoided if the demand-side response is deployed to shift EV charging away from times of peak demand.

4.2. Subperiod 2017

As depicted in Figure 3, the research around EVs and their interaction with distribution systems from different perspectives reduced by approximately 29% in 2017, in contrast with 2016. According to Figure 2, the majority of the work decreased in 2017, except the approach with ID8 (related with charging station planning and battery swap stations), which presents the largest number of publications for this year. Harb and Hamdan in [78] take into account that when new modern technology is introduced to the power grid, it should be compatible with the grid in order to improve its operation and ensure stability and reliability. In this work, several subjects are considered in which the EVs are involved with distribution networks, that is, assessment of different insertion levels of EVs in accordance with power quality (in terms of harmonic distortion) as well as voltage and frequency stability. In regards to power quality, the work shown in [79] focuses on experimental evaluation of EVs to reduce voltage unbalances by modulating the charging current according to local voltage measurements. This autonomous control could partially solve voltage quality issues without the need for grid upgrades or costly communication infrastructure, enabling a higher number of EVs to be integrated in the existing power network. The experiment is carried out with EVs that do not have the V2G technology incorporated but are able to modulate the charging current in steps according to the predefined droop control. Some energy market-oriented works, such as that shown in [80], proposes an eVoucher program to encourage participation of parking lots, with a high EV penetration rate, in the retail electricity market at distribution level.

As a vital part of smart grids, demand response supports the restoration of balance between electricity demand and supply. This concept is highlighted by [81], where a real-time charging scheme is proposed to coordinate the EV charging loads based on the dynamic electricity tariff. On the other hand, an optimization problem is formulated to maximize the number of EVs selected for charging at each time period. Two objective functions are in conflict: maximizing the EV owner's convenience in meeting all charging requests and minimizing the total electricity bill for the parking station. Similar contributions are presented in [82], focused on the real time interactions between energy supplier and the EVs users, in a fully distributed system in which the only information available to the end users is the current price. In this sense, a real-time charging pricing algorithm is introduced to maximize the aggregate utility of all the EV users and minimize the electricity cost generated by the energy supplier. In addition, the EV users and the energy supplier interact each other when running the distributed algorithm to find the optimal power consumption level, and the optimal price values to be revealed by the energy supplier, in order to adapt the users' demands constantly and maximize their own utility. Another study in [83] is presented in this context, defining demand response as "voluntary change of demand", proposing an approach to enable EV smart charging technology among residential customers. This proposal incorporates operation and analysis of power transactions between the energy user and the electricity grid, including the concept of power sharing among neighbors in the residential demand response framework. In the context of V2G, power system stability, and energy markets, the efforts done by [84] are highlighted. The introduction of network characteristics (distribution power losses and maximum power limits of the transformers and lines) in the V2G concept upgrades the accuracy of the EV model to participate in the load frequency control. This approach shows that EVs quickly respond during contingency and are very effective in driving the error to zero. From the other side, in [85] a multi-objective mathematical framework has been presented to cater frequency deviations at the grid level using a fleet of EVs. The objective functions of this model are presented below:

- minimization of grid frequency deviations using the available frequency regulation capacity, dealing with the trade-off between fulfilling EV energy demands and providing maximum grid support;
- maximization of V2G support to EVs while minimizing EV's battery degradation—the objective of this problem is to maximize the scheduling of EV participation while considering the trade-off with battery degradation issues;
- optimal regulation signal dispatch among aggregators and charging stations.

4.3. Subperiod 2018

Compared with 2017 and according to Figure 3, in 2018 the number of publications slightly increased, where EV demand management was the most addressed topic in this subperiod, followed by EVCS planning and studies associated with the V2G concept and ancillary services. In regard to EVCS profit, a parking lot management system (PLMS) is proposed in [86], which promotes EV recharging with the energy produced by a set of solar panels attached to the EVCS, instead of drawing energy from the grid. This behavior increases the parking lot owner's profit by selling more energy from the solar panels (avoiding high energy prices of the grid) and reduces the power network congestion in periods where the energy demand is relatively high. A similar approach is addressed by [87], which proposes an algorithm called JoAP (Joint Admission control and Pricing) to maximize the average profit of an EVCS. The profit is defined as the difference between the revenue and a penalty proportional to the average charging waiting time. This latter reflects the impatience of EV owners to wait in a queue for an excessively long time, which affects both the EVCS's reputation and the long-term profit. On the other side, the aggregators (entities that act as the mediators between the users and the utility operator) play an important role in the optimal regulation of the EV fleet charging plan, in order to minimize the overall cost of EV charging considering EV charging constraints. This aspect is widely dealt with by Mediwaththe and Smith in [88], providing a model for the competition among multiple EV aggregators by using a non-cooperative game-theoretic framework. Each aggregator determines the EV charging start time and charging energy profiles to minimize the EV charging energy cost, considering the actions taken by the neighboring EV aggregators. Other concepts have been recently adopted, such as the "internet of energy" addressed in [89], that refers to enhancing and automating the electricity infrastructure, e.g., EVCS, to move forward to a more efficient EV energy management. From the point of view of the EVCSs planning, several works are developed along this subperiod, considering the transportation and power networks. In [90], siting and sizing of fast charging stations is performed in coupled transportation and power grids with heterogeneous EV charging demands. The number of spots to be installed in a charging station is found in order to provide an adequate charging service quality measured by a performance metric. The transportation network is modeled by using a capacitated flow refueling location model, and Kirchhoff's laws are utilized to roughly approximate the electrical constraints of distribution networks. A similar focus is developed by [91] in terms of the EVCS location, by using a binary lightning search algorithm as an optimization technique for fast charging stations. Other works, such as those published in [92,93], have notable technical and economical contributions for EVCS planning. In [92], a two-level optimal planning approach is developed to optimize the siting, sizing, and demand drawn by the fast charging stations, and the number of chargers at each station is found, as part of a comprehensive benefit analysis. Furthermore, a more complex approach is performed in [93] from the cost–benefit analysis point of view. EVCS planning (siting and sizing) considers an economic analysis based on the life cycle cost and net present value, to provide optimal decisions for investors and charging stations operators. Given the probabilistic perspective, a two-stage stochastic programming model to determine the location and capacity of urban EVCSs is developed in [94], incorporating the uncertainties associated with the EV demand flows, charging patterns, arrival and departure times, and preferred walking distances. Due to the complexity of the two-stage stochastic problem, a heuristic is implemented for large-scale instances to obtain near optimal solutions. Novel strategies for EV charging are studied by [95,96] with wireless charging systems, which allow the EV battery to charge remotely while moving over the highways. This technology mitigates the range limitation of EVs by using power tracks as additional sources of electric energy.

4.4. Subperiod 2019

Considering the rules in this paper to develop the power grids and EV-related research classification, it can be observed that the subperiod 2019 saw double the number of works in comparison with 2018 and has been, until now, the year with the largest number of publications, as depicted in Figure 3. According to the database in this research [6], in 2019 scenario studies were the most frequent, followed by works related to demand management, V2G concept, and power system stability. The fifth and sixth places are for EVCSs/BSSs and electricity markets, respectively, and the last two places are held by power quality and demand response research.

On the other hand, and with the purpose of providing a broader context for the current trends in terms of EV interactions with power grids, the IEA (International Energy Agency) published its annual report, the Global EV Outlook 2019 [97], which discusses key challenges in reference to implications of electric mobility for power systems. Particularly, the report emphasizes the potential of controlled EV charging to increase the flexibility in power systems via DSR (Demand-Side Response) services. This includes charging events during low-demand periods for shaping electricity demand, frequency response based on control signals, and supporting the increase of VRE (Variable Renewable Energy) generation in the power systems for reliability purposes. Likewise, the NARUC (National Association of Regulatory Utility Commissioners) reports in [98] the possibility of EVs as time-movable loads to increase the grid flexibility through rate design options, envisioned within the EVs participating in energy markets. This encompasses TOU (Time of Use) rates and dynamic RTP (Real Time Pricing) implementation for an efficient usage of existing assets, instead of implementing expensive upgrades in the distribution system to serve EVs. Similarly, the report performed by the Center on global Energy Policy in [99] highlights the relevance of demand response programs for aggregated EV charging, with pilot projects being developed in the states of California and Vermont. These projects are framed in assuring that EV charging times can respond to grid requests, and EV charging interruption can be triggered at super-peak times.

Considering the DSR context and EV participation in electricity markets, common ground for several works in the 2019 subperiod corresponds to the EV aggregator as a commercial middleman between the power grid operator and the EVs. The concept of EV aggregator is linked to facilitating EV interactions with power grids, in terms of reducing the charging cost, provision of ancillary services, and balancing between supply and demand [100], with a subsequent profit for both the aggregator and EV owners. Participation in day-ahead markets represents, in many cases, an instrument for the EV aggregator to get profit; furthermore, biding and pricing strategies should be considered due to the unpredictable nature of electricity markets [101]. Additionally, under incentive mechanisms an EV charging schedule adjustment is promoted according to the charging price adopted by the EV aggregator [102]. Therefore, improvement in the voltage profile and decrease in the power losses cost can be experimented by the power distribution operator, and customers are expected to make a profit [103]. Likewise, the EV aggregator coordinates charging and discharging strategies, taking into consideration EV driving patterns, features associated with the battery state of health [104], assessment of energy-efficient batteries [105], and unmodeled externalities acting on the energy price [106], with a direct effect on the aggregator's energy bids on the day-ahead market and its profitability [107]. On the other hand, sometimes the term "EV aggregator" is not used, and the DSR programs are directly performed by EV owners. In this sense, the DSR program can utilize a real-time pricing scheme, leading to a suitable option to alleviate network congestion when customers are encouraged to shift their charging process to off-peak periods. Compared with other price-based DSR programs, i.e., time of use, critical peak pricing, and peak time rebate, the real-time pricing scheme represents a more appropriate alternative for bill savings and dampening energy price volatility [108].

4.5. Subperiod 2020: Recent Research

During this subperiod, efforts around EV integration into the power grid have been adopted in terms of their spatial–temporal property [109]. This includes the introduction of traffic network as a remarkable component on the EV charging/discharging events. Additionally, for optimal and feasible operation, potential benefits of a dynamic distribution network reconfiguration have been considered, which complements V2G services and minimizes the total system cost [110]. A similar approach is taken in [111], showing the impact of traffic topological characteristics on EV charging. This results in practical applications such as EV load forecasting, construction of urban traffic networks, and charging and driving strategy optimization. On the other hand, and following the 2019 subperiod related with DSR programs, Bin Duan et al. in [112] propose an adjustment on electricity price via smart contracts between the user and the charging station, and Babar et al. in [113] present a mechanism for computing the EV charging prices using individualized energy consumption patterns of EVs contingent upon the region.

5. Brief Observations

As analyzed in the sections above, the paper database listed in [6] was divided into two ranges of time, the first set encompassed the elapsed time 1973–2015, and the second set of papers covered 2016–2019. This distribution resulted in a fair partition as each time range included almost half of the debugged database list. In this sense, a clear comparison between both divisions can be made, as presented in Figure 6.

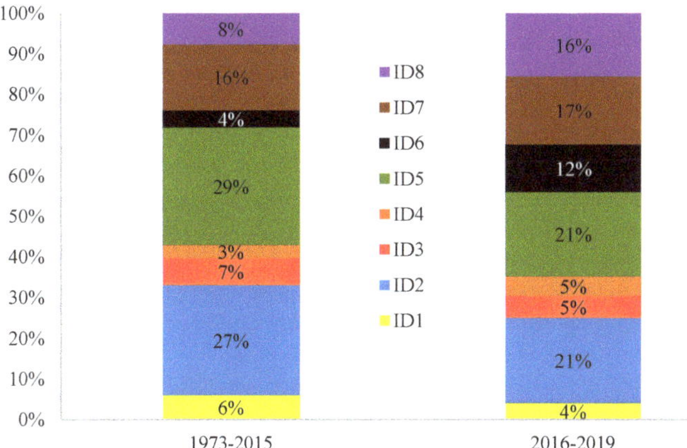

Figure 6. Comparison between the first and second half research periods.

In regards with Figure 6, slight changes in the proportions of all categories can be noticed, except for the ID8 and ID6 corresponding to EVCSs and power system stability categories, respectively. An additional remark is framed within the time taken for both periods in finding almost the same number of works. An evident explanation comes from the fact that, in the second period (2016–2019), the subject of EVs and power grids was mature enough and developed, and many ideas had already been formed. Further, the first period (1973–2015), with a duration of 42 years, had to involve the evolution of the information dynamics and the first stages of research around interaction of EVs and power distribution systems.

6. Conclusions and Future Works

This paper presented a detailed review of the literature related to EVs and their impact on power distribution systems, considering categories that were carefully identified over an exhaustive examination on the IEEE Xplore database. The research encompassed the range of 1973 until the end of 2019, and 2008 was a starting point for increased publication.

Demand management has been identified as the category with a major presence throughout the development of this review. EV technologies are constantly growing, and researchers are always looking for methodologies to optimize several aspects in the EV charging process, i.e., energy drawn from the power grid and the time along the demand curve this process should take place. Moreover, other categories have been largely investigated, such as the Vehicle-to-Grid (V2G) topic and study of scenarios. The battery swap and charging stations topic, which involves the optimal location of these infrastructures throughout the power distribution system, has had notable relevance during the last few years as the spatial problem is solved, in contrast with several papers that only address the temporal and quantity aspects (e.g., when does the EV have to be recharged, and how much energy does the EV draw from the power grid?). The transportation network plays an important role in the optimal location of battery swap stations and EVCSs, despite the low number of publications in IEEE Xplore that also consider the transportation network.

It can be identified that the problems found in the subperiod 1973–1999, in regard to the impact of EVs on the distribution systems, i.e., inaccurate load model and non-desired demand peaks, are in conformity with the typical issues that EV charging may provide to the power grids in the coming years. For the sake of reliable power network operation, demand response programs shall be necessary to be implemented, as suggested by the IEA. On the other side, more research would be needed towards improving the EV load model, by exploration of data-driven EV load models instead of using the conventional ZIP model.

In regards with the classification performed in this review, some of the categories could be combined into one category. Since the V2G concept is the property of an EV to transfer energy to power network, other categories can be intrinsic within this topic, such as "power system stability" and "electricity markets". Furthermore, discrimination in several categories allows future researchers to directly find the corresponding sources of information for their purposes, knowingly that there is a meticulous process to select the papers listed in the database developed in this article.

According to the 2019 outlooks provided by prestigious governmental agencies such as IEA and USDRIVE, and different consultant organizations, the fields of demand management, demand response programs, and strategies for load profile shaving are more attractive rather than investing in expensive upgrades on power distribution infrastructure. Accordingly, it is proposed for future works to develop more specific literature reviews related to EVs and their corresponding involvement in demand response programs and electricity markets. Likewise, elaboration of an extended literature review encompassing the interaction between EVs and power grids can be considered, addressing other prominent databases such as ElSevier, Springer, MDPI, and Taylor and Francis.

7. Final Reflections

Decarbonization of power systems has been a topic under discussion during the last decades due to the imminent depletion of oil reserves all over the world. One of the strategies to pursue this objective is framed within the electrification of transportation. Electric vehicles are considered as a promising alternative to reduce the dependence on fossil fuels and counteract the hibernation effect. On the other hand, the introduction of EVs in power distribution systems may undergo assets, i.e., transformers and conductors, to adverse effects and decrease the power grid performance. As demonstrated in the review in this paper, abundant research has been developed around EVs interaction with power grids, considering different trends that have emerged over the course of time as an effort to discriminate how EVs could be related with power grids. Not all the effects from EVs are dis-satisfactory for power distribution systems. As exhibited by many researchers, EVs may be able to provide ancillary services to power grids and increase its flexibility due to their nature of energy storage and movable loads. In this sense, improvement on load factor, frequency regulation, and support for penetration of renewable generation sources fall within the benefits of EVs on power distribution networks.

To this end, there should be a settled communication between the EVs and power grid operator, coordinated by an entity or an intermediate that in several works in this review has been called "the EV aggregator". This results in a more efficient exchange of information between the interested parties. As presented in AMI (Advanced Metering Infrastructure) system, different key hardware and software components have to be utilized by the aggregator in order to properly track and channel the abundant data coming from the EVs, i.e., charging and discharging profiles, driving patterns, demand response parameters, etc.

We assume that the EV aggregator will control the data exchange between the EVs and power grid utility, for demand response purposes or provision of network ancillary services, but this paradigm can be changed and, instead, the intermediate could be a figure of energy commercialization that negotiates the energy in wholesale or day-ahead market to sell it directly to EVs at consistent price. Furthermore, one can think about the existence of both figures, the "EV energy marketer" and "EV aggregator". The first one would negotiate energy for the EVs by using biding and pricing strategies, and the second one would be in charge of the technical segment. Under this context, the EV owner will have to choose a certain EV aggregator and/or "energy marketer" for operation and profitability purposes, as adopted nowadays with any household electric power installation.

Author Contributions: Conceptualization and writing—review and editing, A.A.-L., O.D.M., L.F.G.-N. All authors have read and agreed to the published version of the manuscript.

Funding: This project was financially supported by own funds of the authors.

Acknowledgments: The authors acknowledge the support provided by MINCIENCIAS, as manager of the Colombian National System of Science, Technology and Innovation (SNCTI).

Conflicts of Interest: The authors declare no conflicts of interest.

References

1. Arias, N.B.; Hashemi, S.; Andersen, P.B.; Træholt, C.; Romero, R. Distribution System Services Provided by Electric Vehicles: Recent Status, Challenges, and Future Prospects. *IEEE Trans. Intell. Transp. Syst.* **2019**, *20*, 4277–4296. [CrossRef]
2. Dubey, A.; Santoso, S. Electric vehicle charging on residential distribution systems: Impacts and mitigations. *IEEE Access* **2015**, *3*, 1871–1893. [CrossRef]
3. Deb, S.; Kalita, K.; Mahanta, P. Review of impact of electric vehicle charging station on the power grid. In Proceedings of the 2017 International Conference on Technological Advancements in Power and Energy (TAP Energy), Kollam, India, 21–23 December 2017; IEEE: Piscataway, NJ, USA, 2017; pp. 1–6.

4. Jia, Q.S. On supply demand coordination in vehicle-to-grid—A brief literature review. In Proceedings of the 2018 33rd Youth Academic Annual Conference of Chinese Association of Automation (YAC), Nanjing, China, 18–20 May 2018; IEEE: Piscataway, NJ, USA, 2018; pp. 1083–1088.

5. Arias, A.; Martínez, L.H.; Hincapie, R.A.; Granada, M. An IEEE Xplore database literature review regarding the interaction between electric vehicles and power grids. In Proceedings of the 2015 IEEE PES Innovative Smart Grid Technologies Latin America (ISGT LATAM), Montevideo, Uruguay, 5–7 October 2015; IEEE: Piscataway, NJ, USA, 2015; pp. 673–678.

6. Papers Classification EVs-Power Grids 1973–2019. Available online: http://academia.utp.edu.co/planeamiento/sistemas-de-prueba/recarga-de-vehiculos-electricos-en-sistemas-de-distribucion-3/papers-classification-evs-power-grids-1973-2019/ (accessed on May 2020).

7. Salihi, J.T. Energy requirements for electric cars and their impact on electric power generation and distribution systems. *IEEE Trans. Ind. Appl.* **1973**, *5*, 516–532. [CrossRef]

8. Patil, P. Prospects for electric vehicles. *IEEE Aerosp. Electron. Syst. Mag.* **1980**, *5*, 15–19. [CrossRef]

9. Heydt, G.T. The impact of electric vehicle deployment on load management straregies. *IEEE Trans. Power Appar. Syst.* **1983**, *5*, 1253–1259. [CrossRef]

10. Orr, J.A.; Emanuel, A.E.; Pileggi, D.J. Current harmonics, voltage distortion, and powers associated with electric vehicle battery chargers distributed on the residential power system. *IEEE Trans. Ind. Appl.* **1984**, *4*, 727–734. [CrossRef]

11. Rahman, S.; Shrestha, G. An investigation into the impact of electric vehicle load on the electric utility distribution system. *IEEE Trans. Power Deliv.* **1993**, *8*, 591–597. [CrossRef]

12. Suggs, C.R. Electric vehicles-driving the way to a cleaner future. In *Conference Record Southcon*; IEEE: Piscataway, NJ, USA, 1994; pp. 28–30.

13. KeKoster, D.; Morrow, K.P.; Schaub, D.A.; Hubele, N.F. Impact of electric vehicles on select air pollutants: A comprehensive model. *IEEE Trans. Power Syst.* **1995**, *10*, 1383–1388. [CrossRef]

14. Staats, P.; Grady, W.; Arapostathis, A.; Thallam, R. A statistical analysis of the effect of electric vehicle battery charging on distribution system harmonic voltages. *IEEE Trans. Power Deliv.* **1998**, *13*, 640–646. [CrossRef]

15. Chan, M.S.; Chau, K.; Chan, C. Modeling of electric vehicle chargers. In Proceedings of the 24th Annual Conference of the IECON'98 Industrial Electronics Society, Aachen, Germany, 31 August–4 September 1998; Volume 1, pp. 433–438.

16. Koyanagi, F.; Inuzuka, T.; Uriu, Y.; Yokoyama, R. Monte Carlo simulation on the demand impact by quick chargers for electric vehicles. In Proceedings of the Power Engineering Society Summer Meeting, Edmonton, AB, Canada, 18–22 July 1999; IEEE: Piscataway, NJ, USA, 1999; Volume 2, pp. 1031–1036.

17. Peres, L.P.; Lambert-Torres, G.; Nogueira, L.H. Electric vehicles impacts on daily load curves and environment. In Proceedings of the Electric Power Engineering, 1999. PowerTech Budapest 99. International Conference, Budapest, Hungary, 29 August–2 September 1999; IEEE: Piscataway, NJ, USA, 1999; p. 55.

18. Chan, C. The past, present and future of electric vehicle development. In Proceedings of the IEEE 1999 International Conference on Power Electronics and Drive Systems, Hong Kong, China, 27–29 July 1999; IEEE: Piscataway, NJ, USA, 1999; Volume 1, pp. 11–13.

19. Brahma, A.; Guezennec, Y.; Rizzoni, G. Optimal energy management in series hybrid electric vehicles. In Proceedings of the American Control Conference, Chicago, IL, USA, 28–30 June 2000; IEEE: Piscataway, NJ, USA, 2000; Volume 1, pp. 60–64.

20. Ceraolo, M.; Pede, G. Techniques for estimating the residual range of an electric vehicle. *IEEE Trans. Veh. Technol.* **2001**, *50*, 109–115. [CrossRef]

21. Gómez, J.C.; Morcos, M.M. Impact of EV battery chargers on the power quality of distribution systems. *IEEE Trans. Power Deliv.* **2003**, *18*, 975–981. [CrossRef]

22. Brooks, A. Integration of electric drive vehicles with the power grid-a new application for vehicle batteries. In Proceedings of the SeventeenthAnnual Battery Conference on Applications and Advances, Long Beach, CA, USA, 18 January 2002; IEEE: Piscataway, NJ, USA, 2002; p. 239.

23. De Breucker, S.; Jacqmaer, P.; De Brabandere, K.; Driesen, J.; Belmans, R. Grid power quality improvements using grid-coupled hybrid electric vehicles pemd 2006. In Proceedings of the 2006 3rd IET International Conference on Power Electronics, Machines and Drives—PEMD 2006, Dublin, Ireland, 4–6 April 2006.

24. Das, T.; Aliprantis, D.C. Small-signal stability analysis of power system integrated with PHEVs. In Proceedings of the Energy 2030 Conference, ENERGY 2008, Atlanta, GA, USA, 17–18 November 2008; IEEE: Piscataway, NJ, USA, 2008; pp. 1–4.
25. Wang, L. Potential impacts of plug-in hybrid electric vehicles on locational marginal prices. In Proceedings of the Energy 2030 Conference, ENERGY 2008, Atlanta, GA, USA, 17–18 November 2008; IEEE: Piscataway, NJ, USA, 2008; pp. 1–7.
26. Guille, C.; Gross, G. Design of a conceptual framework for the V2G implementation. In Proceedings of the Energy 2030 Conference, ENERGY 2008, Atlanta, GA, USA, 17–18 November 2008; IEEE: Piscataway, NJ, USA, 2008; pp. 1–3.
27. Kramer, B.; Chakraborty, S.; Kroposki, B. A review of plug-in vehicles and vehicle-to-grid capability. In Proceedings of the Industrial Electronics, IECON 2008. 34th Annual Conference of IEEE, Orlando, FL, USA, 10–13 Novemer 2008; IEEE: Piscataway, NJ, USA, 2008; pp. 2278–2283.
28. Larsen, E.; Chandrashekhara, D.K.; Ostergard, J. Electric vehicles for improved operation of power systems with high wind power penetration, ENERGY 2008. In Proceedings of the Energy 2030 Conference, Atlanta, GA, USA, 17–18 November 2008; IEEE: Piscataway, NJ, USA, 2008; pp. 1–6.
29. Pina, A.; Ioakimidis, C.S.; Ferrao, P. Introduction of electric vehicles in an island as a driver to increase renewable energy penetration. In Proceedings of the Sustainable Energy Technologies, ICSET 2008. IEEE International Conference, Singapore, 24–27 November 2008; IEEE: Piscataway, NJ, USA, 2008; pp. 1108–1113.
30. Saber, A.Y.; Venayagamoorthy, G.K. Optimization of vehicle-to-grid scheduling in constrained parking lots. In Proceedings of the Power & Energy Society General Meeting, Calgary, AB, Canada, 26–30 July 2009; IEEE: Piscataway, NJ, USA, 2009; pp. 1–8.
31. El Chehaly, M.; Saadeh, O.; Martinez, C.; Joos, G. Advantages and applications of vehicle to grid mode of operation in plug-in hybrid electric vehicles. In Proceedings of the Electrical Power & Energy Conference (EPEC), Montreal, QC, Canada, 22–23 October 2009; IEEE: Piscataway, NJ, USA, 2009; pp. 1–6.
32. Lopes, J.P.; Soares, F.J.; Almeida, P.R. Identifying management procedures to deal with connection of electric vehicles in the grid. In Proceedings of the Powertech, Bucharest, Romania, 28 June–2 July 2009; IEEE: Piscataway, NJ, USA, 2009; pp. 1–8.
33. Clement, K.; Haesen, E.; Driesen, J. Coordinated charging of multiple plug-in hybrid electric vehicles in residential distribution grids. In Proceedings of the Power Systems Conference and Exposition, PSCE'09. IEEE/PES, Seattle, WA, USA, 15–18 March 2009; Piscataway, NJ, USA, 2009; pp. 1–7.
34. Acha, S.; Green, T.C.; Shah, N. Impacts of plug-in hybrid vehicles and combined heat and power technologies on electric and gas distribution network losses. In Proceedings of the Sustainable Alternative Energy (SAE), 2009 PES/IAS Conference, Valencia, Spain, 28 September 2009; IEEE: Piscataway, NJ, USA, 2009; pp. 1–7.
35. Wang, Z.; Liu, P. Analysis on storage power of electric vehicle charging station. In Proceedings of the Power and Energy Engineering Conference (APPEEC), 2010 Asia-Pacific, Chengdu, China, 28–31 March 2010; IEEE: Piscataway, NJ, USA, 2010; pp. 1–4.
36. Singh, M.; Kumar, P.; Kar, I. Analysis of vehicle to grid concept in Indian scenario. In Proceedings of the Power Electronics and Motion Control Conference (EPE/PEMC), 2010 14th International, Ohrid, North Macedonia, 6–8 September 2010; IEEE: Piscataway, NJ, USA, 2010.
37. Acha, S.; Green, T.C.; Shah, N. Effects of optimised plug-in hybrid vehicle charging strategies on electric distribution network losses. In Proceedings of the Transmission and Distribution Conference and Exposition, 2010 IEEE PES, New Orleans, LA, USA, 19–22 April 2010; IEEE: Piscataway, NJ, USA, 2010; pp. 1–6.
38. Devie, A.; Montaru, M.; Pelissier, S.; Venet, P. Classification of duty pulses affecting energy storage systems in vehicular applications. In Proceedings of the 2010 IEEE Vehicle Power and Propulsion Conference, Lille, France, 1–3 September 2010; IEEE: Piscataway, NJ, USA, 2010; pp. 1–6.
39. Mallette, M.; Venkataramanan, G. Financial incentives to encourage demand response participation by plug-in hybrid electric vehicle owners. In Proceedings of the Energy Conversion Congress andExposition (ECCE), 2010, Atlanta, GA, USA, 12–16 September 2010; IEEE: Piscataway, NJ, USA, 2010; pp. 4278–4284.
40. Makasa, K.J.; Venayagamoorthy, G.K. Estimation of voltage stability index in a power system with Plug-in Electric Vehicles. In Proceedings of the Bulk Power System Dynamics and Control (iREP)-VIII (iREP), 2010 iREP Symposium, Rio de Janeiro, Brazil, 1–6 August 2010; IEEE: Piscataway, NJ, USA, 2010; pp. 1–7.

41. Fluhr, J.; Ahlert, K.H.; Weinhardt, C. A stochastic model for simulating the availability of electric vehicles for services to the power grid. In Proceedings of the System Sciences (HICSS), 2010 43rd Hawaii International Conference, Honolulu, HI, USA, 5–8 January 2010; IEEE: Piscataway, NJ, USA, 2010; pp. 1–10.
42. Soares, F.; Lopes, J.P.; Almeida, P.R. A Monte Carlo method to evaluate electric vehicles impacts in distribution networks. In Proceedings of the Innovative Technologies for an Efficient and Reliable Electricity Supply (CITRES), 2010 IEEE Conference, Waltham, MA, USA, 27–29 September 2010; IEEE: Piscataway, NJ, USA, 2010; pp. 365–372.
43. Yamashita, D.; Niimura, T.; Takamori, H.; Yokoyama, R. A dynamic model of plug-in electric vehicle markets and charging infrastructure for the evaluation of effects of policy initiatives. In Proceedings of the Power Systems Conference and Exposition (PSCE), 2011 IEEE/PES, Phoenix, AZ, USA, 20–23 March 2011; IEEE: Piscataway, NJ, USA, 2011; pp. 1–8.
44. Almeida, P.R.; Lopes, J.P.; Soares, F.; Seca, L. Electric vehicles participating in frequency control: Operating islanded systems with large penetration of renewable power sources. In Proceedings of the PowerTech, 2011 IEEE Trondheim, Trondheim, Norway, 19–23 June 2011; IEEE: Piscataway, NJ, USA, 2011; pp. 1–6.
45. Wang, X.; Tian, W.; He, J.; Huang, M.; Jiang, J.; Han, H. The application of electric vehicles as mobile distributed energy storage units in smart grid. In Proceedings of the Power and Energy Engineering Conference (APPEEC), 2011 Asia-Pacific, Wuhan, China, 25–28 March 2011; IEEE: Piscataway, NJ, USA, 2011; pp. 1–5.
46. Kezunovic, M. BEVs/PHEVs as dispersed energy storage in smart grid. In Proceedings of the Innovative Smart Grid Technologies (ISGT), 2012 IEEEPES, Washington, DC, USA, 16–20 January 2012; IEEE: Piscataway, NJ, USA, 2012; pp. 1–2.
47. Feng, L.; Ge, S.; Liu, H. Electric vehicle charging station planning based on weighted voronoi diagram. In Proceedings of the Power and Energy Engineering Conference (APPEEC), 2012 Asia-Pacific, Changchun, China, 16–18 December 2011; IEEE: Piscataway, NJ, USA, 2011; pp. 1–5.
48. Falahati, B.; Fu, Y.; Darabi, Z.; Wu, L. Reliability assessment of power systems considering the large-scale PHEV integration. In Proceedings of the Vehicle Power and Propulsion Conference (VPPC), Chicago, IL, USA, 6–9 September 2011; IEEE: Piscataway, NJ, USA, 2011; pp. 1–6.
49. Rolink, J.; Rehtanz, C. Capacity of low voltage grids for electric vehicles. In Proceedings of the Environment and Electrical Engineering (EEEIC), 2011 10th International Conference, Rome, Italy, 8–11 May 2011; IEEE: Piscataway, NJ, USA, 2011; pp. 1–4.
50. Grahn, P.; Söder, L. The customer perspective of the electric vehicles role on the electricity market. In Proceedings of the Energy Market (EEM), 2011 8th International Conference on the European, Zagreb, Croatia, 25–27 May 2011; IEEE: Piscataway, NJ, USA, 2011; pp. 141–148.
51. Turker, H.; Hably, A.; Bacha, S. Dynamic programming for optimal integration of plug-in hybrid electric vehicles (phevs) in residential electric grid areas. In Proceedings of the IECON 2012-38th Annual Conference on IEEE Industrial Electronics Society, Montreal, QC, Canada, 25–28 October 2012; IEEE: Piscataway, NJ, USA, 2012; pp. 2942–2948.
52. O'Connell, N.; Wu, Q.; Stergaard, J. Efficient determination of distribution tariffs for the prevention of congestion from ev charging. In Proceedings of the Power and Energy Society General Meeting, San Diego, CA, USA, 22–26 July 2012; IEEE: Piscataway, NJ, USA, 2012; pp. 1–8.
53. Zheng, D.; Wen, F.; Huang, J. Optimal planning of battery swap stations. In Proceedings of the International Conference On Sustainable Power Generation And Supply, Hangzhou, China, 8–9 September 2012.
54. Kutt, L.; Saarijárvi, E.; Lehtonen, M.; Mõlder, H.; Niitsoo, J. A review of the harmonic and unbalance effects in electrical distribution networks due to EV charging. In Proceedings of the Environment and Electrical Engineering (EEEIC), 2013 12th International Conference, Wroclaw, Poland, 5–8 May 2013; IEEE: Piscataway, NJ, USA, 2013; pp. 556–561.
55. Kutt, L.; Saarijarvi, E.; Lehtonen, M.; Molder, H.; Niitsoo, J. Current harmonics of EV chargers and effects of diversity to charging load current distortions in distribution networks. In Proceedings of the Connected Vehicles and Expo (ICCVE), 2013 International Conference, Las Vegas, NV, USA, 2–6 December 2013; IEEE: Piscataway, NJ, USA, 2013; pp. 726–731.
56. Tuttle, D.P.; Fares, R.L.; Baldick, R.; Webber, M.E. Plug-In Vehicle to Home (V2H) duration and power output capability. In Proceedings of the Transportation Electrification Conference and Expo (ITEC), Detroit, MI, USA, 16–19 June 2013; IEEE: Piscataway, NJ, USA, 2013; pp. 1–7.

57. Andrés, A.L.; Geovanny, M.G.; Camila, P.G.; Hincapié, R.A.; Mauricio, G.E. Optimal charging schedule of electric vehicles considering variation of energy price. In Proceedings of the Transmission & Distribution Conference and Exposition-Latin America (PES T&D-LA), 2014 IEEE PES, Medellin, Colombia, 10–13 September 2014; IEEE: Piscataway, NJ, USA, 2014; pp. 1–5.

58. Su, J.; Marmaras, C.E.; Xydas, E.S. Technical and environmental impact of electric vehicles in distribution networks. In Proceedings of the Green Energy for Sustainable Development (ICUE), 2014 International Conference and Utility Exhibition, Pattaya, Thailand, 19–21 March 2014; IEEE: Piscataway, NJ, USA, 2014; pp. 1–9.

59. Zheng, Y.; Dong, Z.Y.; Xu, Y.; Meng, K.; Zhao, J.H.; Qiu, J. Electric vehicle battery charging/swap stations in distribution systems: Comparison study and optimal planning. *IEEE Trans. Power Syst.* **2014**, *29*, 221–229. [CrossRef]

60. Xu, N.; Chung, C. Reliability evaluation of distribution systems including vehicle-to-home and vehicle-to-grid. *IEEE Trans. Power Syst.* **2016**, *31*, 759–768. [CrossRef]

61. Riviere, E.; Venet, P.; Sari, A.; Meniere, F.; Bultel, Y. LiFePO4 battery state of health online estimation using electric vehicle embedded incremental capacity analysis. In Proceedings of the 2015 IEEE Vehicle Power and Propulsion Conference (VPPC), Montreal, QC, Canada, 19–22 October 2015; IEEE: Piscataway, NJ, USA, 2015; pp. 1–6.

62. Illing, B.; Warweg, O. Analysis of international approaches to integrate electric vehicles into energy market. In Proceedings of the European Energy Market (EEM), 2015 12th International Conference, Lisbon, Portugal, 19–22 May 2015; IEEE: Piscataway, NJ, USA, 2015; pp. 1–5.

63. Vagropoulos, S.I.; Kyriazidis, D.K.; Bakirtzis, A.G. Real-time charging management framework for electric vehicle aggregators in a market environment. *IEEE Trans. Smart Grid* **2016**, *7*, 948–957. [CrossRef]

64. Gough, B.; Rowley, P.; Khan, S.; Walsh, C. The value of electric vehicles in the context of evolving electricity markets. In Proceedings of the European Energy Market (EEM), 2015 12th International Conference, Lisbon, Portugal, 19–22 May 2015; IEEE: Piscataway, NJ, USA, 2015; pp. 1–6.

65. González-Romera, E.; Barrero-González, F.; Romero-Cadaval, E.; Milanés-Montero, M.I. Overview of plug-in electric vehicles as providers of ancillary services. In Proceedings of the Compatibility and Power Electronics (CPE), 2015 9th International Conference, Costa da Caparica, Portugal, 24–26 June 2015; IEEE: Piscataway, NJ, USA, 2015; pp. 516–521.

66. Izadkhast, S.; Garcia-Gonzalez, P.; Frías, P. An aggregate model of plug-in electric vehicles for primary frequency control. *IEEE Trans. Power Syst.* **2015**, *30*, 1475–1482. [CrossRef]

67. Bayat, M.; Sheshyekani, K.; Rezazadeh, A. A unified framework for participation of responsive end-user devices in voltage and frequency control of the smart grid. *IEEE Trans. Power Syst.* **2015**, *30*, 1369–1379. [CrossRef]

68. Hussain, M.N.; Agarwal, V. A new control technique to enhance the stability of a DC microgrid and to reduce battery current ripple during the charging of plug-in electric vehicles. In Proceedings of the Environment and Electrical Engineering (EEEIC), 2015 IEEE 15th International Conference, Rome, Italy, 10–13 June 2015; IEEE: Piscataway, NJ, USA, 2015; pp. 2189–2193.

69. Poornazaryan, B.; Abedi, M.; Gharehpetian, G.; Karimyan, P. Application of PHEVs in controlling voltage and frequency of autonomous microgrids. In Proceedings of the Power System Conference (PSC), 2015 30th International, Tehran, Iran, 23–25 November 2015; IEEE: Piscataway, NJ, USA, 2015; pp. 60–66.

70. Arias, A.; Martínez, L.H.; Hincapie, R.A.; Granada, M. An efficient approach to solve the combination between Battery Swap Station Location and CVRP by using the MTZ formulation. In Proceedings of the Innovative Smart Grid Technologies Latin America (ISGT LATAM), 2015 IEEE PES, Montevideo, Uruguay, 5–7 October 2015; IEEE: Piscataway, NJ, USA, 2015; pp. 574–578.

71. Zhang, B.; Kezunovic, M. Impact on power system flexibility by electric vehicle participation in ramp market. *IEEE Trans. Smart Grid* **2016**, *7*, 1285–1294. [CrossRef]

72. Sarker, M.R.; Dvorkin, Y.; Ortega-Vazquez, M.A. Optimal participation of an electric vehicle aggregator in day-ahead energy and reserve markets. *IEEE Trans. Power Syst.* **2016**, *31*, 3506–3515. [CrossRef]

73. Johal, R.; Jain, D. Demand response as a load shaping tool integrating electric vehicles. In Proceedings of the Power Systems (ICPS), 2016 IEEE 6th International Conference, New Delhi, India, 4–6 March 2016; IEEE: Piscataway, NJ, USA, 2016; pp. 1–6.

74. Hafez, O.; Bhattacharya, K. Integrating ev charging stations as smart loads for demand response provisions in distribution systems. *IEEE Trans. Smart Grid* **2016**, *9*, 1096–1106

75. Behboodi, S.; Crawford, C.; Djilali, N.; Chassin, D.P. Integration of price-driven demand response using plug-in electric vehicles in smart grids. In Proceedings of the Electrical and Computer Engineering (CCECE), 2016 IEEE Canadian Conference, Vancouver, BC, Canada, 15–18 May 2016; IEEE: Piscataway, NJ, USA, 2016; pp. 1–5.

76. Catalão, J.; Osorio, G.; Gil, F.; Aghaei, J.; Barani, M.; Heydarian Forushani, E. Optimal Behavior of Electric Vehicle Parking Lots as Demand Response Aggregation Agents. 2016. *IEEE Trans. Smart Grid* **2016**, *7*, 2654–2665

77. Cross, J.; Hartshorn, R. My Electric Avenue: Integrating Electric Vehicles into the Electrical Networks. In Proceedings of the 6th Hybrid and Electric Vehicles Conference (HEVC 2016), London, UK, 2–3 November 2016.

78. Harb, A.; Hamdan, M. Power quality and stability impacts of Vehicle to grid (V2G) connection. In Proceedings of the Renewable Energy Congress (IREC), 2017 8th International, Amman, Jordan, 21–23 March 2017; IEEE: Piscataway, NJ, USA, 2017; pp. 1–6.

79. Martinenas, S.; Knezović, K.; Marinelli, M. Management of power quality issues in low voltage networks using electric vehicles: Experimental validation. *IEEE Trans. Power Deliv.* **2017**, *32*, 971–979. [CrossRef]

80. Chen, T.; Pourbabak, H.; Liang, Z.; Su, W.; Yu, P. Participation of electric vehicle parking lots into retail electricity market with evoucher mechanism. In Proceedings of the Transportation Electrification Asia-Pacific (ITEC Asia-Pacific), 2017 IEEE Conference and Exp, Harbin, China, 7–10 August 2017; IEEE: Piscataway, NJ, USA, 2017; pp. 1–5.

81. Yao, L.; Lim, W.H.; Tsai, T.S. A real-time charging scheme for demand response in electric vehicle parking station. *IEEE Trans. Smart Grid* **2017**, *8*, 52–62. [CrossRef]

82. Lu, Z.; Qi, J.; Zhang, J.; He, L.; Zhao, H. Modelling dynamic demand response for plug-in hybrid electric vehicles based on real-time charging pricing. *Iet Gener. Transm. Distrib.* **2017**, *11*, 228–235. [CrossRef]

83. Pal, S.; Kumar, R. Electric Vehicle Scheduling Strategy in Residential Demand Response Programs with Neighbor Connection. *IEEE Trans. Ind. Informatics* **2017**. *14*, 980–988. [CrossRef]

84. Dutta, A.; Debbarma, S. Frequency Regulation in Deregulated Market Using Vehicle-to-Grid Services in Residential Distribution Network. *IEEE Syst. J.* **2017**. *12*, 2812–2820. [CrossRef]

85. Kaur, K.; Singh, M.; Kumar, N. Multiobjective Optimization for Frequency Support Using Electric Vehicles: An Aggregator-Based Hierarchical Control Mechanism. *IEEE Syst. J.* **2017**. *13*, 771–782. [CrossRef]

86. Mathur, A.K.; Yemula, P.K. Optimal Charging Schedule for Electric Vehicles in Parking Lot with Solar Power Generation. In Proceedings of the 2018 IEEE Innovative Smart Grid Technologies-Asia (ISGT Asia), Singapore, 22–25 May 2018; IEEE: Piscataway, NJ, USA, 2018; pp. 611–615.

87. Wang, S.; Bi, S.; Zhang, Y.J.; Huang, J. Electrical Vehicle Charging Station Profit Maximization: Admission, Pricing, and Online Scheduling. *IEEE Trans. Sustain. Energy* **2018**, *9*, 1722–1731. [CrossRef]

88. Mediwaththe, C.P.; Smith, D.B. Game-Theoretic Electric Vehicle Charging Management Resilient to Non-Ideal User Behavior. *IEEE Trans. Intell. Transp. Syst.* **2018**, *19*, 3486–3495. [CrossRef]

89. Lin, C.C.; Deng, D.J.; Kuo, C.C.; Liang, Y.L. Optimal charging control of energy storage and electric vehicle of an individual in the internet of energy with energy trading. *IEEE Trans. Ind. Inform.* **2018**, *14*, 2570–2578. [CrossRef]

90. Zhang, H.; Moura, S.J.; Hu, Z.; Qi, W.; Song, Y. A Second-Order Cone Programming Model for Planning PEV Fast-Charging Stations. *IEEE Trans. Power Syst.* **2018**, *33*, 2763–2777. [CrossRef]

91. Islam, M.M.; Shareef, H.; Mohamed, A. Optimal location and sizing of fast charging stations for electric vehicles by incorporating traffic and power networks. *Iet Intell. Transp. Syst.* **2018**, *12*, 947–957. [CrossRef]

92. Sun, S.; Yang, Q.; Wenjung, Y. A hierarchical optimal planning approach for plug-in electric vehicle fast charging stations based on temporal-SoC charging demand characterization. *Iet Gener. Transm. Distrib.* **2018**. *12*, 4388–4395. [CrossRef]

93. Huang, X.; Chen, J.; Yang, H.; Cao, Y.; Guan, W.; Huang, B. Economic planning approach for electric vehicle charging stations integrating traffic and power grid constraints. *Iet Gener. Transm. Distrib.* **2018**, *12*, 3925–3934. [CrossRef]

94. Faridimehr, S.; Venkatachalam, S.; Chinnam, R.B. A stochastic programming approach for electric vehicle charging network design. *IEEE Trans. Intell. Transp. Syst.* **2018**, *20*, 1870–1882. [CrossRef]

95. Li, C.; Ding, T.; Liu, X.; Huang, C. An Electric Vehicle Routing Optimization Model with Hybrid Plug-in and Wireless Charging Systems. *IEEE Access* **2018**, *6*, 27569–27578. [CrossRef]

96. Hwang, I.; Jang, Y.J.; Ko, Y.D.; Lee, M.S. System optimization for dynamic wireless charging electric vehicles operating in a multiple-route environment. *IEEE Trans. Intell. Transp. Syst.* **2018**, *19*, 1709–1726. [CrossRef]

97. Outlook, I.G.E. *Scaling-Up the Transition to Electric Mobility*; IEA: Paris, France, 2019.

98. Harper, C.; Gregory McAndrews, D.S.B. Electric Vehicles: Key Trends, Issues, and Considerations for State Regulators. In *Proceedings of the Technical Report*; National Association of Regulatory Utility Commissioners NARUC: Washington, DC, USA, 2019.

99. Hove, A.; Sandalow, D. Electric Vehicle Charging in China and the United States. Columbia, School of International and Public Affairs, Center on Global Energy Policy. Available online: https://energypolicy.columbia.edu/sites/default/files/file-uploads/EV_ChargingChina-CGEP_Report_Final.pdf (accessed on 22 December 2019).

100. Doumen, S.; Paterakis, N.G. Economic viability of smart charging EVs in the Dutch ancillary service markets. In Proceedings of the 2019 International Conference on Smart Energy Systems and Technologies (SEST), Porto, Portugal, 9–11 September 2019; IEEE: Piscataway, NJ, USA, 2019; pp. 1–6.

101. Han, B.; Lu, S.; Xue, F.; Jiang, L. Day-ahead electric vehicle aggregator bidding strategy using stochastic programming in an uncertain reserve market. *Iet Gener. Transm. Distrib.* **2019**, *13*, 2517–2525. [CrossRef]

102. Jiang, Z.; Ai, Q.; Hao, R.; Yousif, M.; Zhang, Y. Joint Optimization for Bidding and Pricing of Electric Vehicle Aggregators Considering Reserve Provision and EV Response. In Proceedings of the 2019 IEEE Power & Energy Society General Meeting (PESGM), Atlanta, GA, USA, 4–8 August 2019; IEEE: Piscataway, NJ, USA, 2019; pp. 1–5.

103. Hashemi, B.; Shahabi, M.; Teimourzadeh-Baboli, P. Stochastic-based optimal charging strategy for plug-in electric vehicles aggregator under incentive and regulatory policies of DSO. *IEEE Trans. Veh. Technol.* **2019**, *68*, 3234–3245. [CrossRef]

104. Ren, H.; Zhang, A.; Li, W. Study on Optimal V2G Pricing Strategy under Multi-Aggregator Competition Based on Game Theory. In Proceedings of the 2019 IEEE Sustainable Power and Energy Conference (iSPEC), Beijing, China, 21–23 November 2019; IEEE: Piscataway, NJ, USA, 2019; pp. 1027–1032.

105. Redondo-Iglesias, E.; Venet, P.; Pelissier, S. Efficiency degradation model of lithium-ion batteries for electric vehicles. *IEEE Trans. Ind. Appl.* **2018**, *55*, 1932–1940. [CrossRef]

106. Fele, F.; Margellos, K. Scenario-based Robust Scheduling for Electric Vehicle Charging Games. In Proceedings of the 2019 IEEE International Conference on Environment and Electrical Engineering and 2019 IEEE Industrial and Commercial Power Systems Europe (EEEIC/I&CPS Europe), Genova, Italy, 11–14 June 2019; IEEE: Piscataway, NJ, USA, 2019; pp. 1–6.

107. Gomes, I.; Melicio, R.; Mendes, V. Stochastic Management of Bidirectional Electric Vehicles: The Case of an Electric Vehicles Aggregator. In Proceedings of the 2019 IEEE International Conference on Environment and Electrical Engineering and 2019 IEEE Industrial and Commercial Power Systems Europe (EEEIC/I&CPS Europe), Genova, Italy, 11–14 June 2019; IEEE: Piscataway, NJ, USA, 2019; pp. 1–5.

108. Skolthanarat, S.; Somsiri, P.; Tungpimolrut, K. Contribution of Real-Time Pricing to Impacts of Electric Cars on Distribution Network. In Proceedings of the 2019 IEEE Industry Applications Society Annual Meeting, Baltimore, MD, USA, 29 September–3 October 2019; IEEE: Piscataway, NJ, USA, 2019; pp. 1–5.

109. Chen, C.; Chen, J.; Wang, Y.; Duan, S.; Cai, T.; Jia, S. A Price Optimization Method for Microgrid Economic Operation Considering Across-Time-and-Space Energy Transmission of Electric Vehicles. *IEEE Trans. Ind. Inform.* **2019**, *16*, 1873–1884

110. Guo, Z.; Zhou, Z.; Zhou, Y. Impacts of Integrating Topology Reconfiguration and Vehicle-to-Grid Technologies on Distribution System Operation. *IEEE Trans. Sustain. Energy* **2019**. *11*, 1023–1032

111. Chen, C.; Wu, Z.; Zhang, Y. The Charging Characteristics of Large-Scale Electric Vehicle Group Considering Characteristics of Traffic Network. *IEEE Access* **2020**, *8*, 32542–32550. [CrossRef]

112. Duan, B.; Xin, K.; Zhong, Y. Optimal Dispatching of Electric Vehicles Based on Smart Contract and Internet of Things. *IEEE Access* **2019**, *8*, 9630–9639. [CrossRef]

113. Rasheed, M.B.; Awais, M.; Alquthami, T.; Khan, I. An Optimal Scheduling and Distributed Pricing Mechanism for Multi-Region Electric Vehicle Charging in Smart Grid. *IEEE Access* **2020**, *8*, 40298–40312. [CrossRef]

Article

Optimal Location-Reallocation of Battery Energy Storage Systems in DC Microgrids

Oscar Danilo Montoya [1,2,*], Walter Gil-González [2] and Edwin Rivas-Trujillo [1]

[1] Facultad de Ingeniería, Universidad Distrital Francisco José de Caldas, Bogotá D.C. 11021, Colombia; erivas@udistrital.edu.co

[2] Laboratorio Inteligente de Energía, Universidad Tecnológica de Bolívar, Cartagena 131001, Colombia; wjgil@utp.edu.co

* Correspondence: o.d.montoyagiraldo@ieee.org; Tel.: +57-310-5461-067

Received: 6 April 2020; Accepted: 28 April 2020; Published: 5 May 2020

Abstract: This paper deals with the problem of optimal location and reallocation of battery energy storage systems (BESS) in direct current (dc) microgrids with constant power loads. The optimization model that represents this problem is formulated with two objective functions. The first model corresponds to the minimization of the total daily cost of buying energy in the spot market by conventional generators and the second to the minimization of the costs of the daily energy losses in all branches of the network. Both the models are constrained by classical nonlinear power flow equations, distributed generation capabilities, and voltage regulation, among others. These formulations generate a nonlinear mixed-integer programming (MINLP) model that requires special methods to be solved. A dc microgrid composed of 21-nodes with existing BESS is used for validating the proposed mathematical formula. This system allows to identify the optimal location or reallocation points for these batteries by improving the daily operative costs regarding the base cases. All the simulations are conducted via the general algebraic modeling system, widely known as the General Algebraic Modeling System (GAMS).

Keywords: battery energy storage system; economic dispatch problem; nonlinear programming formulation; optimal reallocation of batteries; mathematical optimization

---

1. Introduction

Electrical networks have progressively transformed from thermal dependent systems to networks with high penetration of renewable energy resources [1,2]. This transformation is promoted to the Paris agreement, where many countries around the world have signed compromises regarding the minimization of greenhouse emissions [3]. This agreement forces conventional power systems to change their fossil fuel-based energy matrices (i.e., coal, natural gas, or diesel) by inserting renewable energy sources [4]. These sources are mainly photovoltaic and wind power plants, since their construction and production costs have decreased significantly in the last years [5]. Nevertheless, the inclusion of renewable energy sources is not a perfect solution since power systems must deal with uncertainties produced by weather variations (solar radiance or wind speed uncertainties). These uncertainties depend on the geographical location of the power system as well as the period of the year (winter or summer seasons). To tackle these uncertainties in renewable power generation, we have developed large-scale energy storage systems that allow to reduce the energy oscillations in the power system by compensating these in a dynamical form. Some of these energy storage devices can be: supercapacitors [6], fly-wheels [7], superconductors [8], compressed air systems [9], pumped-hydro systems [10], or batteries [11], among others. The selection of the energy storage depends on the application, i.e., for voltage and frequency compensation, fly-wheel, supercapacitors, or superconductors are preferred [12], while long-time power supplies are preferred for pumped-hydro and battery energy storage systems [13].

The inclusion of renewable energy sources and energy storage devices in power systems is not the only paradigm shift since other remarkable transformations have occurred, especially in distribution voltage levels [14]. This change is caused by the transition from classical alternating current (ac) networks to direct current (dc) systems [15]. The main advantage of using dc grids in comparison to ac is that the reactive power and frequency concepts disappear [16], which makes dc grids easily controllable, with low energy losses and higher voltage profiles [17]. In general, dc grids are more efficient and reliable than their ac counterparts [16–18].

Regarding the integration of renewable energy sources and energy storage systems in dc grids, there is a clear advantage, since some of them, such as photovoltaic sources and batteries, can operate directly in dc systems, which reduces the number of power conversion stages for integrating these devices into the grids [19]. Additionally, wind turbines or superconductors are only required in dc conversion stages since the inversion ones are unnecessary. This implies fewer electronic power converters can reduce the probability of failure and the costs of investment, maintenance, and operation [20].

It is important to note that when distribution levels are highly influenced by renewable and energy storage technologies, they integrate into smart grids with the capability of self-management regarding control and optimization [21,22]. Nevertheless, in the case of DC networks, smart grids can be reduced to microgrids since dc networks are an emerging concept in distribution levels and their current size and loadability are currently constrained to small areas (i.e., buildings or data centers) [23,24]. Here, we wish to analyze the problem of battery location and reallocation in dc microgrids from the point of economical and technical goals that focus on proposing a new mathematical model for representing this problem. However, it is important to note that the proposed model will be extended to a large-scale dc distribution feeder when it becomes a reality in the near future without modifications [25].

There are three different approaches to operate dc networks with high penetration of renewable energy resources and batteries, which are condensed in hierarchical control methods. The first two approaches are related to primary and secondary control strategies, which deal with power current and voltage controls [26], and measure local state variables to maintain voltage under nominal operative conditions [27]. Some of these controllers are designed via the sliding mode control [28], passivity based control [29], model predictive approach [30], and droop control [27], which are directly applied to the power electronic converters. The third approach is related to the optimization stage, where there exists specialized literature on tertiary control methods [31]. This stage defines the set points regarding power and voltage to be assigned for all the active devices (i.e., power electronic converters that interfaces batteries and distributed generators) to minimize some objective function, such as, typically, grid energy losses or energy purchase costs [18,31,32].

This study seeks to understand the tertiary control stage regarding optimization methods for the optimal operation of battery energy storage systems in dc networks with high penetration of renewable energy resources. In specialized literature, the optimal operation of batteries in dc grids has been addressed via economic dispatch formulation in three recent references as follows: Authors of [33] present a semi-definite programming model to operate batteries and renewables in dc grids. The authors of [33] proposed a relaxation of the power balance equations via semi-definite matrices, which guarantee the uniqueness of the global solution regarding the problem. The authors of [18] proposed a second-order cone programming model to optimally dispatch renewables and batteries in dc grids. Both relaxations coincide numerically to the exact nonlinear model. The main disadvantage of this model is that the number of variables increase in square form compared to the number of nodes in the dc grid. In [11], a nonlinear non-convex model for the optimal operation of batteries and renewable energies in dc grids considering voltage-dependent load models has been presented. This mathematical model is solved using the General Algebraic Modeling System (GAMS). It should be noted that in the previous models, the location of the batteries was predefined, and the data of the network was given by the utility. Nevertheless, there is no guarantee these locations are optimal.

Regarding the optimal location and operation of batteries in an ac electrical network, different approaches have been proposed in scientific literature, some of which are described below: the authors in [32] presented a methodology based on genetic algorithms for the optimal location battery energy storage in ac microgrids considering the performance indicator, the net present value of investments, and the costs of the energy losses. Different simulation cases considering sunny, cloudy, and rainy days were included to analyze the interdependence between batteries and distributed generators. In addition, the numerical results reported allowed to identify the best set of renewables and batteries to provide high-quality service to grid users. The authors of [34] proposed a methodology to increase the flexibility of microgrids with renewables that could be affected by the winter season. The authors of this research achieved a mathematical formulation with a mixed-integer programming form that can solve the battery scheduling problem efficiently via the CPLEX solver. Numerical results in a large-scale power system demonstrate the efficiency of the proposed approach for the management of heat demands in power systems, which are drastically affected by seasons through the optimal scheduling of batteries. In [35], an optimal economic dispatch of batteries in ac distribution networks was proposed considering the minimization of the energy purchase in conventional sources. The results were obtained via GAMS implementation by considering renewables as constant inputs and batteries in fixed points. The authors of [36] presented an optimization methodology based on genetic algorithms for the optimal location of batteries in radial distribution feeders. This approach allowed to reduce the daily operation losses of the network. In the case of batteries, the authors of [36] proposed a binary strategy to dispatch them, which can make the implementation of the model in conventional solvers difficult; nevertheless, the results are interesting for utilities since the grid is absent of renewable energy resources, making it the main contributor.

Note that the previous approaches in ac and dc grids demonstrate that the problem of optimal location and reallocation needs more research since this is an important issue in power system analysis. For this purpose, this paper proposes an optimization model for the location-reallocation batteries, focusing on dc networks. This problem has not yet been reported or analyzed for dc networks. The proposed model has a mixed-integer nonlinear programming (MINLP) structure because of binary and nonlinear variables. The binary variables appear due to location and reallocation of batteries while the nonlinear variables are the products between the voltage variables in the power balance constraint. The proposed model analyzes two objective functions, where the first corresponds to the minimization of the energy purchase costs in the spot market by conventional generators and the second to the minimization of the daily energy losses. In addition, it also employs artificial neural networks to forecast the power generated by wind and solar generators to increase the effectiveness of the proposed model. The main contributions of this study are as follows:

- To propose an MINLP model for the location-reallocation batteries in dc networks, which considers two objective functions independently or a linear combination of them. This problem has not been previously proposed in the scientific literature to the best of the knowledge of the authors. In addition, the proposed model allows to understand the compromise between the location of the batteries as a function of the perforce indicator, i.e., objective function, which demonstrates the interdependence of the batteries' location/operation regarding energy costs and renewable energy availability.
- To include in the proposed MINLP model the economic dispatch problem to maximize the use of the batteries during the day and, thus, obtain a suitable location and reallocation for them.
- Three simulation cases were analyzed for the proposed model to evaluate different objective functions according to the location-reallocation of the batteries. These simulations allow to identify the best trade-offs between the final positioning of the batteries and the daily behavior of the grid, which can help the distribution grid make the best decision as a function of its goals, i.e., technical or economic objectives.

This study is organized as follows: Section 2 presents the mathematical formulation for the optimization model, Section 3 formulates a strategy to solve the proposed optimization model, and Section 4 explains the test systems and proposed scenarios. The computational validation and results are analyzed in Section 5. Lastly, the main conclusions derived from this study and possible future works are presented in Section 6.

2. Mathematical Formulation

The problem of the optimal location-reallocation of batteries in dc microgrids with high penetration of renewable energy resources is a discrete (binary) version of the multi-period economic dispatch models for BESS proposed in [11,18,33]. The model for locating-reallocating batteries is indeed at nonlinear and strong non-convex (also integer) due to the hyperbolic relations between power and voltages in the power-balance equations. Here, we consider two possible objective functions that can be used as indicators to define the best location of the batteries: the first is related to the energy purchase costs in the spot market by the conventional generator and the second to the costs of the daily energy losses in all the branches of the network.

2.1. Objective Functions

$$\min z_1 = \sum_{t \in \mathcal{T}} \sum_{i \in \mathcal{N}} CoE_{i,t} p_{i,t} \Delta t \tag{1}$$

$$\min z_2 = \sum_{t \in \mathcal{T}} \sum_{i \in \mathcal{N}} CoE_{i,t} v_{i,t} \left(\sum_{j \in \mathcal{N}} G_{ij} v_{j,t} \Delta t \right) \tag{2}$$

where z_1 is the objective function value related to energy buying costs, z_2 is the objective function associated to the costs of the daily energy losses, $CoE_{i,t}$ is the cost of buying energy (spot market purchase) at node i in period t, $p_{i,t}$ corresponds to the power bought (generated) at node i during period t, and Δt is the length of the time period under analysis (e.g., 1 h, 30 min or 15 min); $v_{i,t}$ is the voltage value at node i during the period of time t; G_{ij} is the value of the conductance that relates nodes i and j. Observe that $\mathcal{T}$ and $\mathcal{N}$ are the sets that contain all periods of time of the dispatch planning and the total number of nodes in the dc microgrid, respectively.

It should be noted that the mathematical formulation of the objective functions z_1 and z_2 originate from convex functions since the energy purchase costs are a linear function, and the daily energy costs are a positive definite quadratic form based on the properties of the conductance matrix [37].

2.2. Set of Constraints

$$p_{i,t} + p_{i,t}^{dg} + \sum_{b \in \mathcal{B}} p_{i,t}^b - p_{i,t}^d = v_{i,t} \sum_{j \in \mathcal{N}} G_{ij} v_{j,t}, \ \{\forall i \in \mathcal{N} \ \& \ \forall t \in \mathcal{T}\} \tag{3}$$

$$SoC_{i,t}^b = SoC_{i,t-1}^b - \varphi_i^b p_{i,t}^b \Delta t, \ \{\forall b \in \mathcal{B}, \ \forall i \in \mathcal{N} \ \& \ \forall t \in \mathcal{T}\} \tag{4}$$

$$SoC_{i,t_0}^b = x_i^b SoC_i^{b,ini}, \ \{\forall b \in \mathcal{B} \ \& \forall i \in \mathcal{N}\} \tag{5}$$

$$SoC_{i,t_f}^b = x_i^b SoC_i^{b,fin}, \ \{\forall b \in \mathcal{B} \ \& \forall i \in \mathcal{N}\} \tag{6}$$

$$p_{i,t}^{min} \le p_{i,t} \le p_{i,t}^{max}, \ \{\forall i \in \mathcal{N} \ \& \ \forall t \in \mathcal{T}\} \tag{7}$$

$$p_{i,t}^{dg,min} \le p_{i,t}^{dg} \le p_{i,t}^{dg,max}, \ \{\forall i \in \mathcal{N} \ \& \ \forall t \in \mathcal{T}\} \tag{8}$$

$$x_i^b p_i^{b,\min} \leq p_{i,t}^b \leq x_i^b p_i^{b,\max}, \quad \{\forall b \in \mathcal{B}, \ \forall i \in \mathcal{N} \ \& \ \forall t \in \mathcal{T}\} \tag{9}$$

$$v_i^{\min} \leq v_{i,t} \leq v_i^{\max}, \quad \{\forall i \in \mathcal{N} \ \& \ \forall t \in \mathcal{T}\} \tag{10}$$

$$x_i^b SoC_i^{b,\min} \leq SoC_{i,t}^b \leq x_i^b SoC_i^{b,\max}, \quad \{\forall b \in \mathcal{B}, \ \forall i \in \mathcal{N} \ \& \ \forall t \in \mathcal{T}\} \tag{11}$$

$$\sum_{b \in \mathcal{B}} \sum_{i \in \mathcal{N}} x_i^b = N_b^{\max}, \tag{12}$$

where $p_{i,t}^{dg}$, $p_{i,t}^b$, and $p_{i,t}^d$ are the power generation by renewable energy resources (i.e., distributed generation), the power delivered/absorbed by the batteries, and the power demand at node i during the time period t, respectively; $SoC_{i,t}^b$ represents the state-of-charge of the battery in the ith node at the tth time period; x_i^b is a binary variable related to the possibility of locating/reallocating the battery b at node i; $SoC_i^{b,ini}$ and $SoC_i^{b,fin}$ are the initial and final desired states of charge of the batteries, respective;y, while $SoC_i^{b,\min}$ and $SoC_i^{b,\max}$ are the minimum and maximum state-of-charge bounds; $p_{i,t}^{\min}$, $p_{i,t}^{\max}$, $p_{i,t}^{dg,\min}$, and $p_{i,t}^{dg,\max}$ are the minimum and maximum bounds of admissible generation for conventional and renewable generators located in the ith node in time period t, respectively, while $p_i^{b,\min}$ and $p_i^{b,\max}$ represent the minimum and maximum charge/discharge capabilities of a battery connected at node i; $v_i^{\min}$ and $v_i^{\max}$ are the voltage regulation bounds of the dc microgrid. Finally, φ_i^b represents the coefficient of charge of a battery connected at node i. Observe that $N_b^{\max}$ corresponds to the maximum number of batteries available for location or reallocation, and $\mathcal{B}$ is the set that contains all the types of batteries.

The interpretation of the complete mathematical model described from Equation (1) to (12) is the following: Expression (1) presents the objective function related to the minimization of the energy purchase cost in the spot market by conventional sources; Equation (2) defines the objective function related to the total costs of the energy losses in all the branches of the network; Equation (3) presents the power balance constraint per node associated with the combination of Kirchhoff's first law and the first Tellegen's theorem [38]; Expression (4) shows the linear relation between the state-of-charge of the battery and the power injection/absorption [33]; Equations (5) and (6) present the operative consigns for battery operation regarding initial and final state-of-charges, which are defined by the utility. Expressions (7)–(9) are defined as the minimum and maximum power bounds for conventional and distributed generators as well as for batteries, respectively. In Equation (10), the lower and upper bounds admissible for voltage profiles, i.e., voltage regulation limits are presented. Equation (11) shows the minimum and maximum bounds for the states-of-charge in batteries. Expression (12) presents the constraint related to the maximum number of batteries available for location or reallocation in the dc network.

In the mathematical model of Equations (1) and (12) for optimal location-reallocation of batteries, it is important to highlight the following facts:

- This formulation has a mixed-integer nonlinear programming (MINLP) structure due to the presence of binary variables regarding the location-reallocation of batteries as well as products between voltage variables in the power balance constraint, which implies robust optimization methodologies or toolboxes are required to get the optimal solution, even if it is local or global [11].
- The effectiveness of the proposed model highly depends on the weather conditions in renewable generation, since their power injections, i.e., $p_{i,t}^{dg}$, are conditioned to the generation technology. Here we consider these renewables are based on wind and power technologies and their outputs forecast via artificial neural networks, as recommended in [12,19,33].

- The location or reallocation of the batteries in the dc grid will depend on the performance indicator, i.e., objective functions z_1 and z_2, even if they are used independently or as a linear combination.
- Regarding the complexity of the proposed MINLP model, the main difficulty is growing due to the solution space with the number of nodes in relation to the possibilities for locating or reallocating batteries. The size of this part of the solution space can be calculated as follows [39]:

$$C_{n,b} = \binom{n}{r} = \frac{n!}{b!\,(n-b)!}$$

where n is the number of nodes and b the number of batteries. In this sense, if we have a dc grid with 50 nodes and batteries from 1 to 5, then the number of possibilities for their location is 50, 1225, 19,600, 230,300, and 2,118,760, respectively, which demonstrate that the problem of optimal location-reallocation of batteries is highly complex. Note that each possible combination of batteries is needed to solve the resulting economic dispatch problem, which is also nonlinear and strong non-convex.

In this paper, the main interest is regarding the formulation of the MINLP formulation presented from Equation (1) to (12) since, for dc grids, after a careful revision has not found evidence of previous formulations, as most of the works are related with economic dispatch analysis considering the fixed location of the batteries, as reported in [11,12,18,33]. For this reason, we have employed the General Algebraic Modeling System (GAMS) for reaching the solution of the proposed model, since it has previously been used in [11] for battery dispatch in dc grids considering voltage-dependent load models. The next section presents the main characteristics of the GAMS software as a solution methodology.

It is important to mention that the optimization model proposed in this research (see Equations (1)–(12)) considers the basic relation between state-of-charge in batteries and the active power injections, as reported in [40]; nevertheless, in the future, for the purpose of analysis, it is highly recommended to make additional improvements regarding batteries, such as life cost analysis, self-discharge phenomena, or efficiency in power electronic converters that interface them [26], to identify other relevant aspects that can affect the short-term economic dispatch analysis in this research.

3. Solution Strategy

To deal with the mathematical model Equations (1)–(12) that describe the optimal location-reallocation of batteries in dc networks as an MINLP model, we select the GAMS optimization package as the solution strategy. This software is selected as it has been largely used in specialized literature to address complex optimization problems with hundreds of variables. Some of the most relevant works where GAMS has been used are the optimal locations of distributed generators in ac and dc networks considering daily load behaviors [19,41], the optimal design of osmotic power plants for electricity generation and desalinization processes [42,43], the optimization of the pump and valve schedules in complex, large-scale water distribution systems [44], the optimal dispatch of batteries in dc and ac networks [11,12,18,35], the multi-objective optimization of the stack of thermoacoustic engines [45], the multi-objective optimization in power systems with renewable sources [46], and economic dispatch approaches in thermal power systems [47], etc.

3.1. Software Implementation

To illustrate an implementation of an optimization model in GAMS, let us consider the simple MINLP model that represents the line selection for a transmission system using the transportation model, as depicted in Figure 1.

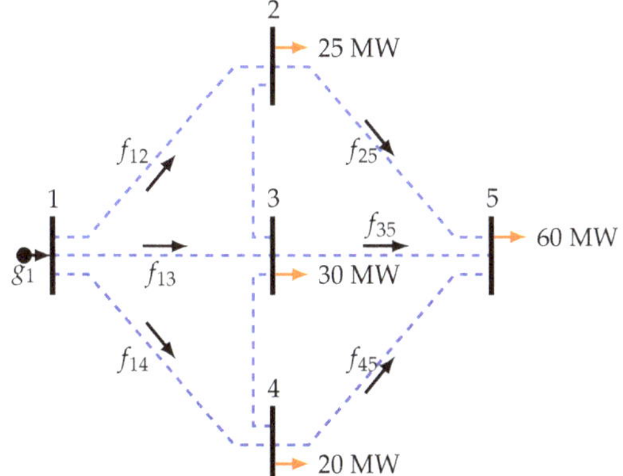

Figure 1. Small transmission system for the illustrative example.

This problem was proposed as an illustrative example of planning distribution networks in [48]. The formulation of this problem is presented from Equation (13a) to (13g).

$$\min z = \sum_{i=1}^{n} \sum_{j=1}^{n} f_{ij}^2,\tag{13a}$$

$$p_i^g - p_i^d = \sum_{i=1}^{n} \left(f_{ij} - f_{ji} \right),\ i = 1, 2, ..., n,\tag{13b}$$

$$- x_{ij} F_{ij}^{\max} \leq f_{ij} \leq x_{ij} F_{ij}^{\max},\ i, j = 1, 2, ..., n,\tag{13c}$$

$$0 \leq p_i^g \leq p_i^{g,\max},\ i = 1, 2, ..., n,\tag{13d}$$

$$x_{i,j} \in \{0, 1\},\ i = 1, 2, ..., n,\tag{13e}$$

$$x_{i,i} = 0,\ i = 1, 2, ..., n,\tag{13f}$$

$$x_{i,j} = 0,\ i, j = \{1, 5;\ 5, 1;\ 2, 4;\ 4, 2\}\tag{13g}$$

where f_{ij} is the power flow through the line that connected nodes i and j; p_i^g is the total power generation at node i, p_i^d is the total power consumption at node i; x_{ij} is a binary variable that defines the line between nodes i and j is built; z is the objective function defined as the sum of the square power flows (sensitivity index without physical interpretation). Note that $p_i^{g,\max}$ is the maximum power generation at node i, $F_{ij}^{\max}$ is the maximum power flow through the line that connects nodes i and j, and n is the number of nodes.

It is important to note that the main idea of this optimization problem is to determine the set of lines that need to be built to supply power to all the nodes. An important aspect in this formulation is that it is not unique since [48] has used a set of lines to reach the solution.

Figure 2 presents the GAMS implementation of the mathematical model Equations (13a)–(13g). Some important aspects about GAMS implementation can be extracted from this interface.

- It works with plain text using five main steps: (*i*) definition of all the sets related to the variables domain; (*ii*) definition of scalars (constant numbers), parameters (constant vectors), and matrices (constant matrices); (*iii*) definition of variables and their nature, i.e., continuous, binaries, or integers; (*iv*) declaration of the names of equations and their mathematical structures; and (*v*) solution of the model with the appropriate structure, i.e., maximization or minimization.
- The mathematical structure is pretty similar to the symbolic formulation (see model Equations (13a)–(13g)).

- This is an ideal toolbox that introduces to mathematicians and engineering students mathematical optimization since it allows to concentrate on the development of well-structured mathematical models to solve physical problems without focusing on the solution techniques.
- The implementation of any mathematical model in GAMS requires only a few concepts on computer programming, which is an advantage in comparison with metaheuristic approaches in MINLP optimization.

Once the optimization is solved by GAMS, we can visualize the solution variables, as depicted in Figure 3.

```
1    SETS
2    i Set that contains all the nodes /N1*N5/
3    g Set that contains all the generators /G1/
4    map(g,i) Relation between generators and nodes /G1.N1/
5    alias(i,j);
6    SCALARS
7    Fijmax Maximum flow allowed in the branch i-j /100/
8    Pgmax Maximum power generation /150/;
9    PARAMETER LOAD(i)
10   /N1 0,N2 25,N3 30,N4 20,N5 60/;
11   VARIABLES
12   Z Obj.function
13   f(i,j) Power flow in lines
14   p(g) Power generation;
15   BINARY VARIABLE
16   x(i,j) Construction of the line between node i and j;
17   * Minimum and maximum bounds and fixed variables
18   p.lo(g) = 0; p.up(g) = Pgmax;
19   x.fx('N1','N5') = 0;  x.fx('N5','N1') = 0; x.fx('N2','N4') = 0;
20   x.fx('N4','N2') = 0;  x.fx('N1','N1') = 0; x.fx('N2','N2') = 0;
21   x.fx('N3','N3') = 0;  x.fx('N4','N4') = 0; x.fx('N5','N5') = 0;
22   EQUATIONS
23   OBJ Objective funtion
24   Balance(i) Power balance at each node
25   MaxFij(i,j) Maximum flow
26   MinFij(i,j) Minimum flow
27   Radial Radial topology (tree);
28   OBJ.. z =E= sum(i,sum(j,sqr(f(i,j))));
29   Balance(i).. sum(g$map(g,i),p(g)) - LOAD(i) =E= sum(j,f(i,j) - f(j,i));
30   MaxFij(i,j).. f(i,j) =L= FijMax*x(i,j);
31   MinFij(i,j).. f(i,j) =G= -FijMax*x(i,j);
32   Radial.. sum(i,sum(j,x(i,j))) =E= card(i)-1;
33   MODEL INTERCONECTION /all/;
34   SOLVE INTERCONECTION US MINLP min z;
35   DISPLAY z.l,p.l,x.l,f.l;
```

Figure 2. Example of the implementation of a nonlinear mixed-integer programming (MINLP) model by General Algebraic Modeling System (GAMS).

```
1    VARIABLE Z.L =  11525.000  Obj.function
2    VARIABLE p.L Power generation
3    G1 135.000
4    VARIABLE x.L  Construction of the line between node i and j
5    N1            N5
6    N2      1.000
7    N3      1.000
8    N4      1.000           1.000
9    VARIABLE f.L  Power flow in lines
10   N1            N5
11   N2     -25.000
12   N3     -30.000
13   N4     -80.000         60.000
```

Figure 3. Solution of the MINLP model by GAMS.

Note that the objective function reaches an optimum value of about 11525 MW^2, and this system has building lines $x_{1,2}$, $x_{1,3}$, $x_{1,4}$, and $x_{1,5}$ with the following power flows: $f_{1,2} = 25$ MW, $f_{1,3} = 30$ MW, $f_{1,4} = 80$ MW, and $f_{4,5} = 60$ MW (this is the same solution reported in [48] using the AMPL software). In addition, the remainder of variables is zero, which implies that this is the optimal solution of the model since it fulfills all the constraints in the mathematical model Equations (13a)–(13g) used in this example to present the main characteristics of any mathematical optimization via the GAMS package. For more details about the software, please refer to [46,47].

3.2. Definition of the Renewable Generation Profiles

The forecast of the renewable generation profiles has been carried out by implementing the methodology described in [33]. This methodology works with artificial neural network (ANN) and combined receding horizon control. The receding horizon control works as a moving time horizon that calculates the economic dispatch model in each period using the forecast of the renewable generation profiles estimated by the ANN approach. The implemented methodology has the following steps:

- The ANN predicts the renewable generation profiles in periods n, where n is the prediction horizon.
- Solving the economic dispatch model with GAMS, the reallocation of battery energy storage systems is achieved.
- Employing the previous data of the renewable generation profiles from $t - m$ to t, the forecast of the profiles is recomputed for the $t + 1$ period.

For more details of this methodology, see [33].

Table 1 lists the ANN settings for each type of renewable energy source, which has implemented in MATLAB using *ntstool*. The ANN was configured with 70%, 15%, and 15% of the data for training, adjustment, and validation processes, respectively. Figure 4 illiterates the ANN scheme for the wind power forecasting.

Table 1. Parameters for wind and solar generation forecasting.

	Wind	Solar Power
Inputs	Temperature, humidity, pressure, and time	Temperature and time
Output	Wind speed	Solar radiation
Delay number	4	6
Hidden neurons	12	18
Training optimizer	Levenberg–Marquardt algorithm	Levenberg–Marquardt algorithm

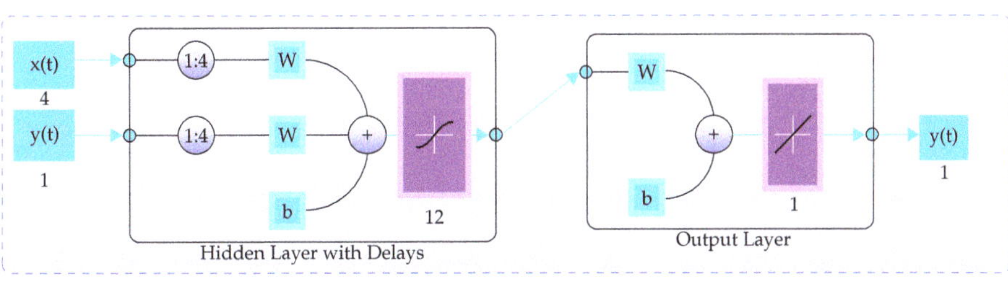

Figure 4. Artificial neural network (ANN) scheme for wind speed prediction [49].

3.3. Flow Diagram of the Proposed Approach

To summarize the application of the proposed methodology in a dc distribution network for location-reallocation of batteries the flow chart depicted in Figure 5 was implemented.

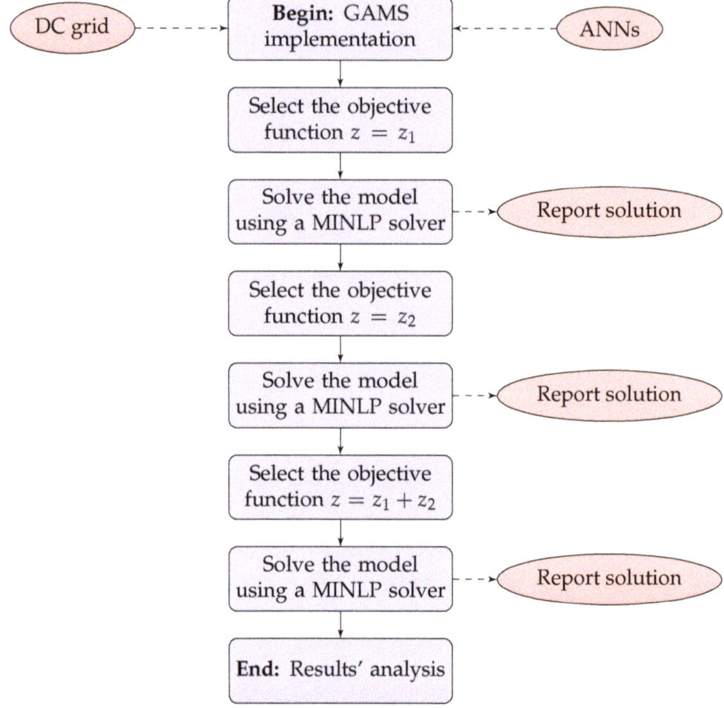

Figure 5. Flow chart of the proposed optimization approach for optimal location-reallocation of batteries in dc grids.

4. Test Systems

The validation of the proposed mathematical model for the optimal location-reallocation of batteries in dc distribution networks via GAMS implementation is made inside a 21-node test feeder [18]. All the information about this test feeder is presented below.

The 21-node test system is a radial test feeder composed of 21 nodes and 20 branches in radial connection, where the slack node is located at node 1. The configuration of this dc network is presented in Figure 6.

The information about loads, energy purchase cost, demand variation, and batteries for this test system is reported in Tables 2–4.

Table 2. Parameters for the 21-node test feeder.

From i	To j	R_{ij} [p.u.]	P_j [p.u.]	From i	To j	R_{ij} [p.u.]	P_j [p.u.]	From i	To j	R_{ij} [p.u.]	P_j [p.u.]
1 (slack)	2	0.0053	0.70	7	9	0.0072	0.80	15	16	0.0064	0.23
1	3	0.0054	0.00	3	10	0.0053	0.00	16	17	0.0074	0.43
3	4	0.0054	0.36	10	11	0.0038	0.45	16	18	0.0081	0.34
4	5	0.0063	0.04	11	12	0.0079	0.68	14	19	0.0078	0.09
4	6	0.0051	0.36	11	13	0.0078	0.10	19	20	0.0084	0.21
3	7	0.0037	0.00	10	14	0.0083	0.00	19	21	0.0081	0.21
7	8	0.0079	0.32	14	15	0.0065	0.22	–	–	–	–

All parameters are in per unit considering as bases $P_{\text{base}} = 100$ kW and $V_{\text{base}} = 1$ kV.

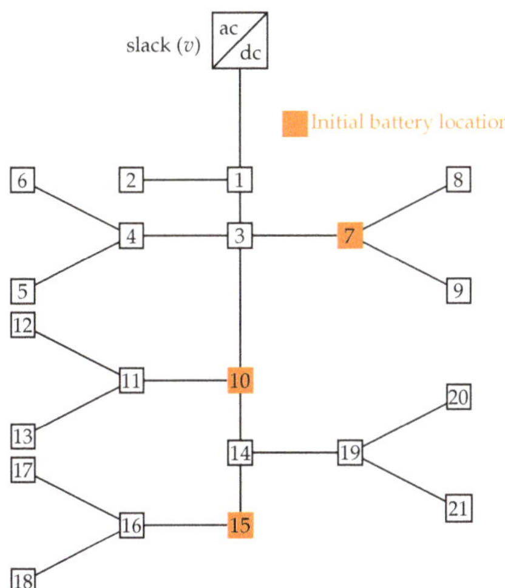

Figure 6. Electrical configuration for the 21-nodes test system.

Table 3. Energy purchasing cost and hourly demand.

Time [h]	CoE [p.u.]	Demand Variation [%]	Time [h]	CoE [p.u.]	Demand Variation [%]	Time [h]	CoE [p.u.]	Demand Variation [%]
0.5	0.8105	34	8.5	0.9263	62	16.5	0.9737	90
1.0	0.7789	28	9.0	0.9421	68	17.0	1	90
1.5	0.7474	22	9.5	0.9579	72	17.5	0.9947	90
2.0	0.7368	22	10.0	0.9579	78	18.0	0.9895	90
2.5	0.7263	22	10.5	0.9579	84	18.5	0.9737	86
3.0	0.7316	20	11.0	0.9579	86	19.0	0.9579	84
3.5	0.7368	18	11.5	0.9579	90	19.5	0.9526	92
4.0	0.7474	18	12.0	0.9526	92	20.0	0.9474	100
4.5	0.7579	18	12.5	0.9474	94	20.5	0.9211	98
5.0	0.8000	20	13.0	0.9474	94	21.0	0.8947	94
5.5	0.8421	22	13.5	0.9421	90	21.5	0.8684	90
6.0	0.8789	26	14.0	0.9368	84	22.0	0.8421	84
6.5	0.9158	28	14.5	0.9421	86	22.5	0.7947	76
7.0	0.9368	34	15.0	0.9474	90	23.0	0.7474	68
7.5	0.9579	40	15.5	0.9474	90	23.5	0.7211	58
8.0	0.9421	50	16.0	0.9474	90	24.0	0.6947	50

The energy-based cost is COP\$/kWh 479.3389 based on the prices in May 2019 for the CODENSA utility [11].

Table 4. Parameters associated with batteries and their initial locations.

Node	φ^b	$p^{b,\max}$	$p^{b,\min}$	Node	φ^b	$p^{b,\max}$	$p^{b,\min}$	Node	φ^b	$p^{b,\max}$	$p^{b,\min}$
7	0.0625	4	−3.2	10	0.0813	3.2	−2.4616	15	0.0813	3.2	−2.4616

In the case of renewable generation, we consider the wind power plant is connected at node 12 with a maximum power capability of about 221.52 kW, and the photovoltaic source is connected at node 21 with a maximum power capability about 281.58 kW. Note that these multiply the maximum capacities of the normalized generation curves reported in Table 5.

Table 5. Normalized power generation curve.

Time [h]	P_{WT} [p.u.]	P_{PV} [p.u.]	Time [h]	P_{WT} [p.u.]	P_{PV} [p.u.]	Time [h]	P_{WT} [p.u.]	P_{PV} [p.u.]
0.5	0.6303	0	8.5	0.8271	0.0403	16.5	0.9892	0.4193
1.0	0.6194	0	9.0	0.8523	0.1344	17.0	0.9652	0.2784
1.5	0.6098	0	9.5	0.8788	0.2710	17.5	0.9244	0.1373
2.0	0.6050	0	10.0	0.9064	0.3673	18.0	0.8607	0.0374
2.5	0.6122	0	10.5	0.9328	0.4584	18.5	0.7743	0.0007
3.0	0.6411	0	11.0	0.9520	0.6125	19.0	0.7251	0
3.5	0.6927	0	11.5	0.9640	0.8134	19.5	0.7167	0
4.0	0.7395	0	12.0	0.9700	0.9122	20.0	0.7167	0
4.5	0.7779	0	12.5	0.9748	0.9633	20.5	0.7251	0
5.0	0.7887	0	13.0	0.9784	1.0000	21.0	0.7263	0
5.5	0.7671	0	13.5	0.9832	0.9582	21.5	0.7179	0
6.0	0.7479	0	14.0	0.9880	0.8791	22.0	0.7095	0
6.5	0.7287	0	14.5	0.9940	0.7308	22.5	0.6987	0
7.0	0.7371	0	15.0	0.9988	0.7645	23.0	0.6915	0
7.5	0.7731	0	15.5	1.0000	0.6866	23.5	0.6867	0
8.0	0.8031	0.0016	16.0	0.9964	0.5893	24.0	0.6831	0

5. Computational Validation

All simulations were carried out on a desktop computer running on INTEL(R) Core(TM) i7-7700, 3.60 GHz, 8 GB RAM with 64-bit Windows 10 Pro using GAMS 25.1.3 with the nonlinear large-scale solver BONMIN licensed by Universidad Tecnológica de Bolívar in Colombia.

5.1. Simulation Conditions and Initial Function Values

To simulate the 21-node test feeder, we considered the following facts:

- The batteries begin and end the day with a total charge of about 50%. During the day, this state-of-charge can vary between 10% and 90%, as recommended for Ion-Lithium batteries in [11].
- Both objective functions are evaluated with the initial position of the batteries reported in the previous section, to identify the base cases and the possible improvements when they are reallocated.
- A linear combination of both objective functions was made to identify the effect of adding energy purchase costs with energy losses costs.

Once the minimization of the energy purchase cost of energy and the minimization of the cost of the daily energy losses were performed, we found the objective function z_1 took COP\$/day 1,139,524.00 (with $z_2 = 131,198.70$) (see Equation (1)), and the objective function z_2 took COP\$/day 52,957.92 (with $z_1 = 1,941,395.00$) (see Expression (2)). These values are considered the base cases for each objective function.

5.2. Optimal Reallocation of Batteries

Here, we present the optimal location-reallocation of batteries considering both objective functions. Table 6 reports each objective function and its corresponding BESS' location.

Table 6. Optimal location-reallocation of batteries considering different objective functions.

Minimization of z_1 [COP\$/Day]		Minimization of z_2 [COP\$/Day]	
z_1 :	**1,089,974.00** ($z_2 = 87426.51$)	z_2 :	**47,209.95** ($z_1 = 1843467.00$)
Battery type 1:	1	Battery type 1:	13
Battery type 2:	{2, 3}	Battery type 2:	{20, 21}

Minimization of the Linear Combination $z_1 + z_2$ [COP\$/Day]			
z_1 :	**1,188,233.00**	z_2 :	**94,347.07**
Battery type 1:	13	Battery type 2:	{9, 21}

To understand the optimal reallocation of batteries as a function of the performance indicator, let us plot the locations of the batteries in the test feeder, as depicted in Figure 7.

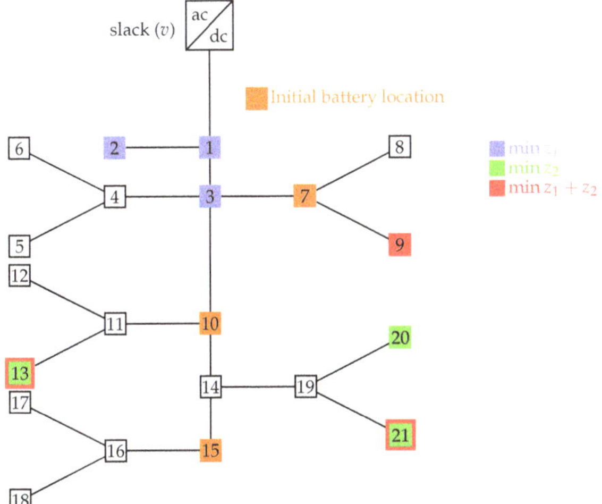

Figure 7. Reallocation of batteries in the 21-node test feeder.

Based on the location-reallocation of the batteries in Table 6 (see Figure 7), the following facts can be highlighted:

- When the objective of the optimization is to minimize the energy purchase costs at the conventional sources, i.e., z_1, the system reduces the daily operation cost of about 4.35% passing from COP\$/day 1,139,524.00 to COP\$/day 1,089,974.00, which implies a reduction per day about COP\$/day 49,550. In addition, when we observe the total costs of the daily losses, it passes from COP\$/day 131,198.70 to COP\$/day 87,426.51; which is traduced in 33.36% of the energy losses reduction. These results confirm that the optimal reallocation of the batteries from nodes 7, 10, and 15 to nodes 1, 2, and 3 has a positive effect on both the objective functions.

- When the objective of the optimization model is to minimize the daily energy losses, i.e., z_2, this function is reduced from COP\$/day 52,957.92 to COP\$/day 47,209.95, this corresponds a reduction about 10.85%. This reduction is achieved by reallocating batteries from nodes 7, 10, and 15 to nodes 13, 20, and 21. In addition, the costs of the energy purchase pass from COP\$/day 1,941,395.00 to COP\$/day 1,843,467.00, which corresponds to a reduction of about 5.04%

- When the objective function is the linear combination of z_1 and z_2, i.e., $z_1 + z_2$, the objective function is COP\$/day 1,282,580.07; this implies an increment regarding the base of the case of the energy purchase about COP\$/day 11,857.37 per day of operation. In the case of daily energy losses reduction, the linear combination reaches a reduction of COP\$/day 711,772.85 per day of operation.

The previous analyses regarding different objective functions allowed us to conclude that the 21-node test feeder is always positive, considering the energy purchase costs as the objective function since it has the most important effect on the daily operation cost for the test feeder. In addition, reducing z_1 also allows to reduce the daily energy losses with respect to the base case. On the other hand, when we minimize the daily energy losses, this is reduced. Nevertheless, it also produces a negative effect on the energy purchase cost with an important increment in this objective function. For this reason, here, the minimization of the daily energy loss reduction as an adequate indicator for the optimal operation of batteries in dc networks is discarded.

Regarding the reallocation of the batteries, we observe that when we reduce the energy purchase costs, all the batteries are positioned near the slack node, as these locations allow to charge all the batteries with minimum losses when the daily energy cost is lower to inject this power when this cost increases.

5.3. Complementary Analysis

To understand whether the reallocation of the batteries satisfy the operative conditions imposed on the mathematical model, i.e., begin and end the day with 50% considering possible variations between 10% and 90%, Figure 8 presents the behaviors for both base cases and the three possible locations, reported in Table 6 for the battery type 1.

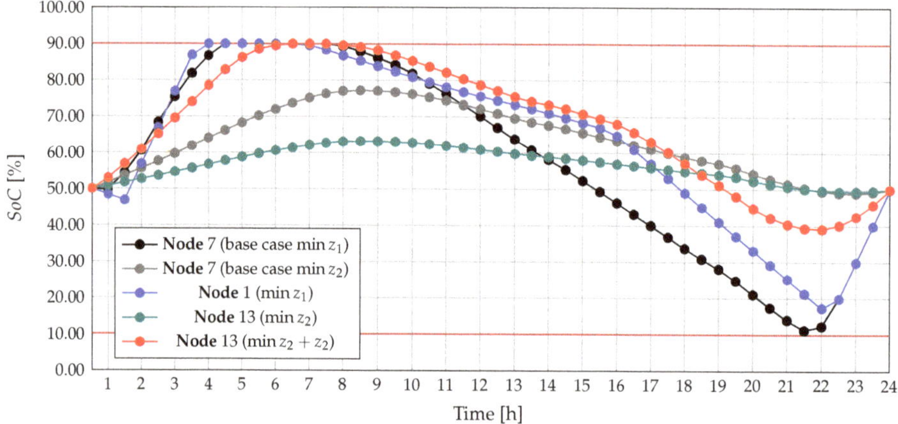

Figure 8. Behavior of battery type 1 in its different locations.

The behavior of the battery type 1 when it is located at different nodes considering variations in the objective function allows to identify the following important facts:

- When the objective function is minimizing the total energy purchase cost, i.e., z_1, this battery charges at its maximum admissible value (see periods comprehended between 0.5 h to 6). In addition, when energy cost is expensive, this battery begins to discharge continuously during fourteen hours (see periods comprehended between 8 h to 22). Finally, this battery recovers this charge in the final part of the time.

- Regarding the minimization of energy losses, i.e., z_2, we observe that the batteries charge between 50% and 80% during the day. This behavior is explained by the fact that the battery works as a generator or load; it is modifying the power flowing through the lines, which implies that it is also affecting the power losses. For this reason, it has small changes in its state-of-charge to help and minimize the total power losses during the day of operation.

- In the case of combining both objective functions linearly, battery type 1 experiences both behaviors, as previously reported, at the same time, i.e., soft state-of-charge variations overpass values lower than 50% at the end of the day.

Note that the main message of the battery behavior is regarding the different possibilities of having state-of-charges behavior as a function of the performance indicator (i.e., objective function) as well as the possible location of it. Nevertheless, in all the simulation cases, the proposed optimization model (Equations (1)–(12)) is feasible and allows to determine the best operation practice for batteries depending on the operative consigns imposed by the utility, which becomes the proposed optimization model the main contribution of this research.

Figure 9 reports the power generation in the slack node for the base cases as well as for the different battery locations. From this plot, we can observe that:

- When the objective function is minimizing z_1, the conventional generator is used to charge all the batteries to reach their maximum admissible values at the beginning of the day. This conventional generator is also used to recover the final state of the charge imposed, i.e., 50% at the end of the day. In the rest of the periods, the energy provided by the conventional generator is zero, which minimizes the total purchase costs in the spot market and also allows to maximize the use of renewable sources.
- In the case of minimization of z_2, we can see that the conventional source generates during all the periods since this generation allows redistributing line power flows, which helps minimize the total cost of the energy losses.
- Regarding the linear combination of the objective functions, it is possible to observe that at the beginning and the end of the day, the conventional generator is used for charging all the batteries while in the intermediate times, it is used for redistributing power flows; in other words, the behavior of the conventional generator is a linear combination of both analyses mentioned above.
- The behavior of the conventional generators shows that in relation to the energy purchase cost minimization, its generation appears mainly in periods of time where solar energy is absent, which can be considered as an indicator for the utility to introduce additional renewable energy resources (i.e., small-hydro power [50]) to complement solar and wind sources in order to reduce to zero the conventional generation in normal operating conditions. This also can help to indirectly reduce greenhouse emissions in isolated grids power supply by diesel or in predominantly thermal interconnected systems.

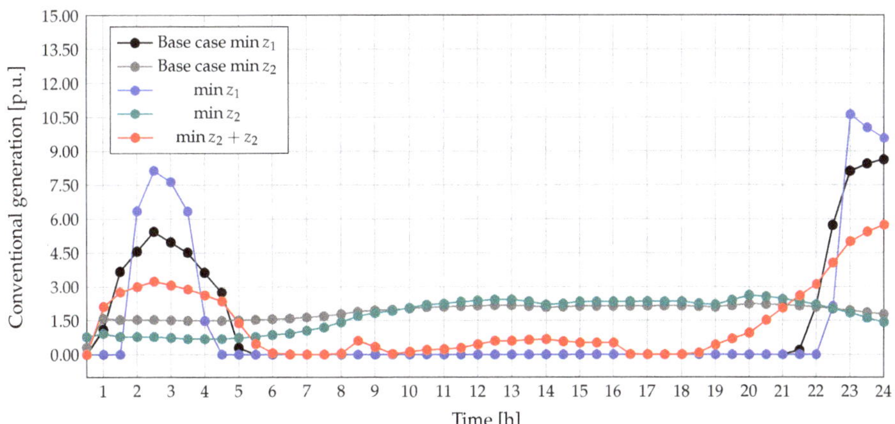

Figure 9. Behavior of the conventional generation (e.g., slack node) for all the cases studied.

In conclusion, it is important to keep in mind that the behavior of all the variables in the proposed mathematical MINLP model highly depends on the objective function selected since it guides them to minimize or maximize the benefit established by the electricity company. In addition, all the solutions reported in this research can be taken as indicators of the grid performance; nevertheless, these are not taken as absolute results, since they also depend on the renewable generation availability, grid operative conditions, and demand behavior.

Regarding the computational effort of the proposed approach for optimal locating-reallocating batteries in dc grids, it is important to highlight that the GAMS and its SCIP solver takes about 10 min to define the optimal position of the batteries, which can be considered as a pretty small time considering the complexity of the problem (3 batteries in 21-nodes generates 1330 possible locations). In addition, if the batteries are considered in fixed locations, then, the optimization problem is transformed from an MINLP into a nonlinear programming one, reducing the processing times to 2 s for knowing the economic dispatch output, which allows access to the utilities having multiple scenarios of simulation before defining the final day-ahead economic dispatch.

5.4. Scalability of the Proposed Model

To demonstrate the scalability of the proposed approach, we present an additional simulation case regarding batteries' location and reallocation in dc grids by using the 30-node test feeder reported in [11]. In this test system, there are three batteries in nodes 3, 15, and 22. In addition, the objective function reported in that paper corresponds to the daily energy losses cost, which takes a value of COP\$/day 254,539.39. Once we apply our proposed reallocation approach, this cost moves to COP\$/day 244,595.39 by reallocating these batteries to nodes 18, 23, and 24, respectively, which implies a daily reduction of about 3.91% regarding power loss minimization, demonstrating that the proposed optimization approach is applicable to dc grids with a different number of nodes, i.e., it is scalable.

It is important to point out that the computational cost of the proposed approach for this test feeder is 14 and takes minutes to decide where the batteries must be reallocated; nevertheless, when the batteries are fixed, this time is reduced to 3 s, which implies that this methodology is perfectly for day-ahead analysis for dc grids with batteries, since it allows to create multiple simulation scenarios in less time.

6. Conclusions and Future Works

The problem of the optimal location-reallocation of batteries in dc distribution networks has been analyzed in this paper through an MINLP formulation. This mathematical model has binary variables associated with the position of the batteries that are modified as a function of the performance indicator, i.e., minimization of the energy purchase costs in the conventional generators or minimization of the total costs of the daily energy losses. Numerical results confirm that in all the analyzed scenarios, the re-positioning of batteries allows to achieve better objective function values by modifying the daily state-of-charge performances on them. In addition, the behavior of the conventional power generator is highly conditioned by the performance indicator since this variable defines the total energy purchase costs in the spot market and also has an influence over the power flow redistribution in lines regarding power loss minimization. This means that this variable is highly sensitive to the proposed MINLP model.

For future works, we state the following: (*i*) to propose a mixed-integer convex optimization model to deal with the non-linearities of the power balance equations to guarantee the uniqueness of the global optimum solution via branch and bound methods added with a second-order cone or semi-definite programming methods, (*ii*) to employ heuristic methods to avoid the usage of specialized software in the solution of the MINLP model to develop free applications for engineering students and small electricity companies, (*iii*) to extend the proposed MINLP formulation to alternating current networks considering active and reactive power capabilities in batteries via optimal control of power electronic converters that interface them, (*iv*) to consider different battery's aspects in the model such as lifecycle, self-discharge phenomenon, or power losses analysis to have more realistic models that can affect the grid behavior in the short-term horizon, and (*v*) to propose a hybrid optimization problem based on the convex optimization for the nonlinear optimization part of the MINLP model (economic dispatch problem) and metaheuristics that can help address the integer part (location of batteries and renewables) by conforming a master-slave optimization methodology.

Energies **2020**, *13*, 2289

Author Contributions: Conceptualization, O.D.M. and W.G.-G.; Methodology, O.D.M. and W.G.-G.; Investigation, O.D.M. and W.G.-G.; Writing—review and editing, O.D.M., W.G.-G., and E.R.-T. All authors have read and agreed to the published version of the manuscript.

Funding: This work was partially supported by the Universidad Tecnológica de Bolívar under grant CP2019P011 associated with the project: "Operación eficiente de redes eléctricas con alta penetración de recursos energéticos distribuidos considerando variaciones en el recurso energético primario".

Conflicts of Interest: The authors declare no conflicts of interest.

Abbreviations

The following abbreviations and nomenclature are used in this manuscript:

Acronyms

AC	Alternating current.
COP	Colombian pesos.
ANN	Artificial neural network.
BESS	Battery energy storage system.
DC	Direct current.
GAMS	General algebraic modeling system.
MINLP	Mixed-integer nonlinear programming.
MCOP	Millions of Colombian pesos.
PV	Photovoltaic.
WT	Wind turbine.

Sets and subscripts

$\mathcal{B}$	Set of batteries.
$\mathcal{N}$	Set of nodes.
$\mathcal{T}$	Set of periods of time.
b	Type of batteries.
t	Periods of time.
i, j	Nodes.

Parameters

$CoE_{i,t}$	Energy purchase cost in the conventional source in the node i at the period of time t.
G_{ij}	Conductance value that relates nodes i and j.
$N_b^{\max}$	Maximum number of batteries available.
$p_{i,t}^{\max}$	Maximum power generation in the conventional source in the i at the period of time t.
$p_{i,t}^{\min}$	Minimum power generation in the conventional source in the i at the period of time t.
$p_{i,t}^{b,\max}$	Minimum power in the battery b connected at i at the period of time t.
$p_{i,t}^{b,\min}$	Maximum power in the battery b connected at i at the period of time t.
$p_{i,t}^{dg,\max}$	Minimum power generation in the distributed generator in the i at the period of time t.
$p_{i,t}^{dg,\min}$	Maximum power generation in the distributed generator in the i at the period of time t.
$p_{i,t}^{d}$	Power consumption in the i at the period of time t.
$SoC_i^{b,fin}$	Final state of charge in the battery b connected to the node i.
$SoC_i^{b,ini}$	Initial state of charge in the battery b connected to the node i.
$v_i^{\max}$	Maximum voltage bound at node i.
$v_i^{\min}$	Minimum voltage bound at node i.
φ_i^b	Charge coefficient of the battery b connected at node i.
Δt	Length of the period of time.

Variables

$p_{i,t}$	Power generation in the conventional source at node i in the period of time t.
$p_{i,t}^b$	Power input/output in battery b connected at node i in the period of time t.
$p_{i,t}^{dg}$	Renewable power generation at node i in the period of time t.
$SoC_{i,t}^b$	State of charge in battery b connected at node i in the period of time t.
$SoC_{i,t-1}^b$	State of charge in battery b connected at node i in the period of time $t-1$.
$v_{i,t}$	Voltage profile in the node i at the period of time t.
x_i^b	Binary decision variable regarding the installation of the battery b at node i.
z	Objective function.
z_1	Objective function related to the energy purchase costs in conventional sources.
z_2	Objective function related to the cost of the daily energy losses.

References

1. Council, G.W.E. Global Status of Wind Power. 2018. Available online: http://gwec.net/global-figures/windenergy-global-status (accessed on 20 March 2020).
2. Strunz, K.; Abbasi, E.; Huu, D.N. DC microgrid for wind and solar power integration. *IEEE Trans. Emerg. Sel. Top. Power Electron.* **2013**, *2*, 115–126. [CrossRef]
3. United Nations Framework Convention on Climate Change. (UNFCCC) *Adoption of the Paris Agreement*; I: Proposal by the President (Draft Decision); United Nations Office: Geneva, Switzerland, 2015.
4. Mahabir, R.; Shrestha, R.M. Climate change and forest management: Adaptation of geospatial technologies. In Proceedings of the 2015 Fourth International Conference on Agro-Geoinformatics (Agro-geoinformatics), Istanbul, Turkey, 20–24 July 2015; pp. 209–214.
5. Ray, S. *Construction Cost Data for Electric Generators Installed in 2013*; US Energy Information Administration (EIA): Washington, DC, USA, 2016. Available online: http://www.eia.gov/electricity/generatorcosts (accessed on 22 March 2020).
6. de Carvalho, W.C.; Bataglioli, R.P.; Fernandes, R.A.; Coury, D.V. Fuzzy-based approach for power smoothing of a full-converter wind turbine generator using a supercapacitor energy storage. *Electr. Power Syst. Res.* **2020**, *184*, 106287. [CrossRef]
7. Elbouchikhi, E.; Amirat, Y.; Feld, G.; Benbouzid, M.; Zhou, Z. A Lab-scale Flywheel Energy Storage System: Control Strategy and Domestic Applications. *Energies* **2020**, *13*, 653. [CrossRef]
8. Gil, W.; Montoya, O.D.; Garces, A. Direct power control of electrical energy storage systems: A passivity-based PI approach. *Electr. Power Syst. Res.* **2019**, *175*, 105885.
9. Zhao, P.; Xu, W.; Zhang, S.; Wang, J.; Dai, Y. Technical feasibility assessment of a standalone photovoltaic/wind/adiabatic compressed air energy storage based hybrid energy supply system for rural mobile base station. *Energy Convers. Manag.* **2020**, *206*, 112486. [CrossRef]
10. Javed, M.S.; Zhong, D.; Ma, T.; Song, A.; Ahmed, S. Hybrid pumped hydro and battery storage for renewable energy based power supply system. *Appl. Energy* **2020**, *257*, 114026. [CrossRef]
11. Montoya, O.D.; Gil-González, W.; Grisales-Noreña, L.; Orozco-Henao, C.; Serra, F. Economic Dispatch of BESS and Renewable Generators in DC Microgrids Using Voltage-Dependent Load Models. *Energies* **2019**, *12*, 4494. [CrossRef]
12. Gil-González, W.; Montoya, O.D.; Garces, A. Control of a SMES for mitigating subsynchronous oscillations in power systems: A PBC-PI approach. *J. Energy Storage* **2018**, *20*, 163–172. [CrossRef]
13. Essallah, S.; Khedher, A.; Bouallegue, A. Integration of distributed generation in electrical grid: Optimal placement and sizing under different load conditions. *Comput. Electr. Eng.* **2019**, *79*, 106461. [CrossRef]
14. Montoya, O.D.; Gil-González, W.; Garrido, V. Voltage Stability Margin in DC Grids with CPLs: A Recursive Newton–Raphson Approximation. *IEEE Trans. Circuits Syst. II Express Briefs* **2020**, *67*, 300–304. [CrossRef]
15. Garces, A. Uniqueness of the power flow solutions in low voltage direct current grids. *Electr. Power Syst. Res.* **2017**, *151*, 149–153. [CrossRef]
16. Lotfi, H.; Khodaei, A. AC versus DC microgrid planning. *IEEE Trans. Smart Grid* **2015**, *8*, 296–304. [CrossRef]
17. Nojavan, S.; Pashaei-Didani, H.; Mohammadi, A.; Ahmadi-Nezamabad, H. Energy management concept of AC, DC, and hybrid AC/DC microgrids. In *Risk-Based Energy Management*; Elsevier: Amsterdam, The Netherlands, 2020; pp. 1–10.
18. Gil-González, W.; Montoya, O.D.; Grisales-Noreña, L.F.; Cruz-Peragón, F.; Alcalá, G. Economic Dispatch of Renewable Generators and BESS in DC Microgrids Using Second-Order Cone Optimization. *Energies* **2020**, *13*, 1703. [CrossRef]
19. Montoya, O.D.; Grisales-Noreña, L.F.; Gil-González, W.; Alcalá, G.; Hernandez-Escobedo, Q. Optimal Location and Sizing of PV Sources in DC Networks for Minimizing Greenhouse Emissions in Diesel Generators. *Symmetry* **2020**, *12*, 322. [CrossRef]
20. Hu, J.; Shan, Y.; Xu, Y.; Guerrero, J.M. A coordinated control of hybrid ac/dc microgrids with PV-wind-battery under variable generation and load conditions. *Int. J. Elec. Power* **2019**, *104*, 583–592. [CrossRef]
21. Kazmi, S.A.A.; Shahzad, M.K.; Khan, A.Z.; Shin, D.R. Smart Distribution Networks: A Review of Modern Distribution Concepts from a Planning Perspective. *Energies* **2017**, *10*, 501. [CrossRef]

22. Siano, P.; Rigatos, G.; Piccolo, A. Active Distribution Networks and Smart Grids: Optimal Allocation of Wind Turbines by Using Hybrid GA and Multi-Period OPF. In *Atlantis Computational Intelligence Systems*; Atlantis Press: Paris, France, 2012; pp. 579–599. [CrossRef]
23. Becker, D.J.; Sonnenberg, B.J. DC microgrids in buildings and data centers. In Proceedings of the 2011 IEEE 33rd International Telecommunications Energy Conference (INTELEC), Amsterdam, The Netherlands, 9–13 October 2011; pp. 1–7.
24. Noritake, M.; Yuasa, K.; Takeda, T.; Hoshi, H.; Hirose, K. Demonstrative research on DC microgrids for office buildings. In Proceedings of the 2014 IEEE 36th International Telecommunications Energy Conference (INTELEC), Vancouver, BC, Canada, 28 September–2 October 2014; pp. 1–5.
25. Mackay, L.; van der Blij, N.H.; Ramirez-Elizondo, L.; Bauer, P. Toward the Universal DC Distribution System. *Electr. Power Compon. Syst.* **2017**, *45*, 1032–1042. [CrossRef]
26. Jing, W.; Lai, C.H.; Wong, S.H.W.; Wong, M.L.D. Battery-supercapacitor hybrid energy storage system in standalone DC microgrids: A review. *IET Renew. Power Gener.* **2017**, *11*, 461–469. [CrossRef]
27. Weaver, W.W.; Robinett, R.D.; Parker, G.G.; Wilson, D.G. Energy storage requirements of dc microgrids with high penetration renewables under droop control. *Int. J. Electr. Power Energy Syst.* **2015**, *68*, 203–209. [CrossRef]
28. Li, Y.; Meng, K.; Dong, Z.Y.; Zhang, W. Sliding Framework for Inverter-Based Microgrid Control. *IEEE Trans. Power Syst.* **2020**, *35*, 1657–1660. [CrossRef]
29. Azimi, S.M.; Hamzeh, M. Adaptive Interconnection and Damping Assignment Passivity-Based Control of Interlinking Converter in Hybrid AC/DC Grids. *IEEE Syst. J.* **2020**. [CrossRef]
30. Guo, Z.; Li, S.; Zheng, Y. Feedback linearization based distributed model predictive control for secondary control of islanded microgrid. *Asian J. Control* **2020**. [CrossRef]
31. Garcés, A. Convex Optimization for the Optimal Power Flow on DC Distribution Systems. In *Handbook of Optimization in Electric Power Distribution Systems*; Springer: Cham, Switzerland, 2020; pp. 121–137.
32. Chen, C.; Duan, S.; Cai, T.; Liu, B.; Hu, G. Optimal Allocation and Economic Analysis of Energy Storage System in Microgrids. *IEEE Trans. Power Electron.* **2011**, *26*, 2762–2773. [CrossRef]
33. Gil-González, W.; Montoya, O.D.; Holguín, E.; Garces, A.; Grisales-Noreña, L.F. Economic dispatch of energy storage systems in dc microgrids employing a semidefinite programming model. *J. Energy Storage* **2019**, *21*, 1–8. [CrossRef]
34. Li, Y.; Wang, C.; Li, G.; Wang, J.; Zhao, D.; Chen, C. Improving operational flexibility of integrated energy system with uncertain renewable generations considering thermal inertia of buildings. *Energy Convers. Manag.* **2020**, *207*, 112526. [CrossRef]
35. Montoya, O.D.; Grajales, A.; Garces, A.; Castro, C.A. Distribution Systems Operation Considering Energy Storage Devices and Distributed Generation. *IEEE Latin Am. Trans.* **2017**, *15*, 890–900. [CrossRef]
36. Grisales-Noreña, L.; Montoya, O.D.; Gil-González, W. Integration of energy storage systems in AC distribution networks: Optimal location, selecting, and operation approach based on genetic algorithms. [CrossRef]
37. Garcés, A.; Montoya, O.D. A Potential Function for the Power Flow in DC Microgrids: An Analysis of the Uniqueness and Existence of the Solution and Convergence of the Algorithms. *J. Control. Autom. Electr. Syst.* **2019**, *30*, 794–801. [CrossRef]
38. Zia, M.F.; Elbouchikhi, E.; Benbouzid, M.; Guerrero, J.M. Energy management system for an islanded microgrid with convex relaxation. *IEEE Trans. Ind. Appl.* **2019**, *55*, 7175–7185. [CrossRef]
39. Montoya, O.D.; Gil-González, W.; Grisales-Noreña, L. Relaxed convex model for optimal location and sizing of DGs in DC grids using sequential quadratic programming and random hyperplane approaches. *Int. J. Electr. Power Energy Syst.* **2020**, *115*, 105442. [CrossRef]
40. Luna, A.C.; Diaz, N.L.; Andrade, F.; Graells, M.; Guerrero, J.M.; Vasquez, J.C. Economic power dispatch of distributed generators in a grid-connected microgrid. In Proceedings of the 2015 9th International Conference on Power Electronics and ECCE Asia (ICPE-ECCE Asia), Seoul, Korea, 1–5 June 2015; pp. 1161–1168.
41. Montoya, O.D.; Gil-González, W.; Grisales-Noreña, L. An exact MINLP model for optimal location and sizing of DGs in distribution networks: A general algebraic modeling system approach. *Ain Shams Eng. J.* **2019**. [CrossRef]

42. Naghiloo, A.; Abbaspour, M.; Mohammadi-Ivatloo, B.; Bakhtari, K. GAMS based approach for optimal design and sizing of a pressure retarded osmosis power plant in Bahmanshir river of Iran. *Renew. Sustain. Energy Rev.* **2015**, *52*, 1559–1565. [CrossRef]
43. Du, Y.; Liang, X.; Liu, Y.; Xie, L.; Zhang, S. Exergo-economic analysis and multi-objective optimization of seawater reverse osmosis desalination networks. *Desalination* **2019**, *466*, 1–15. [CrossRef]
44. Skworcow, P.; Paluszczyszyn, D.; Ulanicki, B.; Rudek, R.; Belrain, T. Optimisation of Pump and Valve Schedules in Complex Large-scale Water Distribution Systems Using GAMS Modelling Language. [CrossRef]
45. Tartibu, L.; Sun, B.; Kaunda, M. Multi-objective optimization of the stack of a thermoacoustic engine using GAMS. *Appl. Soft Comput.* **2015**, *28*, 30–43. [CrossRef]
46. Soroudi, A. *Power System Optimization Modeling in GAMS*, 1st ed.; Springer International Publishing: Cham, Switzerland, 2017. [CrossRef]
47. Montoya, O.D. Solving a Classical Optimization Problem Using GAMS Optimizer Package: Economic Dispatch Problem Implementation. *Ingeniería y Ciencia* **2017**, *13*, 39–63. [CrossRef]
48. Lavorato, M.; Franco, J.F.; Rider, M.J.; Romero, R. Imposing Radiality Constraints in Distribution System Optimization Problems. *IEEE Trans. Power Syst.* **2012**, *27*, 172–180. [CrossRef]
49. Gil-González, W.; Montoya, O.D.; Grisales-Noreña, L.F.; Perea-Moreno, A.J.; Hernandez-Escobedo, Q. Optimal Placement and Sizing of Wind Generators in AC Grids Considering Reactive Power Capability and Wind Speed Curves. *Sustainability* **2020**, *12*, 2983. [CrossRef]
50. Gil-González, W.; Montoya, O.D.; Garces, A. Modeling and control of a small hydro-power plant for a DC microgrid. *Electr. Power Syst. Res.* **2020**, *180*, 106104. [CrossRef]

Article

Optimal Design of a Stand-Alone Residential Hybrid Microgrid System for Enhancing Renewable Energy Deployment in Japan

Yuichiro Yoshida [1] and Hooman Farzaneh [1,2,*]

[1] Interdisciplinary Graduate School of Engineering Sciences, Kyushu University, Fukuoka 816-8580, Japan; yoshida.yuichiro.764@s.kyushu-u.ac.jp

[2] Platform of Inter/Transdisciplinary Energy Research, Kyushu University, Fukuoka 819-0395, Japan

* Correspondence: farzaneh.hooman.961@m.kyushu-u.ac.jp; Tel.: +81-92-583-8626

Received: 3 March 2020; Accepted: 31 March 2020; Published: 5 April 2020

Abstract: This paper aims at the optimal designing of a stand-alone microgrid (PV/wind/battery/diesel) system, which can be utilized to meet the demand load requirements of a small residential area in Kasuga City, Fukuoka. The simulation part is developed to estimate the electrical power generated by each component, taking into account the variation of the weather parameters, such as wind, solar irradiation, and ambient temperature. The optimal system design is then based on the Particle Swarm Optimization (PSO) method to find the optimal configuration of the proposed system, using the least-cost perspective approach.

Keywords: renewable energy; microgrids; simulation; optimization

1. Introduction

Japan's energy self-sufficiency rate was as low as 9.6% in 2017, indicating the energy security issues in this country [1]. The energy self-sufficiency rate indicates the proportion of primary energy required for daily life and economic activities that can be secured in the country [2]. Japan's electric power industry faces a wide range of challenges, including the reliance on imports of fossil fuels through the immediate nuclear power phase-out and also further focusing on reducing and decarbonizing its energy system [3,4]. Initiatives are underway to decentralize the power sector in Japan from the centralized fossil fuel-based systems to distributed ones. The evolution of the electric utility system has many drawbacks because it is vulnerable to disasters due to extreme concentration. One way to avoid this problem is to use Distributed Energy Resources (DER), which enables the decentralization of the electric power sector in Japan. The deployment of DER involves both generators and energy storage technologies. A microgrid is a combination of various interconnected DER and loads that can operate as a grid-tied (connected to the grid) or a stand-alone (disconnected from the grid) controllable system. The stand-alone microgrids are considered as the most appropriate and cost-effective ways to electrify off-grid communities. Since the stand-alone microgrids operate as the off-grid systems, matching the quantity of the supplied electricity with the load requirements is an important issue, particularly when they are used for providing reliable power in small communities or remote areas. However, the integration and hybridization of various energy sources into the microgrid system increases the complexity of the system.

Although microgrids have several advantages, including reduced maintenance costs, emissions, and increased reliability and flexibility, their initial investment costs are higher than the other conventional power systems. Therefore, finding the optimal size and configuration of a microgrid in a cost-effective way has been the main focal point of recent research activities in this field of study.

Research that demonstrates the importance of optimizing microgrid systems with hybrid power supplies has gained more attention from scholars worldwide. Many scholars have developed optimization techniques to find the optimal operating point and configurations of microgrid systems. The main methods include minimizing the total cost and emissions or maximizing the reliability of the system. Table 1 shows the different optimization methods used in microgrid modeling based on the various approaches.

Table 1. Optimization methods used in microgrid modeling.

Authors	Year	System Components								Objective Function	Optimization Approach	Model Period
		Wind	PV	FC	Biomass	Hydro	Storage	Diesel	Other			
Zhang et al. [5]	2019	•	•	•		•		•		Total cost	CS-HS-SA-ANN [2]	20 years
Farzaneh [6,7]	2019		•	•	•		•			Total cost	NLP [3]	1 year
Bukar et al. [8]	2019	•	•				•	•		Total cost	GOA [4]	1 year
Angelopoulos et al. [9]	2019	•	•				•		Tidal	Total cost	DP [5]	1 year
Jing et al. [10]	2015	•	•					•		Total cost or CO_2 emissions	NSGA-II [6]	1 year
Sharafi et al. [11]	2014	•	•	•			•	•		Total cost and CO_2 emissions	PSO [7]	1 year
Kuzunia et al. [12]	2013	•					•	•		Total cost	SMIP [8]	1 year
Khatib et al. [13]	2012	•	•				•			Total cost	GA [9]	1 year
Ahmarinezhad et al. [14]	2012	•	•	•	•		•	•		Total cost	PSO	20 years
Giannakoudis et al. [15]	2010	•	•	•			•	•		Total cost	SA	10 years
Kashefi et al. [16]	2009	•	•	•			•			Total cost	PSO	20 years
Cai et al. [17]	2009	•	•			•		•		Total cost	ISITSP [10]	15 years
Dufo-López et al. [18]	2007	•	•	•		•	•	•		Total cost	GA	1 day
Garcia et al. [19]	2006	•		•			•	•		LEC [1]	LP [11]	1 year
Koutroulis et al. [20]	2006	•	•							Total cost	GA	20 years

[1]. Levelized Energy Cost. [2]. CS-HS-SA-ANN: chaotic search-harmonic search-simulated annealing (CS-HS-SA) using an artificial neural network (ANN). [3]. Non-Linear Programming. [4]. Grasshopper Optimization Algorithm. [5]. Dynamic Programing. [6]. Non-dominated Sorting Genetic Algorithm. [7]. Particle Swarm Optimization. [8]. Stochastic Mixed-Integer Program. [9]. Genetic Algorithm. [10]. Interval Parameter Superiority–Inferiority-based Two-Stage Programming. [11]. Linear Programming.

Following previous studies, this paper addresses a detailed modeling approach that is used to find the optimal configuration of a typical stand-alone microgrid system consisting of solar panels, wind turbines, battery storage and diesel generators, in order to satisfy the demand load of a residential area in Kasuga city in Japan (Figure 1). The proposed research methodology is based on a cost-effectiveness approach, which aims at finding the optimal configuration of the microgrid together with addressing the uncertainties related to the impact of variable weather conditions on the overall performance of the system and its optimal operation. The analytical framework consists of simulation and optimization models. The simulation model is based on developing a detailed power control strategy that is used to match the supplied electricity with the hourly demand load requirements in the different operating conditions. The optimization model uses the Particle Swarm Optimization (PSO) method to find the optimal configuration (optimal capacity of each DER) of the microgrid. The optimality criterion is satisfied at the minimum total cost of the system.

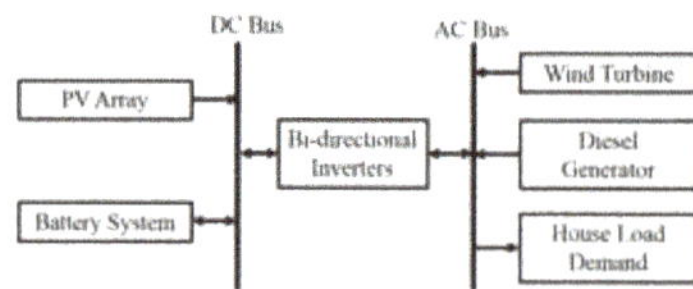

Figure 1. The proposed microgrid system in this study.

2. Simulation Model

2.1. Wind Turbine

In this simulation model, the following equations are used to quantify the amount of power output from a wind turbine [21]:

$$P_w(V) = \begin{cases} \frac{P_r(V - V_{CIN})}{V_{rat} - V_{CIN}}, & V_{CIN} \le V \le V_{rat} \\ P_r, & V_{rat} \le V \le V_{CO} \\ 0, & V \le V_{CIN} \text{ and } V \ge V_{CO} \end{cases} \tag{1}$$

$$V = V_{ref}\left(\frac{H}{H_{ref}}\right)^{\alpha} \tag{2}$$

where V_{ref} (m/s) refers to the measured wind speed at the reference height, H_{ref} (m); α is the power-law exponent; V refers to the wind speed at the height of H(m) [21]; and P_r, V_{CIN}, V_{rat} and V_{CO} refer to the constant power (kW), cut-in speed (m/s), rated wind speed (m/s), and cutout speed (m/s), respectively.

2.2. Solar Photovoltaic (PV)

The Duffie and Beckman principle model was used to calculate the global radiation incident on the PV array, using the following equation [22]:

$$\tau\alpha G_T = \eta_c G_T + U_L(T_c - T_a) \tag{3}$$

where $\tau\alpha$ is the effective transmittance-absorptance of the PV panel (%); G_T is the incident solar radiation on the PV surface (kW/m^2); η_c is the conversion efficiency of the PV array (%); U_L is the overall heat transfer coefficient of the PV (kW/m^2 °C); and T_c and T_a are the PV cell temperature (°C) and the ambient temperature (°C). The equation above states that a balance exists between the solar energy absorbed by the PV array and the amount of electrical output and heat which is transferred to the surroundings. The following formula gives the cell temperature:

$$\frac{\tau\alpha}{U_L} = \frac{T_{c,NOCT} - T_{a,NOCT}}{G_{T,NOCT}} \tag{4}$$

Assuming that $\tau\alpha/UL$ is constant, this equation is substituted into the Equation (3) to calculate T_c:

$$T_c = T_a + G_T\left(\frac{T_{c,NOCT} - T_{a,NOCT}}{G_{T,NOCT}}\right)\left(1 - \frac{\eta_c}{\tau\alpha}\right) \tag{5}$$

Finally, the output of the PV array is calculated by the following equation:

$$P_{pv} = G_{pv} f_{pv}\left(\frac{G_T}{G_{T,STC}}\right)[1 + \alpha_p(T_C - T_{C,STC})] \tag{6}$$

where in the above equations, *STC* and *NOCT* refer to the standard test and nominal operating cell temperature conditions, respectively; G_{pv} is the rated capacity of the PV panel (kW); f_{pv} is the PV derating factor (%); and α_p is the temperature coefficient of power (%/°C).

2.3. Diesel Generator

The diesel generator will be used as the backup system in the proposed microgrid. The rate of the fuel consumption of a diesel generator may be estimated by using the following formula [23]:

$$D_F(t) = A_N \cdot D_R + B_O \cdot D_O(t) \tag{7}$$

where D_F is the rate of fuel consumption (L/h); D_R is the rated power of the diesel generator (kW); D_O is the power output of the diesel generator (kW); A_N and B_O are the coefficients that are set at 0.2461 (L/kWh) and 0.081451 (L/kWh) [18]; and t is the time period in hours.

2.4. Power Converter

The proposed system utilizes a bi-directional converter to link the AC and DC buses to each other. The amount of the converted power by the converter is calculated by the following equation:

$$P_{out}(t) = P_{in}(t) \cdot \eta_{Conv} \tag{8}$$

where P_{in} and P_{out} are the input and output power from the inverter (kW), respectively; and η_{conv} is the converter efficiency, which is assumed to be 90%.

2.5. Battery Storage

In this study, a lead–acid battery is considered as the storage system in the proposed microgrid. The state of charge (SOC) of a lead–acid battery system should be controlled within the following range [24]:

$$SOC(t) = SOC(t-1) \pm \frac{E_{CD}(t) \cdot \eta_B}{P_R} \cdot 100 \tag{9}$$

where η_B is discharge and charging (round trip) efficiency (%); $E_{CD}(t)$ is the amount of electricity that is charged to or discharged from the battery (kWh); and P_R refers to the rated capacity of the battery(kWh).

2.6. Power Control Strategy

The control strategy flowchart of electricity generation and storage in the proposed microgrid system is represented in Figure 2. If the amount of electricity generated by the renewable generators (PV and wind turbine) exceeds the load requirement, the surplus energy is sent to the battery. If the SOC of the battery reached its maximum level and there is still excess electricity, the extra electricity will be sent to a dummy load. If the amount of generated electricity is less than the load requirement, then the battery storage can be discharged to meet the demand. If the battery discharge was insufficient, a diesel generator would be added to the system as a backup.

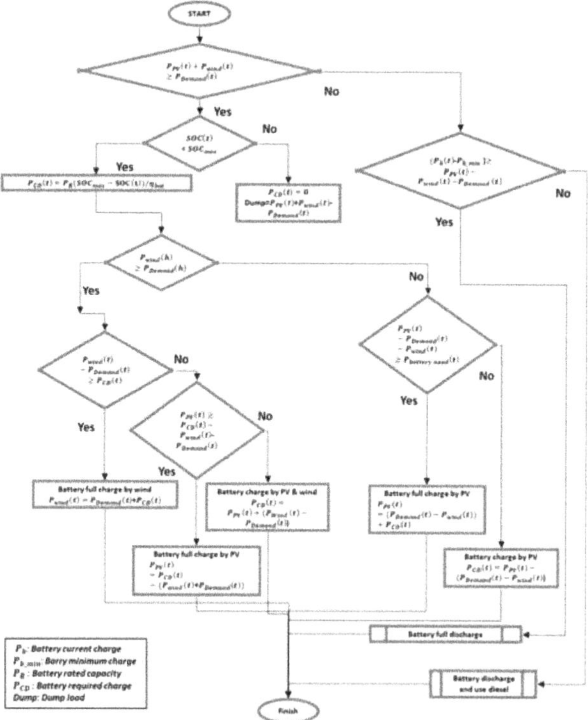

Figure 2. Power control strategy flowchart.

2.7. Demand Load Calculation

The area of study was selected among the Japanese standard residential buildings located in Kasuga city, Fukuoka prefecture, Japan. The total energy consumption (electrical and thermal loads) of the selected building was estimated using EnergyPlus software developed by the National Renewable Energy Laboratory (NREL), Denver, CO, USA [25]. The 3D model of the targeted Japanese standard house was developed using Sketchup 2019, which is shown in Figure 3. This 3D model includes all walls, ceilings, floors, doors, and windows. A standard size family, including four inhabitants, such as a father, a mother, a son, and a daughter, is supposed to live in the targeted building. An activity schedule was set for each person who should be taken into account, in order to calculate each person's internal heat gain. The occupancy schedule of each family member in this building is shown in Figure 4.

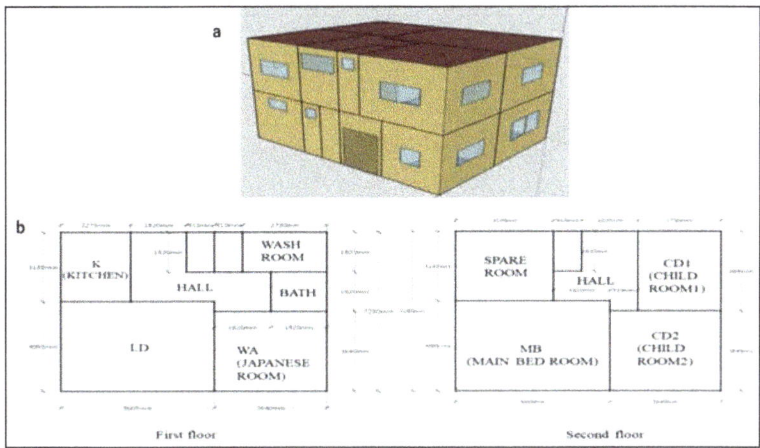

Figure 3. Layouts of the selected building: (**a**) 3D layout; (**b**) 2D Layout.

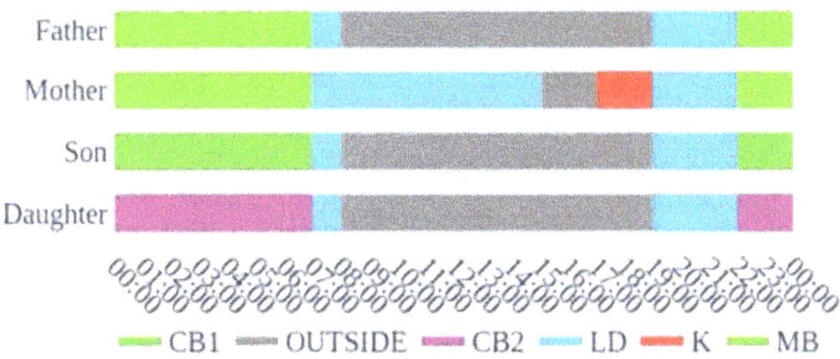

Figure 4. Occupancy schedule of family members in the selected building.

To meet this schedule, the usage plan of the electrical appliances, lightings, cooling, and heating loads are given in Figures 5–8. The internal heat generated from electrical equipment and lighting was not included in the simulation. The variable refrigerant flow (VRF) air conditioning system was considered to provide cooling and heating loads. The values of the cooling and heating Coefficient of Performance (COP) for this air conditioning system was set at 3.4 and 3.3, respectively. The Heating, Ventilation, and Air Conditioning (HVAC) operation strategy is based on the heating and cooling schedule in each room. When the HVAC is ON, cooling occurs when the temperature is higher than the cooling setpoint temperature (26 °C), and heating is performed when the temperature is lower than the heating setpoint (18 °C).

The assumed hourly electricity consumption by the electrical appliances and lighting is represented in Figures 9 and 10.

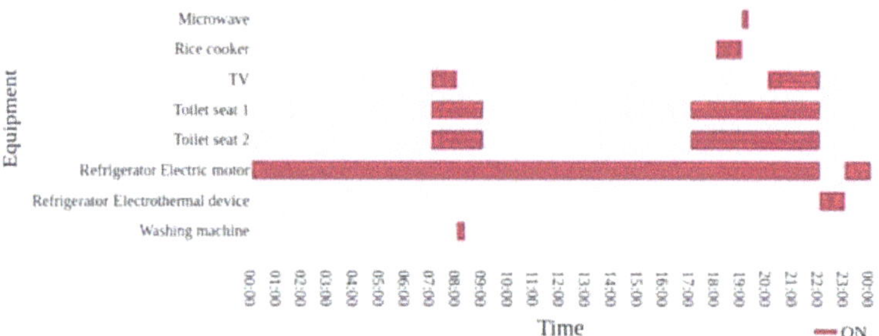

Figure 5. Appliances usage plan.

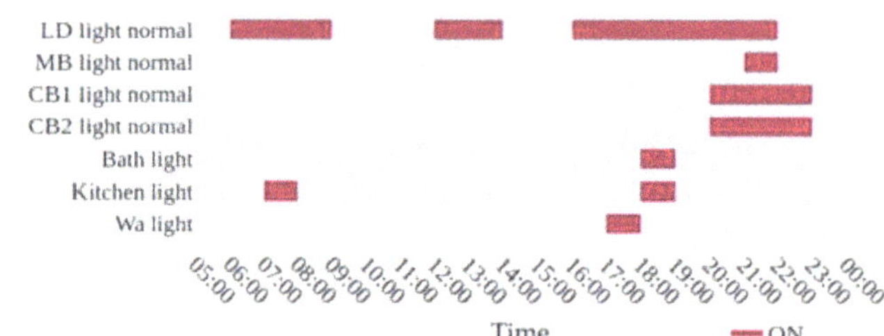

Figure 6. Lighting usage plan.

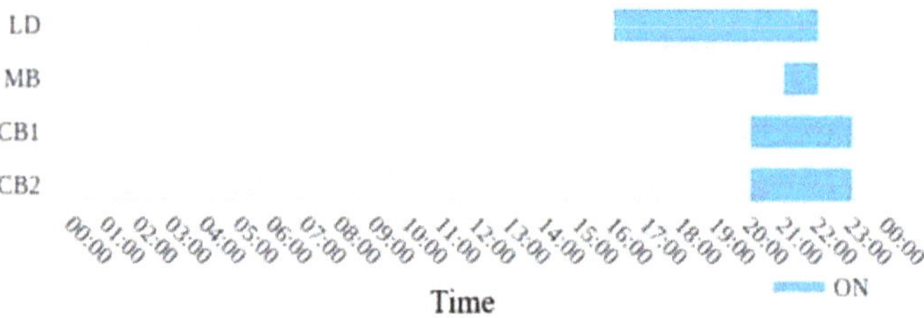

Figure 7. Cooling load usage plan (hot seasons).

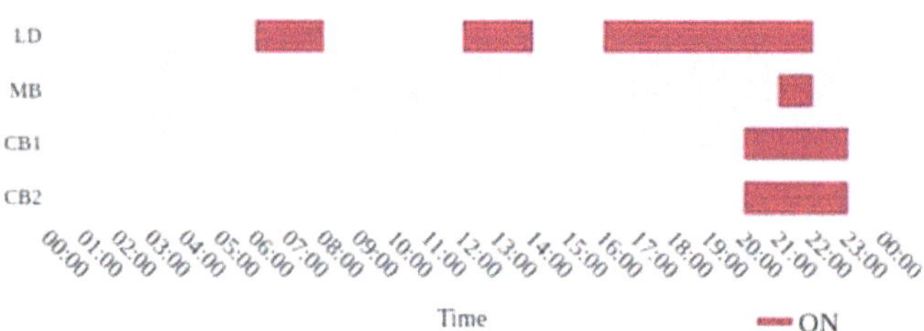

Figure 8. Heating load usage plan (cold seasons).

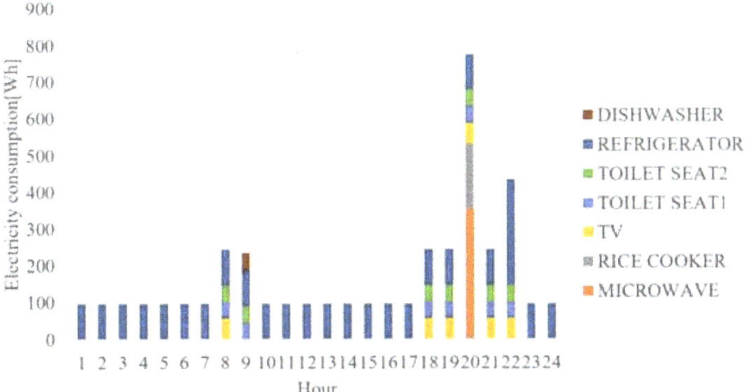

Figure 9. Hourly electricity consumption of the electrical appliances.

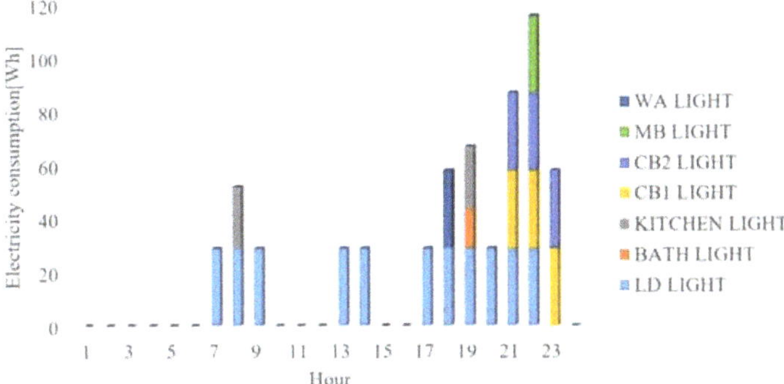

Figure 10. Hourly electricity consumption of lighting.

3. Weather Data

Weather data was used in this research for the following two purposes:

1) Optimal design of the system: The optimal configuration of the system can be found based on the amount of hourly electricity generation from solar and wind power systems. To this aim, the real measured meteorological data on solar irradiation, wind speed, and ambient temperature in Fukuoka were collected from the Japan Meteorological Agency (JMA) throughout the entire period in 2019, which are shown in Figures 11–13 [26].

2) Optimal day-ahead operation: In order to be able to understand the uncertainties related to the impact of variable weather conditions on the day-ahead optimal operation and electric power dispatching of the proposed system, the GPV–MSM weather forecast provided by JMA was used in this study. The GPV–MSM weather forecasting system reproduces atmospheric phenomena using the mesoscale modeling approach, which can be applied to the selected areas in Japan and its neighboring seas, including a horizontal grid of 5 km. Hourly forecasts of seven weather-related variables are provided eight times a day (00:00, 03:00, 06:00, 09:00, 12:00, 15:00, 18: 00, and 21:00; UTC time zone). The forecast period depends on the forecast time and can be up to 39 hours or 51 hours ahead. In this research, we used the first 24 hours of forecast weather data ahead, updated at UTC 00:00 hours (JST 9:00). Temperature and wind speed are the actual meteorological data that were collected from JMA [27]. The comparison between real and forecasting meteorological data used in this study is shown in Figure 14.

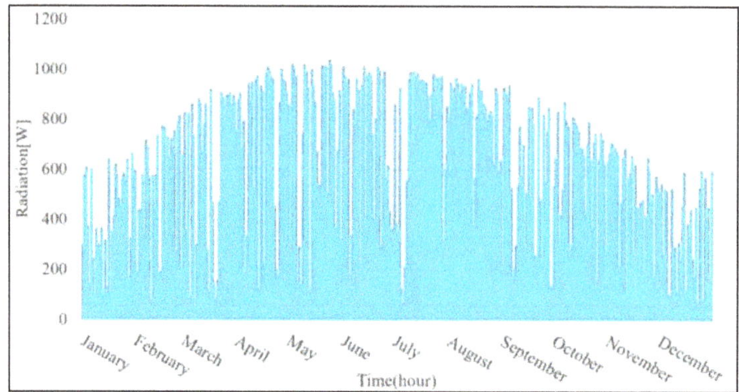

Figure 11. Hourly solar irradiation in Fukuoka (Source: JMA).

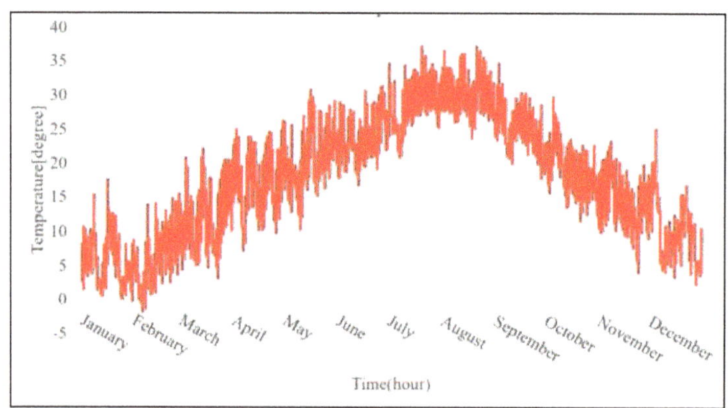

Figure 12. Hourly ambient temperature in Fukuoka (Source: JMA).

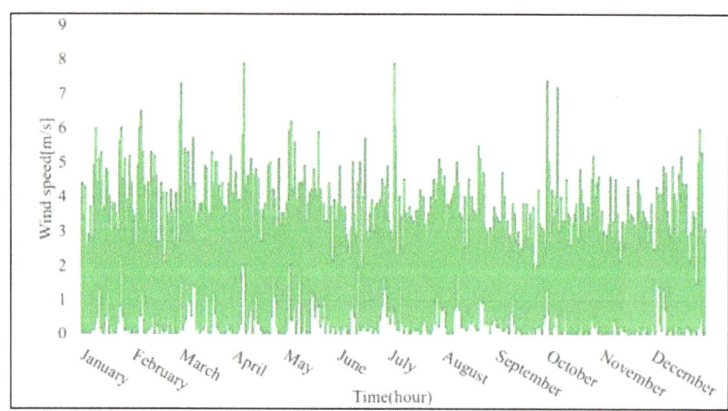

Figure 13. Hourly wind speed in Fukuoka (Source: JMA).

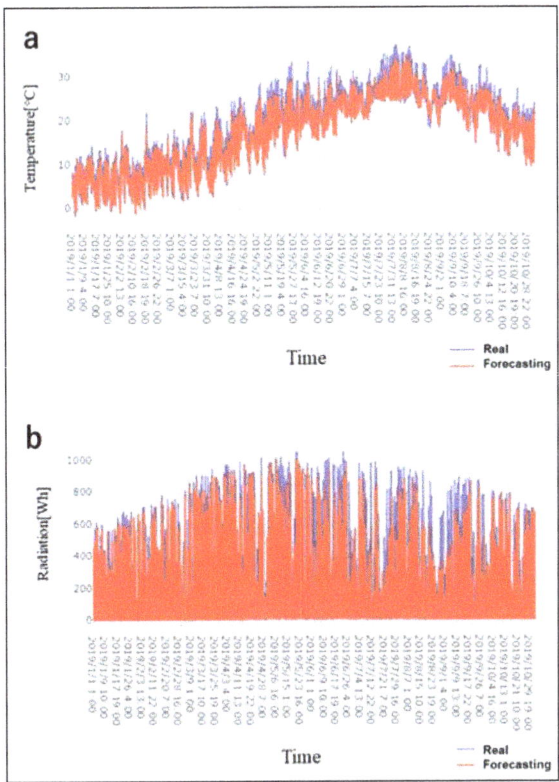

Figure 14. Comparison between the real and forecasting meteorological data used in this study: (**a**) Temperature; (**b**) Solar radaition.

4. Optimization Model

4.1. Objective Function

The optimal sizing of the power generation units in a microgrid system is essential for the efficient utilization of renewable resources. To this aim, the optimization technique was mainly founded on the basis of the minimization of the total cost of the system, subject to satisfying the technical, economic, and environmental constraints. The objective function of the optimization model can be expressed as follows:

$$TC = \sum_{i=1}^{n} \frac{C_t + O_t + F_t}{(1+r)^t} \tag{10}$$

where TC is the total cost of the microgrid system over its lifetime (\$); C_t, O_t and F_t are investment expenditures, operation costs, and fuel costs in the year t, respectively; and r is the discount rate. The total lifetime of the system, n, is considered to be 20 years.

4.2. Demand–Supply Constraint

The main objective of the model is to find the optimal value of the vector of the decision variables $P = (P_{PV}, P_{WG}, P_{bat}, P_{DG})$ which includes the installed capacities of the PV, wind power generator, battery and diesel generator subject to satisfying the following demand–supply equality:

$$P_{PV}(t) + P_{wind}(t) + P_{diesel}(t) + P_{battery\ discharge}(t) = P_{Demand}(th) + P_{battery\ charge}(t) \tag{11}$$

The above concept was applied to the different dispatching modes, which is visualized in Figure 15.

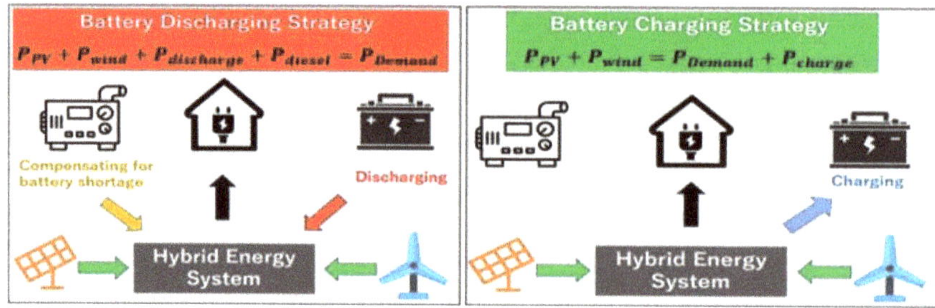

Figure 15. Demand–supply matching based on the battery charging/discharging modes.

4.3. Solving Method

In this research, the Particle Swarm Optimization algorithm was used to find the least expensive combination of the decision variables. This method consists of a constant search of the best solution by moving the particles at a specific speed calculated in each iteration, which is represented as follows [28]:

$$v_{id}(t+1) = \omega \cdot v_{id}(t) + c_1 \cdot \phi_1 \cdot (P_{id}(t) - x_{id}(t)) + c_2 \cdot \phi_2 \cdot (g_{id}(t) - x_{id}(t)) \tag{12}$$

$$x_{id}(t+1) = x_{id}(t) + v_{id}(t+1) \tag{13}$$

where P_{id} and $v_{id}(t)$ represent the particle's best candidate position and the velocity of inertia, respectively; x_{id} is the particle position; and u is the coefficient of inertia. The parameters c_1 and c_2 are positive weighting constants, described as "self-confidence" and "swarm confidence", respectively. The random values of φ_1 and φ_2 are between 0 and 1. ω indicates the inertia weight, which is set in the range (0.5, 1), and near 1 facilitates the global search. The first iteration ends by adjusting the speed and position of the next time step $t + 1$. Consistently, this process is performed until the best value of the objective function. In this paper, for all variants, fixed values considered as defaults for the PSO parameters were used as $c_1 = c_2 = 1.5$, $\omega = 0.8$, iterations = 100, population size = 20 [29,30]. To relocate the wrong particles in the adequate solution space and evaluate the fitness function, as well as provide a valid solution to the optimization problem, the attenuation technique was used to represent the boundary condition of the proposed PSO model, as shown in Figure 16 [30].

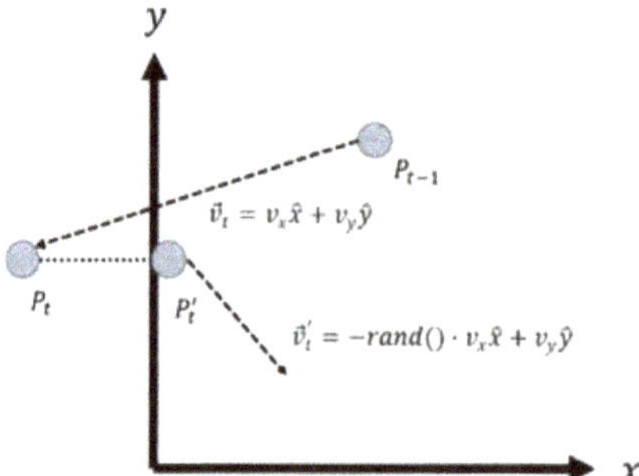

Figure 16. Boundary conditions for the Particle Swarm Optimization (PSO) developed in this research.

Figure 17 shows the interrelationship between the optimization and simulation models. The global solution of the PSO model, the best place that every individual in the flock has ever achieved, is adopted for all particles.

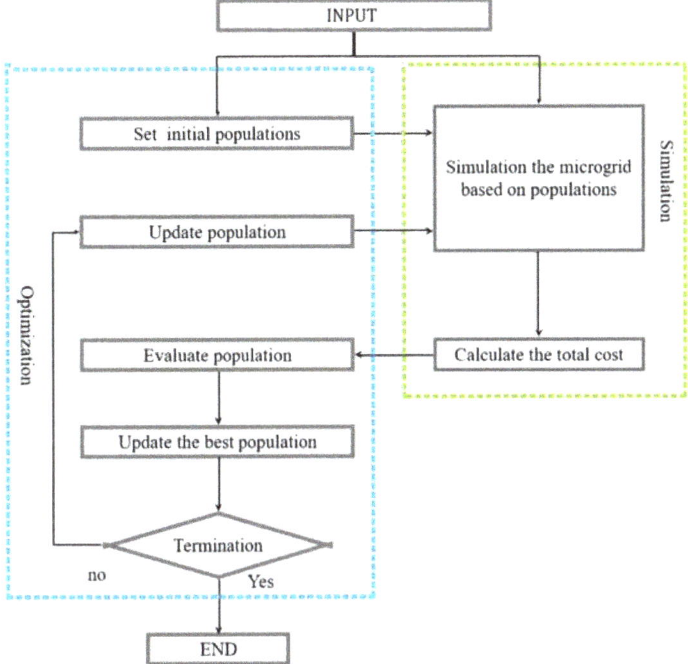

Figure 17. The interrelationship between the simulation and optimization models.

5. Results and Discussion

5.1. Estimation of the Annual Electricity Demand

EnergyPlus software was used to perform a load simulation by computing an hourly energy balance in the building. The amount of required electricity for providing necessary cooling/heating loads to maintain the building at the desired temperature was estimated based on the given scheduled plan of cooling/heating and the difference between the outdoor and indoor temperatures. The total electricity consumption for cooling and heating purposes in the selected residential building in Kasuga city was estimated at 600 kWh/year. This estimated value was summed up with the total electricity consumption by electrical appliances, considering their usage plans and rated power to calculate the total electricity consumption in the building. Figure 18 shows the hourly electricity demand load. The total electricity consumption in this building was estimated at 2303 kWh/year. The annual electricity consumption in the selected residential building is represented in Figure 18.

According to the Japan Agency for National Resource and Energy, the average annual electricity consumption per household in Japan is approximated at 4618 kWh, and the electricity demand of the main electrical appliances accounts for 57% of the total, which is about 2632 kWh [31]. The comparison between this value and the estimated value of total electricity consumption indicates the good agreement between the results of the simulation model and the standard data.

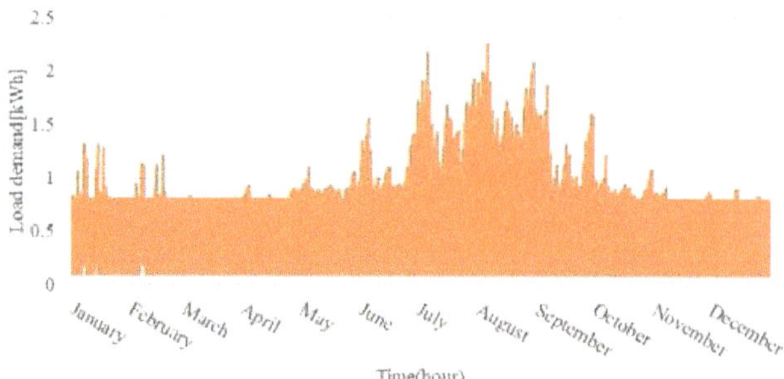

Figure 18. Estimated annual electricity demand in the selected residential building in Kasuga city.

5.2. Optimal Design of the Proposed Microgrid

The technical specifications of the main components of the microgrid system are given in Tables 2–4. The hourly power output per unit of wind generator and the solar panel is shown in Figure 19.

Table 2. Main input data used in the solar panel simulation [32].

Rated capacity of the PV array power under standard test conditions (kW)	G_{pv}	0.245
Ambient temperature at which the NOCT is defined (°C)	$T_{a,NOCT}$	20
Nominal operating cell temperature (°C)	$T_{c,NOCT}$	44
PV cell temperature under standard test conditions (°C)	$T_{C,STC}$	25
Incident solar radiation incident on the PV array (kW/m^2)	G_T	1
Temperature coefficient of power (%/°C)	α_p	-0.258
PV derating factor (%)	f_{pv}	0.8
Effective transmittance-absorptance of the PV panel (%)	$\tau\alpha$	0.9

Table 3. Main input data used in the battery storage simulation [24].

Battery Type		**Lead–Acid**
Nominal capacity (kWh)	P_R	1
SOC_{max} (%)	SOC	100
SOC_{min} (%)	SOC	40
Round-trip efficiency (%)	η_B	80

Table 4. Maim input data used in the wind turbine simulation [33].

Constant power (kWh)	P_r	0.3
Cut-in wind speed (m/s)	V_{CIN}	3
Cut out wind speed (m/s)	V_{CO}	20
Height (m)	H	40
Reference height (m)	H_{ref}	10

Table 5 shows the cost analysis of the proposed system. The optimal size of each component is given in Table 6. Figure 20 demonstrates the pathways towards reaching the optimal solution by each element (particle in the PSO model) based on satisfying the minimum total cost of the system. The total cost of the proposed microgrid is estimated at USD 42,300.

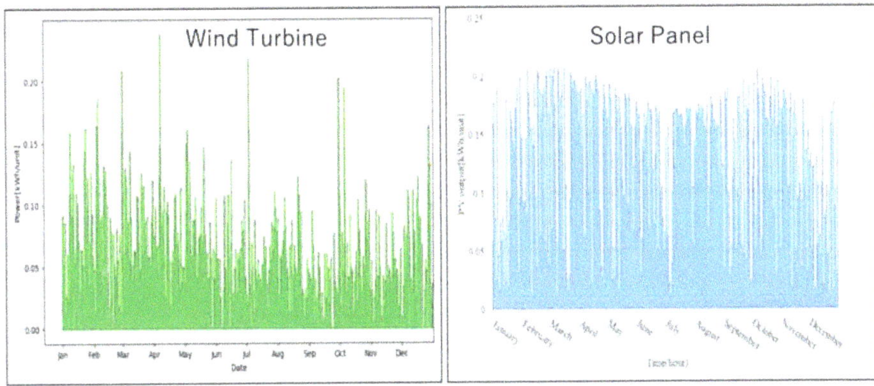

Figure 19. Estimated power output per each unit of PV panel and wind turbine.

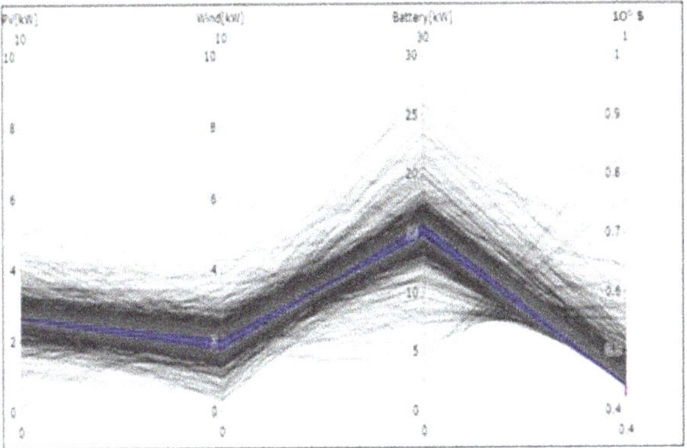

Figure 20. Pathways towards the optimal solution by the PSO model.

Table 5. Cost analysis of the system.

Components	Capital Cost [1] ($/kW)	O&M Cost ($/KW)	Fuel Cost ($/KW)	Lifetime
Wind Turbine [20]	2300	2	0	20 years
PV [34]	5100	10	0	20 years
Diesel [23]	300	0.5	1.3	15,000 h
Battery [23]	120	10	0	4 years
Converter [23]	127	1	0	20 years

[1] Including both purchase and installation costs.

Table 6. Optimal size and cost of each component.

	PV	WG	Battery	Diesel	Converter
Optimal capacity (kW)	2.65	2.01	14.86	3.6	2.8

As can be observed from Figure 21, the solar panel represents the largest share in the total cost of the system, followed by the battery storage and diesel generator. The levelized cost of electricity (LCOE) of the proposed microgrid is estimated at 0.88 $/kWh, which is much higher than the average electricity tariff in Japan (0.2 $/kWh ≈ 22 JPY/kWh).

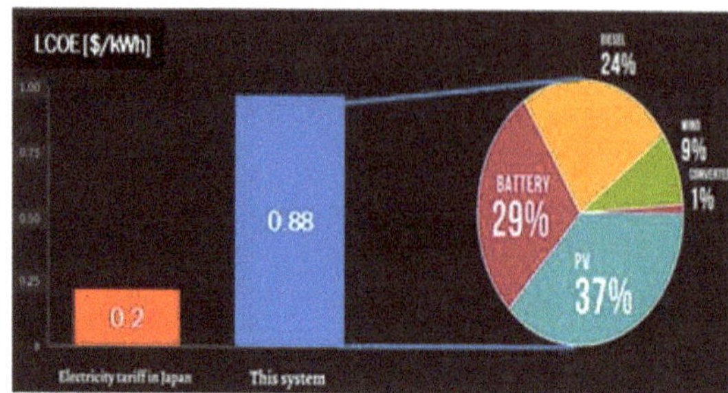

Figure 21. Estimated levelized cost of electricity (LCOE) of the proposed microgrid.

The comparison between the estimated LCOE in this study and other similar off-grid residential microgrids is given in Table 7.

Table 7. Comparison between the estimated LCOE by the model and other references.

System	LCOE ($/kWh)
The proposed system in this paper (PV + wind + battery + diesel)	0.88
Typical off-grid microgrid in Japan: 4kW of PV + 4kWh of battery [35]	0.55–0.72
Typical off-grid microgrid in Pacific Island: PV + diesel [36]	1–1.7

The average monthly electricity generation by the system is shown in Figure 22. The battery SOC is represented in Figure 23.

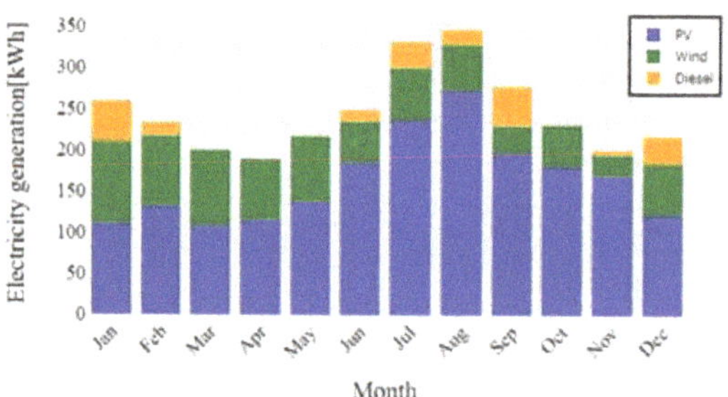

Figure 22. Monthly average electricity generation by the microgrid.

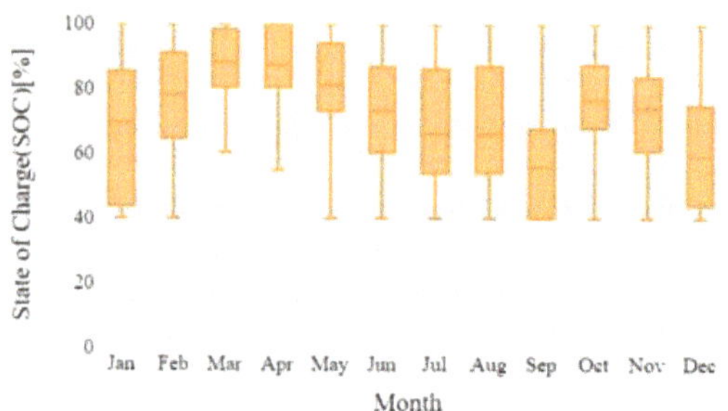

Figure 23. State of charge (SOC) of the battery in the microgrid.

The model results revealed that the power loss due to the charging and discharging efficiency of the battery is considerable, which is estimated at 719 kWh per year, which indicates that improving the round-trip efficiency of the battery is necessary for the effective utilization of the proposed microgrid system (Figure 24).

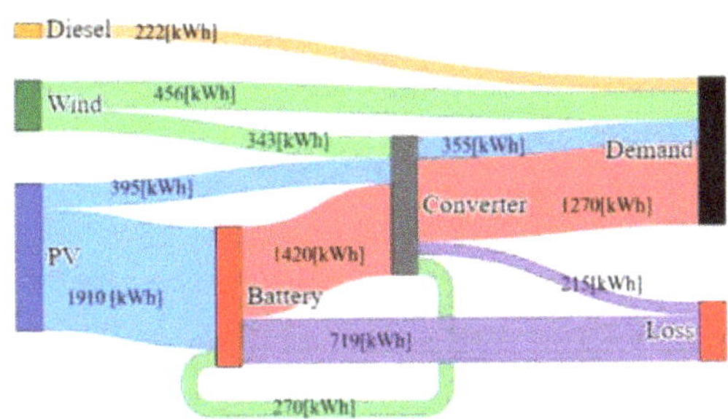

Figure 24. Annual energy balance in the proposed microgrid.

5.3. Assessing the Impact of Weather Conditions on the Optimal Performance and Power Dispatching of the Proposed System

The microgrid's fuel consumption and energy storage requirements are examined as a function of the atmospheric conditions. Weather data is thus necessary to establish optimal operating and dispatching plans according to the operational objective of the microgrid. Day-ahead weather forecasts are also responsible for deviations from these plans, thereby being a valuable source of uncertainty in the scheduling process. Figure 25 represents the weather satellite images taken by the satellite Himawari-8 at 12:00 on 3 July 2018 (real data) and 2019 (forecasts). On 3 July 2018, Typhoon No. 7 went north over the East China Sea and approached Kyushu, and strong winds and heavy rain took Fukuoka. Therefore, there was insufficient solar electricity generation, but since the strong wind blew in the afternoon, a sufficient amount of electricity was generated by the wind power generator. However, on the same day in 2019, Fukuoka was covered with heavy clouds, and the average wind speed was very weak at 2–3 (m/s). Therefore, there was no power generation from the wind turbine on this day. Besides, since the sky had been covered with dark clouds for a long time, there was no sufficient power output from the PV panels, which resulted in reducing the SOC of the battery.

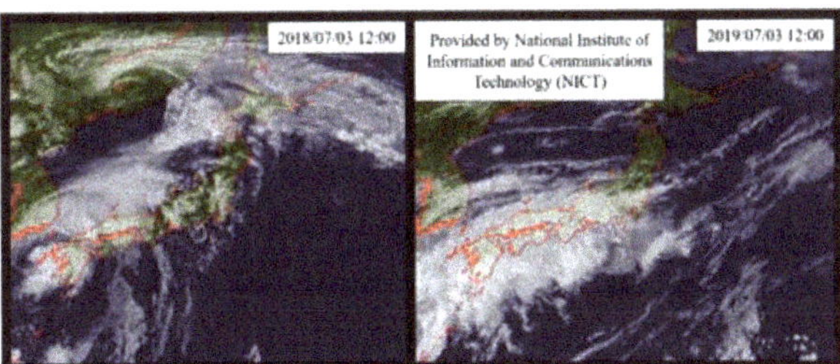

Figure 25. Comparison of satellite images taken on 2018/07/03 at 12:00 and on 2019/07/03 at 12:00.

As shown in Figure 26, the diesel generator was used to offset the shortage of battery discharge. The comparison between the historical (2018) and forecasting data (2019) highlights the remarkable impact of weather conditions on both fuel consumption and energy storage requirement of the proposed microgrid. Based on the forecasting data, the stationary battery storage's SOC decreases to 40%, since it cannot be charged by solar and wind in the morning. Therefore, the diesel generator should be used in order to meet the demand load during the evening. It highlights that forecasts of weather conditions at the site location would be required to know in advance the amount of power that the wind turbine or the PV will feed into the battery over the next hours and days. Day-ahead forecasting of the weather data will help in managing the battery operation through monitoring its SOC condition and lowering the usage of the diesel generator to reduce its cost and environmental impacts on the system.

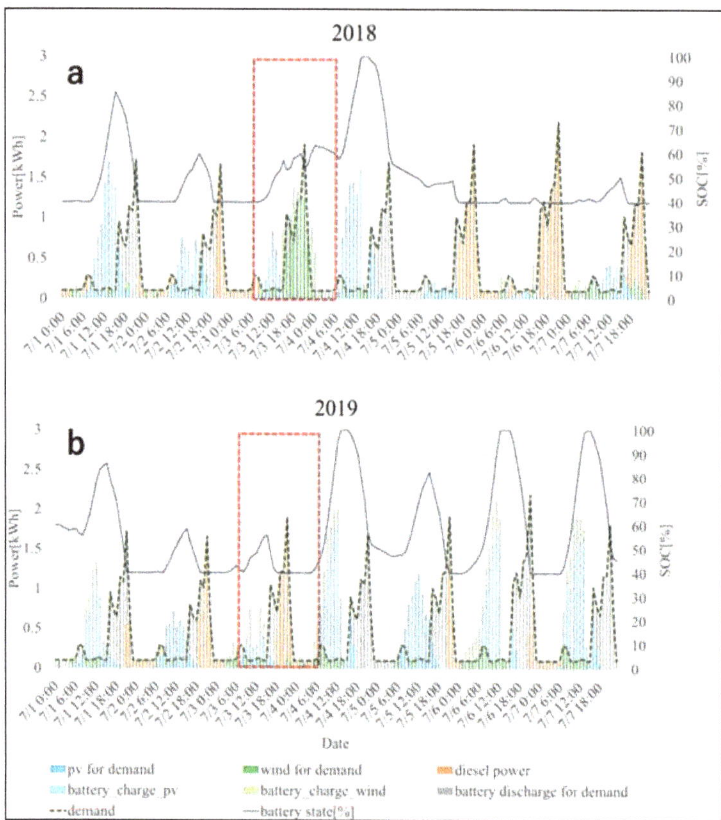

Figure 26. Optimal operation of the microgrid within the same period in 2018 and 2019: (**a**) based on the measured weather data; (**b**) based on the weather forecast.

6. Conclusions

This research addressed the optimal design of a stand-alone microgrid system that can be used in order to meet the electrical load requirement in a selected Japanese standard building in Kasuga city, Fukuoka prefecture, Japan. Based on the results, the optimal size of the main components of the system was estimated for the PV as 2.65 kW; for wind power as 2.01 kW; for the battery as 14.86 kW; for the diesel generator as 3.6 kW; and for the converter as 2.8 kW. The total cost of the proposed system was estimated at USD 42,300. The LCOE of the proposed system was estimated at 0.88 $/kWh, which is much higher than the average electricity rate in Japan. The percentage of power provided by each power unit was estimated at 43.4% by the solar PV, 16.7% by the wind power, 4.9% by the diesel generator, and 35% by the battery discharge. The model results show that the operation of the proposed microgrid system is highly dependent on batteries and solar power, due to the high potential of solar energy in Kasuga city. Furthermore, the results revealed the remarkable impact of weather conditions on the optimal operation of the proposed microgrid, especially during the windy and rainy seasons. Due to the issue of the high LCOE of the proposed microgrid in this study, the investment demand of the system might be insufficient in the early stages, which makes customers lack the motivation to participate in the project. In order to promote the development of such a system, the corresponding incentive mechanism should be designed through an optimal subsidy from the Japanese government and an optimal cooperation incentive from the energy supplier. Another solution would be the introduction of the community microgrid, including some prosumers and consumers, all with access to a local grid, which allows its participants to achieve a greater outcome than they would individually. In this scheme, participants who generate excess electricity are able to share their generation with participants of their choosing. They are also able to take advantage of shared energy-storage systems in the community to improve the operational reliability and the economy. However, it demands extra efforts regarding a detailed cost analysis of the system, which can be considered as the future work.

Author Contributions: Conceptualization, methodology and investigation: Y.Y., Writing—review, editing and supervising: H.F. Both authors have read and agreed to the published version of the manuscript.

Funding: This research was supported by the Kurata grant of the Hitachi Global Foundation.

Acknowledgments: The author wishes to thank the editor and the reviewers for their contributions on the paper.

Conflicts of Interest: The author declares no conflict of interest.

References

1. Agency for Natural Resources and Energy Energy White Paper 2019. Available online: https://www.enecho. meti.go.jp/about/whitepaper/2019html/ (accessed on 14 December 2019).
2. Farzaneh, H.; McLellan, B.; Ishihara, K.N. Toward a CO_2 zero emissions energy system in the Middle East region. *Int. J. Green Energy* **2016**, *13*, 682–694. [CrossRef]
3. McLellan, B.C.; Zhang, Q.; Utama, N.A.; Farzaneh, H.; Ishihara, K.N. Analysis of Japan's post-Fukushima energy strategy. *Energy Strategy Rev.* **2013**, *2*, 190–198. [CrossRef]
4. Esteban, M.; Portugal-Pereira, J.; Mclellan, B.C.; Bricker, J.; Farzaneh, H.; Djalilova, N.; Ishihara, K.N.; Takagi, H.; Roeber, V. 100% renewable energy system in Japan: Smoothening and ancillary services. *Appl. Energy* **2018**, *224*, 698–707. [CrossRef]
5. Zhang, W.; Maleki, A.; Rosen, M.A.; Liu, J. Sizing a stand-alone solar-wind-hydrogen energy system using weather forecasting and a hybrid search optimization algorithm. *Energy Convers. Manag.* **2019**, *180*, 609–621. [CrossRef]
6. Farzaneh, H. Design of a Hybrid Renewable Energy System Based on Supercritical Water Gasification of Biomass for Off-Grid Power Supply in Fukushima. *Energies* **2019**, *12*, 2708. [CrossRef]
7. Farzaneh, H. *Energy Systems Modeling: Principles and Applications*; Springer Nature: Singapore, 2019.
8. Bukar, A.L.; Tan, C.W.; Lau, K.Y. Optimal sizing of an autonomous photovoltaic/wind/battery/diesel generator microgrid using grasshopper optimization algorithm. *Sol. Energy* **2019**, *188*, 685–696. [CrossRef]
9. Angelopoulos, A.; Ktena, A.; Manasis, C.; Voliotis, S. Impact of a Periodic Power Source on a RES Microgrid. *Energies* **2019**, *12*, 1900. [CrossRef]
10. Jing, R.; Wang, M.; Zhang, Z.; Liu, J.; Liang, H.; Meng, C.; Shah, N.; Li, N.; Zhao, Y.Y.; Lu, Y.; et al. Renewable energy system optimization of low/zero energy buildings using single-objective and multi-objective optimization methods. *Energy Build.* **2019**, *89*, 123–139. [CrossRef]
11. Sharafi, M.; ELMekkawy, T.Y. Multi-objective optimal design of hybrid renewable energy systems using PSO-simulation based approach. *Renew. Energy* **2014**, *68*, 67–79. [CrossRef]
12. Kuznia, L.; Zeng, B.; Centeno, G.; Miao, Z. Stochastic optimization for power system configuration with renewable energy in remote areas. *Ann. Oper. Res.* **2013**, *210*, 411–432. [CrossRef]
13. Khatib, T.; Mohamed, A.; Sopian, K. Optimization of a PV/wind micro-grid for rural housing electrification using a hybrid iterative/genetic algorithm: Case study of Kuala Terengganu, Malaysia. *Energy Build.* **2012**, *47*, 321–331. [CrossRef]
14. Ahmarinezhad, A.; Tehranifard, A.A.; Ehsan, M.; Firuzabad, M.F. Optimal sizing of a stand alone hybrid system for Ardabil area of Iran. *IJTPE* **2012**, *4*, 118–125.
15. Giannakoudis, G.; Papadopoulos, A.I.; Seferlis, P.; Voutetakis, S. Optimum design and operation under uncertainty of power systems using renewable energy sources and hydrogen storage. *Int. J. Hydrog. Energy* **2010**, *35*, 872–891. [CrossRef]
16. Kashefi Kaviani, A.; Riahy, G.H.; Kouhsari, S.M. Optimal design of a reliable hydrogen-based stand-alone wind/PV generating system, considering component outages. *Renew. Energy* **2009**, *34*, 2380–2390. [CrossRef]
17. Cai, Y.P.; Huang, G.H.; Tan, Q.; Yang, Z.F. Planning of community-scale renewable energy management systems in a mixed stochastic and fuzzy environment. *Renew. Energy* **2009**, *34*, 1833–1847. [CrossRef]

18. Dufo-López, R.; Bernal-Agust, J.L.; Contreras, J. Optimization of control strategies for stand-alone renewable energy systems with hydrogen storage. *Renew. Energy* **2007**, *32*, 1102–1126. [CrossRef]
19. Garcia, R.S.; Weisser, D. A wind-diesel system with hydrogen storage: Joint optimisation of design and dispatch. *Renew. Energy* **2006**, *31*, 2296–2320. [CrossRef]
20. Koutroulis, E.; Kolokotsa, D.; Potirakis, A.; Kalaitzakis, K. Methodology for optimal sizing of stand-alone photovoltaic/wind-generator systems using genetic algorithms. *Sol. Energy* **2006**, *80*, 1072–1088. [CrossRef]
21. Hiendro, A.; Kurnianto, R.; Rajagukguk, M.; Simanjuntak, Y.M. Junaidi Techno-economic analysis of photovoltaic/wind hybrid system for onshore/remote area in Indonesia. *Energy* **2013**, *59*, 652–657. [CrossRef]
22. HOMER Pro 3.12 User Manual. Available online: https://www.homerenergy.com/products/pro/docs/ (accessed on 13 December 2019).
23. Jamshidi, M.; Askarzadeh, A. Techno-economic analysis and size optimization of an off-grid hybrid photovoltaic, fuel cell and diesel generator system. *Sustain. Cities Soc.* **2019**, *44*, 310–320. [CrossRef]
24. Dhundhara, S.; Verma, Y.P.; Williams, A.; Ehsan, M.; Kuznia, L.; Zeng, B.; Centeno, G.; Miao, Z.; Agüera-Pérez, A.A.; Palomares-Salas, J.C.; et al. Techno-economic analysis of the lithium-ion and lead-acid battery in microgrid systems. *Renew. Energy* **2017**, *145*, 304–317. [CrossRef]
25. Shabunko, V.; Lim, C.M.; Mathew, S. EnergyPlus models for the benchmarking of residential buildings in Brunei Darussalam. *Energy Build.* **2018**, *169*, 507–516. [CrossRef]
26. Japan Meteorological Agency. Available online: https://www.data.jma.go.jp/obd/stats/etrn/view/monthly_ s3_en.php?block_no=47401&view=11 (accessed on 14 December 2019).
27. Japan Meteorological Agency. Available online: https://www.jma.go.jp/en/week/346.html (accessed on 14 December 2019).
28. Parsopoulos, K.; Vrahatis, M. Particle Swarm Optimization Method for Constrained Optimization Problem. In *Intelligent Technologies-Theory and Applications: New Trends in Intelligent Technologies*; IOS Press (Frontiers in Artificial Intelligence and Applications): Fairfax, VA, USA, 2002; Volume 76, pp. 214–220.
29. Zhang, W.J.; Xie, X.F.; Bi, D.C. Handling boundary constraints for numerical optimization by particle swarm flying in periodic search space. In Proceedings of the 2004 Congress on Evolutionary Computation, Portland, OR, USA, 19-23 June 2004; Volume 2, pp. 2307–2311.
30. Xu, S.; Rahmat-Samii, Y. Boundary conditions in particle swarm optimization revisited. *IEEE Trans. Antennas Propag.* **2007**, *55*, 760–765. [CrossRef]
31. Japan Agency for National Resource and Energy. Available online: https://www.enecho.meti.go.jp/index.html (accessed on 14 December 2019).
32. Panasonic Panasonic Photovoltaic Module HIT VBHN245SJ25 VBHN240SJ25. Available online: https://panasonic. net/lifesolutions/solar/download/pdf/VBHN245_240SJ25_ol_190226.pdf (accessed on 14 January 2020).
33. Intelligent Energy—Europe, Catalogue of European Urban Wind Turbine Manufacturers, (2011) 61. Available online: http://123doc.org/document/1227748-catalogue-of-european-urban-wind-turbine-manufacturers-potx.htm (accessed on 21 March 2020).
34. Panasonic, Panasonic Residential Catalog. Available online: https://sumai.panasonic.jp/catalog/solarsystem. html (accessed on 10 December 2018).
35. Shibata, Y. How Can "Solar PV + Battery System" Be Economically Competitive and Reliable Power Generation? *IEEJ* **2017**, 1–23. Available online: https://eneken.ieej.or.jp/data/7457.pdf (accessed on 21 March 2020).
36. IRENA—International Renewable Energy Agency. Solar photovoltaic Summary Charts. Available online: https://www.irena.org/costs/Charts/Solar-photovoltaic (accessed on 21 March 2020).

 *energies*

Article

Structural Condition for Controllable Power Flow System Containing Controllable and Fluctuating Power Devices

Saher Javaid *, Mineo Kaneko and Yasuo Tan

Graduate School of Advanced Science and Technology, Japan Advanced Institute of Science and Technology, 1-1 Asahidai, Nomi City 923-1296, Japan; mkaneko@jaist.ac.jp (M.K.); ytan@jaist.ac.jp (Y.T.)
* Correspondence: saher@jaist.ac.jp

Received: 5 March 2020; Accepted: 20 March 2020 ; Published: 2 April 2020

Abstract: This paper discusses a structural property for a power system to continue a safe operation under power fluctuation caused by fluctuating power sources and loads. Concerns over global climate change and gas emissions have motivated development and integration of renewable energy sources such as wind and solar to fulfill power demand. The energy generated from these sources exhibits fluctuations and uncertainty which is uncontrollable. In addition, the power fluctuations caused by power loads also have the same consequences on power system. To mitigate the effects of uncontrollable power fluctuations, a power flow control is presented which allocates power levels for controllable power sources and loads and connections between power devices. One basic function for the power flow control is to balance the generated power with the power demand. However, due to the structural limitations, i.e., the power level limitations of controllable sources and loads and the limitation of power flow channels, the power balance may not be achieved. This paper proposes two theorems about the structural conditions for a power system to have a feasible solution which achieves the power balance between power sources and power loads. The discussions in this paper will provide a solid theoretical background for designing a power flow system which proves robustness against fluctuations caused by fluctuating power devices.

Keywords: power flow control; power fluctuations; renewable energy sources; demand uncertainty; augmenting path

1. Introduction

The awareness of depletion of fossil fuels, increase of power demand, and global warming have promoted the development of renewable energy sources. These energy sources such as wind turbines and photovoltaic (PV) generation system play important roles because of low impact against the environment. However, the generated power from these energy sources varies greatly, resulting in a risk of power fluctuations which is uncontrollable [1]. Renewable energy sources are often connected to the medium-and-low-voltage grid in smaller unit sizes. With the rapid increase of renewable energy sources, the increase of power fluctuation in the power system gives cause for anxiety [2]. The power fluctuations phenomenon caused by power loads also have the same consequences. The power demand is continuously growing due to the development of the smart consumer electronics equipped with communication and control units [3–5]. Moreover, the introduction of heat pumps and the electric cars are also contributing to a higher power demand. The mixture of renewable energy generation attached to grid and ever-growing power demand have increased the threats of stability and quality of power of the national wide power grid. As a result, the power system needs new strategies for the management and operation of the electricity to maintain balance between changing power supply patterns and consumption patterns [6].

Recently, the interest in the demand side management has been increased [7–11]. This is particularly because residential and commercial domains represent a major part of electricity consumption and carbon gas emissions [12] and partly because small scale distributed power resources such as photovoltaic, wind turbines, fuel cells, and storage batteries are introducing into houses, buildings, offices, and factories etc. The power structure of these facilities is changing dynamically due to the integration of distributed power sources [13,14]. One example of such changing structure is a Nano-grid (NG) [15–17]. It includes numerous power generating sources, an in-house or building power distribution system, and energy storage functions as well as a variety of consumer devices such as lighting, TV, heating/ventilation/air-conditioning, and cooking, i.e., such power systems need to have a sophisticated power control function that can manage dynamically changing power consumption and power supply conditions. We believe this is a new power control function to be developed for these changing power structures.

Due to the uncontrollability of generated power and demand, the power imbalance is a big challenge to consider [18,19]. In the fluctuating environment where power supply and demand both change dynamically and uncontrollably, a real-time power flow control is required, whose main function is to keep the balance between generated power and demand. There have been proposed several implementation models that can be classified into power switching and power packetization. Okabe et al. in [20] proposed a power switch between a multiple power sources and loads. The proposed system architecture for power switching is comprise of power sources, loads, power switch/router and power transmission lines. A physical power line connection is created between a power source and a load. The power control introduced is very simple and support only one-to-one and one-to-many connections types between power sources and loads. However, the power control cannot be enhanced when number of devices increased without changing the existing grid structure to manage many to one connection type. The power fluctuation management and control is out of the scope for this method. Abe et al. [21] and Hikihara et al. [22] proposed power packetization methods. The system design consists of group of power sources, loads, storage battery, power routers, and power lines. They introduced power routers which are able to receive, store, and transmit power, in power packets, a power packet is associated with source ID and destination ID. This method also implements a simple power control; however, explicit power control for uncertainty due to fluctuating power devices is not considered. Moreover, since they emphasis on the power transmission, no mechanism to control distributed multiple power sources and loads is introduced.

To achieve this power balance, cooperation with controllable power sources and loads and their cooperative control seem to be a promising technology [23–25]. The cooperative control of controllable power devices can accommodate power fluctuations caused by fluctuating generators and loads. The absorption of the fluctuation will be achieved by controlling the power (supply/consumption) of controllable power sources/loads.

Information technology is currently being used throughout power grids, where embedded smart power sensors, power actuators, and controllers are used for continuous power monitoring, control, and management. Based on the extreme control-ability of smart power devices, the quantity and direction of power stream at each power source and load can be exactly controlled by the power user, which provides the technical foundation for the realization of the sophisticated power flow control. The concept of Power Flow Coloring proposed in [23,24] can be one example of information-technology-supported power flow control. In this concept, individual power flow between a pair of power source and load can be managed with a unique identification attached to each power flow, which enables us to manage versatile power flow patterns between distributed power sources and loads. Figure 1 illustrates the power flow coloring in a household environment.

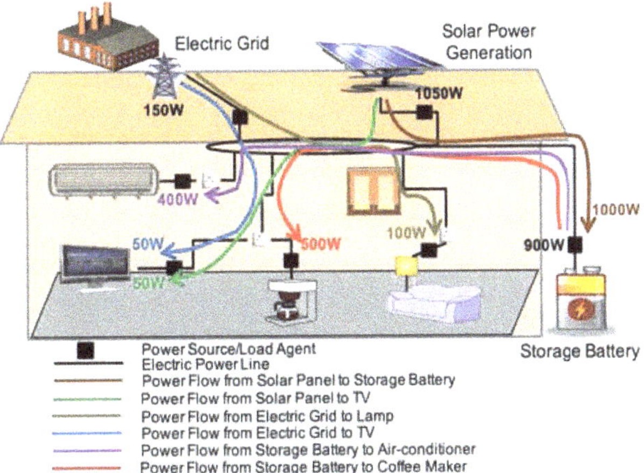

Figure 1. The concept of The Power Flow Coloring.

As mentioned above, one of the most important functions for the power flow control for a power system which contains uncontrollable fluctuating devices and controllable devices is to balance the generated power with the power demand. However, due to the structural limitations, i.e., the power level limitations of controllable sources and loads and the limitation of power flow channels, the power balance may not be achieved. To operate a power system safely under fluctuating environment, the power system must have such property that there always exists a feasible solution, i.e., the power levels for controllable power devices and power flows on connections between power devices which achieve the power balance between power sources and power loads, in any situation of fluctuating devices. We will call such property as "robustness against fluctuation". This paper discusses structural characteristics for a power system to possess the robustness against fluctuation. This paper proposes two theorems about the structural conditions for a power system to possess the robustness against fluctuation. The first one is a preliminary theorem to the second one, which provides the structural conditions for a power system with given power levels of fluctuating devices to have a feasible control solution. The second one is the main theorem of this paper, which provides the structural conditions for a power system to possess the robustness against fluctuation. The first theorem seems to be more relevant to a control problem which a power flow control needs to solve, while the second and main theorem is more relevant to a power system design, i.e., our main theorem will provide how power sources and power loads should be connected and how large the ability of individual controllable power sources and loads should be, in order to achieve the robustness against power fluctuations caused by fluctuating power devices. The proposed theorem can be applied to any level of power flow system e.g., nano-grid or micro-grid if the connections are incomplete between power sources and loads which is the key point of our research.

The power flow management has been discussed in past with respect to different objectives and optimization techniques [25–28]. In [29] authors proposed a flexible distributed multi-energy power generation system considering uncertainty for long term. Another research work in [30] studied the multi-energy microgrids considering long term and short-term uncertainty. The main objective of both papers is to enhance demand side flexibility by efficiently using distributed multi-energy generation system. The authors presented a novel approach for power flow management on wide scale using distributed sources and formulate a Multi-Integer Linear Programming (MILP) problem for optimization. However, the system design guidelines and system robustness in the presence of uncertainty caused by fluctuating power devices in real time is not considered which is the main focus of our study. One of the big challenges for integration of renewable energy sources remains in the matching of the intermittent energy generation/production with the dynamic power demand. On the other hand, there are many software packages which can simulate the behavior of a power system, find solutions to constrained optimization problems, etc. However, most of these approaches are based on numerical computations, and it needs intuition and experience of experts as well as repetition of simulation to understand the relation between cause and effect. Compared with these approaches, our approach is not a numerical approach to show the behavior of a given system, rather it explains the reason of existence/inexistence of a solution under power supply/consumption limitation constraints, which will provide us with firm understanding of a power flow system.

This paper is organized as follows: Section 2 describes our structural characterization problem in terms of our system model and power flow control problem is introduced. Section 3 explains our system model which includes representation, and categorization of power devices and connections. Power flow control problem on our system model with fluctuating uncontrollable devices and controllable devices is introduced also in this section. Section 4, is devoted to our first theorem and its mathematical proof, which describes the structural conditions for a power system with given power levels of fluctuating devices to have a feasible control solution. The second and main theorem which describes the structural condition for a power system to possess the robustness against fluctuation is shown in Section 5. Numerical results with illustrative examples of considered system for the application of proposed theorems have been explained in Section 6, Finally, Section 7 gives concluding remarks.

2. Structural Condition Issues and Reasoning

2.1. Incomplete Graph and Its Advantages

Due to the increase in power load demand on the consumer side, fossil energy reserve of electric power system are being exhausted rapidly and resulting in higher energy prices [31]. Despite the intermittent nature, renewable power sources are gaining much consideration than non-renewable energy sources [32]. As for the future energy support, the use of renewable energy technologies is increasing. Hence, it is essential to maximize their profits without losing the stability of the power system.

On the other hand, deregulation of power system is introduced in many countries of the world [33]. Since power demand and its form is changing rapidly, therefore, recognition and determination of most accurate power sources and efficient power transmission are important aspects to consider and long-distance transmission of power from one place to another through multiple buses is another critical point to focus as mentioned in [34]. Undoubtedly, power transmission from one point to another point through multiple power buses causes power loss. Also, the power system stability decreases constantly when the power transmission lines become longer for the long-distance transmission. For the power loss due to a long power transmission line (and conversion loss from one bus to another bus with different characteristics), we need to avoid long-distance (and/or via many different buses) power transmission (except for the case of emergency). For example, if we impose some limitation on the distance (or the number of intermediate buses) of power line transmission, the connection graph is no longer a complete graph (i.e., it must be an incomplete bipartite graph), and a normal operation (except for emergency) is maintained based on this bipartite graph model.

With the increase of power loads of different types (AC loads or DC loads), the DC power network is separated from the AC power network called hybrid AC-DC power network [35–37]. In a standard renewable energy connected to AC only power network, the power need to be converted not only once but twice to supply power to DC loads. The DC power supply is first converted to AC power, then transformed to DC power for DC power loads/equipment, which leads to power transformation losses. Another motivation for this type of system is to supply the right power to the right equipment. Separating the power network into DC and AC power network allows the native power to be supplied to the devices optimizing efficiency at every level. Moreover, the complexity of the power flow control problem is high with complete connection graph. In order to mitigate the complexity of power flow control problem, we need to limit the possible connections between power sources and loads.

In [23,24], the concept of the Power Flow Coloring is presented, which attaches the unique ID to each connection between a power source and a load. The concept of the power flow coloring is implemented using real physical devices to show that how power fluctuations are managed with the cooperation of controllable power devices. The power generation excess and shortages are handled by efficiently controlling terminal power devices. In the implementation phase, connections are restricted or reduced to supply power from selected sources to selected loads. This represents the incomplete graph implementation in a household environment. In this research study, the system design guidelines and power device constraints for incomplete connections are studied deeply.

2.2. Problem Discussed in This Paper

We will treat a power system which comprises of power sources, power loads and connections between them. We consider fluctuating power sources/loads as well as controllable power sources/loads, where the last can work for absorbing the fluctuation of power generation/demand in the former and for constructing an entire system robust against the consequences from fluctuating devices.

At each time instance, the power flow control problem with measured information of fluctuating power devices needs to be solved. One of the major objectives of the power flow control is to maintain the balance between generating power and demand. In a real physical situation, the system controller desires to handle power transient behavior, latency of system control, cost efficiency, etc. However, the issue whether the power flow system (i.e., power flow control problem) has a feasible solution in terms of power balance or not is one of the most important issues.

If we consider a complete connection between power sources and power loads, i.e., each power source can provide the power to every individual power load, the condition for a feasible solution in terms of power balance might be trivial as, $ps^f \le p\ell^f + p\ell^{c-max}$, and $ps^f + ps^{c-max} \ge p\ell^f$. The first inequality means that the (total) amount of generated power ps^f by fluctuating power source(s) can be consumed completely by the (total) demand $p\ell^f$ of fluctuating load(s) and by controlling controllable power load(s) with the maximum available consuming power $p\ell^{c-max}$. On the other hand, the second inequality means that the (total) demand $p\ell^f$ of fluctuating loads(s) can be fully satisfied by the (total) generating power ps^f of fluctuating source and by controlling controllable power source(s) with the maximum available power ps^{c-max}. However, if the connection between sources and loads is incomplete, i.e., for some pairs of source and load, there is no transfer mechanism/route by which the power is transmitted from a source to a load, the solvability issue is not trivial. In this paper, we will discuss the issue of power balance under such an incomplete connection between sources and loads.

If we use some numerical tool to compute the solution of problem, we can know existence/inexistence of a feasible solution for each individual instance of the problem. However, our approach discussed in this paper is different from it.

Our concern is the structural condition which makes the power flow control problem solvable in terms of power balancing, i.e., how power sources and power loads should be connected and how large the ability of individual controllable power sources and loads should be in order for a power flow system to have a feasible solution of the power flow control problem. Hence, the major application area of our results in this paper is the design issue of a power flow system, since the allocation of power sources and power loads, the connections (power flow channels) between power sources and loads, and the capacity of individual power sources and loads need to be designed so that the resultant power flow system always has a feasible solution under any situation of fluctuating power devices (this property is called "robustness against fluctuation").

In the following of this paper, we will discuss two types of system conditions. The first one is the structural condition for a power flow system with given power levels of fluctuating devices to have a feasible solution of the power flow control problem, and the second one is the structural condition for a power flow system to possess the robustness against fluctuation. As a first attempt to investigate structural conditions for a power flow system to possess the robustness against fluctuation, the connectivity of connections between power sources and loads and the maximum and the minimum power levels of individual power devices are considered to be structural factors, and the capacity of individual connections is assumed to be large enough so that it does not affect the existence/inexistence of a feasible solution of the power flow control problem. The improved structural conditions regarding the capacity of individual connections remain as a future problem.

In this paper, we do not consider any specific target level of a power network, but aim to provide a general discussion about the power balancing under an incomplete connection between power sources and loads.

3. System Model

This section describes the details of our system model and explains the *Power Flow Control Problem*.

3.1. Representation and Categorization of Power Devices

This subsection shows the representation of power devices with both types and connections between them as given in Figure 2.

A power source (PS) can be defined as an electric device which can supply electric power to electric loads, e.g., photovoltaic, wind turbine, utility grid, etc. A power load (PL) is an electric device which consumes electric power supplied by power sources. All power devices (i.e., sources and loads) are divided into two categories based on their characteristics and functionality, such as *Controllable* PS^c/PL^c and *Fluctuating* PS^f/PL^f. A controllable PS^c/PL^c can control its power (supply/consume), whereas fluctuating PS^f/PL^f cannot control its power.

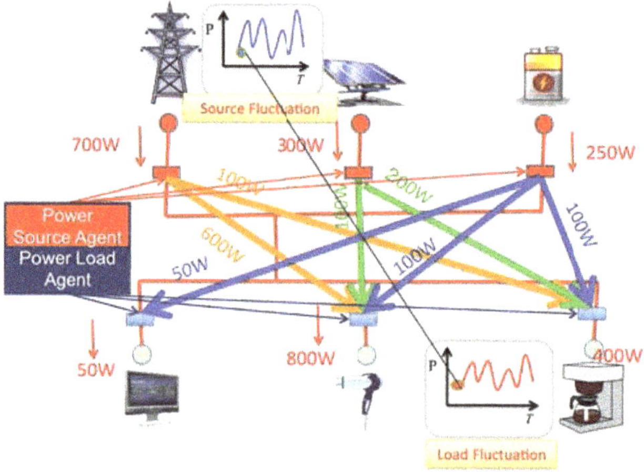

Figure 2. Representation of power sources, power loads, and connections between them.

All power sources with both types will be represented as, $\mathcal{PS} = \{PS_1^c, PS_2^c, \cdots, PS_I^c, PS_1^f, PS_2^f, \cdots, PS_J^f\} = \{PS_1, PS_2, PS_3, \ldots, PS_{I+J}\}$, where I and J show the total numbers of controllable and fluctuating power sources, respectively. Similarly, all power loads will be indexed as, $\mathcal{PL} = \{PL_1^c, PL_2^c, \cdots, PL_K^c, PL_1^f, PL_2^f, \cdots, PL_L^f\} = \{PL_1, PL_2, PL_3, \ldots, PL_{K+L}\}$ where K and L show the total numbers of controllable and fluctuating power loads. For convenience sake, we will use $C(\bullet)$ for representing the set of controllable power devices in a set $\bullet$, and $F(\bullet)$ for representing the set of fluctuating devices in a set $\bullet$.

A power agent is attached to each PS/PL, which measures and controls the power levels of the attached power device. The actual power levels (i.e., generation and consumption) of sources and loads will be represented as ps_i^c, ps_j^f, $p\ell_k^c$ and $p\ell_\ell^f$, respectively for PS_i^c, PS_j^f, PL_k^c and PL_ℓ^f.

Each device PS/PL has a minimum power level and maximum power level limitation, which represents the range of power modes/operation and performance of that particular device. The minimum power supply/generation limit ps_i^{c-min} and maximum limit ps_i^{c-max} show the capacity of a controllable source PS_i^c and the power ps_i^c generated by PS_i^c is assumed to be bounded as,

$$ps_i^{c-min} \leq ps_i^c \leq ps_i^{c-max} \tag{1}$$

Similarly, the minimum and maximum power generation limits will be given as ps_j^{f-min} and ps_j^{f-max} respectively, for PS_j^f and the power generation ps_j^f is limited as,

$$ps_j^{f-min} \leq ps_j^f \leq ps_j^{f-max} \tag{2}$$

For the power demand $p\ell_k^c$ of controllable load PL_k^c with given minimum and maximum levels $p\ell_k^{c-min}$ and $p\ell_k^{c-max}$, and for the power demand $p\ell_\ell^f$ of fluctuating load PL_ℓ^f with given minimum and maximum levels $p\ell_\ell^{f-min}$ and $p\ell_\ell^{f-max}$ are bounded as,

$$p\ell_k^{c-min} \leq p\ell_k^c \leq p\ell_k^{c-max} \tag{3}$$

$$p\ell_\ell^{f-min} \leq p\ell_\ell^f \leq p\ell_\ell^{f-max} \tag{4}$$

3.2. Connections between Power Sources and Power Loads

A connection is a pair of a PS and a PL, (PS_m, PL_n). The real physical arrangement of power devices and connection between them can be modeled with a bipartite graph which is introduced in Figure 3.

The system model considered in this paper thus consists of a set of power sources ($\mathcal{PS}$), a set of power loads ($\mathcal{PL}$), and a set $\mathcal{X}$ of connections between power sources and loads as, $\mathcal{X} \subseteq \mathcal{PS} \times \mathcal{PL}$. The system model represents the incomplete connections between power sources and loads and the connection in this figure shows the power transfer/supply from power source to load. Power devices without connection show that there is no possibility of power transfer/supply between power devices. The key point of the paper is the consideration of between devices.

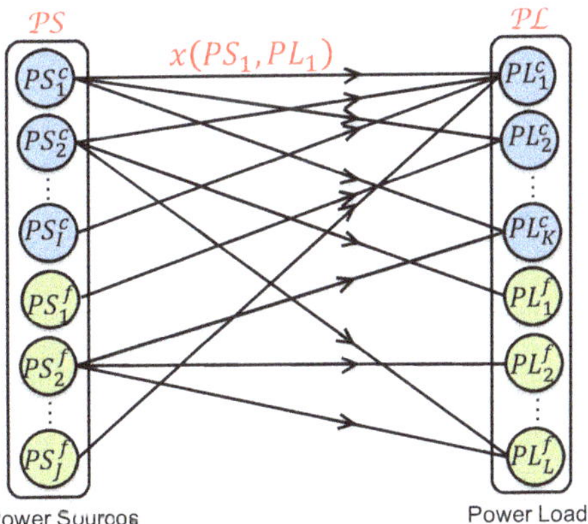

Figure 3. Representation of power sources, power loads, and connections between them using bipartite graph.

In Figure, power devices with each type are represented with different colors. Each connection (PS_m, PL_n) is associated with some power level in Watt $x(PS_m, PL_n)$ to show the amount of power supplied from a source PS_m to a load PL_n via this connection, which is assumed to be always non-negative real number.

3.3. Power Flow Control Problem

As the actual/physical power by a fluctuating power device changes a lot due to its type of device and operation mode, the power stream on each power flow/connection must be altered according to the fluctuating situation. Here, it is supposed that the power levels of fluctuating power devices are noted with smart power sensors for each time instance. In order to adjust power fluctuations triggered by fluctuating power devices, a power flow control is essential. This power flow control method uses measured power levels of fluctuating power devices and calculates power levels for controllable power devices and connections under the power balance restriction such that the total power supplied/generated by all power sources is fully used by power loads, and all power loads take sufficient power from power sources.

Each connection connects a PS to its neighbor on the other side of the connection. The set of neighbors of PS_m is denoted as $N(PS_m)$, which can be separated into $C(N(PS_m))$ and $F(N(PS_m))$, the sets of controllable and fluctuating power devices, respectively. As for the representation of neighboring devices and the power flows, please refer to Figure 4.

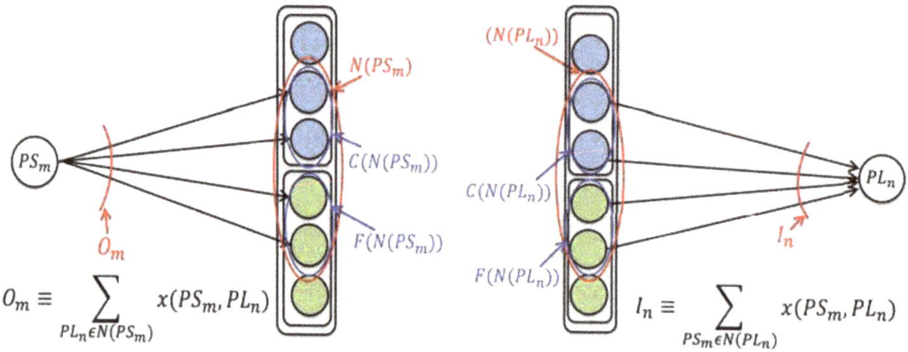

Figure 4. Connections between Power Sources and Power Loads. (**a**) A power source with connections; (**b**) A power load with connections.

The sum of all outgoing power flows, O_m, of power source PS_m can be written as,

$$O_m \triangleq \sum_{PL_n \in N(PS_m)} x(PS_m, PL_n)$$

Similarly, the sum of all incoming flows, I_n, of a power load, PL_n, can be computed as,

$$I_n \triangleq \sum_{PS_m \in N(PL_n)} x(PS_m, PL_n)$$

At the end of power flow control in each time instance, the power generation ps_m of power source PS_m must be equal to the sum of all outgoing flows, O_m as,

$$O_m = ps_m, \tag{5}$$

and the power consumption $p\ell_n$ of power load PL_n must be equal to the sum of all incoming power flows to this PL as,

$$I_n = p\ell_n. \tag{6}$$

Therefore, the ultimate goal of this proposed power control problem is, for given (measured) power levels ps_j^f and $p\ell_\ell^f$ of fluctuating power sources and loads, find the power levels ps_i^c and $p\ell_k^c$ of controllable power sources and loads and power flow assignment $x : \mathcal{X} \to R_+$ such that (5) and (6) are satisfied along with the limitations given by (1) and (3). Since we can control ps_i^c and $p\ell_k^c$ freely while keeping individual minimum and maximum power limitations for each power device, the problem can be considered to be to find $x : \mathcal{X} \to R_+$ such that

$$ps_i^{c-min} \leq O_i^c \leq ps_i^{c-max}, \qquad \forall PS_i^c \in C(\mathcal{PS}) \tag{7}$$

$$O_j^f = ps_j^f, \qquad \forall PS_j^f \in F(\mathcal{PS}) \tag{8}$$

$$p\ell_k^{c-min} \leq I_k^c \leq p\ell_k^{c-max}, \qquad \forall PL_k^c \in C(\mathcal{PL}) \tag{9}$$

$$I_\ell^f = p\ell_\ell^f, \qquad \forall PL_\ell^f \in F(\mathcal{PL}) \tag{10}$$

4. System Condition with Given Power Levels for Fluctuating Power Devices

First we consider a general instance of the power flow control problem where the generated power levels and demand levels for fluctuating power sources and loads, respectively, are given as constant values (values obtained by measurement), and provide the structural condition for this problem instance to have a feasible solution (Theorem 1). The structural conditions described in Theorem 1 can be an important base for our main theorem (Theorem 2 shown in the next section) which provides the structural conditions for a system to possess the robustness against fluctuation, i.e., the conditions for a system to have a feasible solution of the power flow control problem for any power levels of fluctuating power devices.

Theorem 1. *The power flow control problem can find the feasible solution if and only if the following two system conditions are satisfied.*
Condition 1-1:

$$\forall S \subseteq \mathcal{PS}, \quad \sum_{PS_i^c \in C(S)} ps_i^{c-min} + \sum_{PS_j^f \in F(S)} ps_j^f \leq \sum_{PL_k^c \in C(N(S))} p\ell_k^{c-max} + \sum_{PL_\ell^f \in F(N(S))} p\ell_\ell^f$$

Condition 1-2:

$$\forall T \subseteq \mathcal{PL}, \quad \sum_{PS_i^c \in C(N(T))} ps_i^{c-max} + \sum_{PS_j^f \in F(N(T))} ps_j^f \geq \sum_{PL_k^c \in C(T)} p\ell_k^{c-min} + \sum_{PL_\ell^f \in F(T)} p\ell_\ell^f$$

Proof of Theorem 1. First, we introduce the necessity of the system conditions. Let $x : \mathcal{X} \to R_+$ be a feasible solution of power flow control problem and let S be an random subset of power sources, then (5) and (6) are satisfied for every PS and PL, which further yields the following equations.

$$\sum_{PS_j^f \in F(S)} ps_j^f = \sum_{PS_j^f \in F(S)} O_j^f$$

$$\sum_{PS_i^c \in C(S)} ps_i^c = \sum_{PS_i^c \in C(S)} O_i^c$$

and

$$\sum_{PL_\ell^f \in F(N(S))} I_\ell^f = \sum_{PL_\ell^f \in F(N(S))} p\ell_\ell^f$$

$$\sum_{PL_k^c \in C(N(S))} I_k^c = \sum_{PL_k^c \in C(N(S))} p\ell_k^c$$

Since each power source in S is supplying power to power loads in $N(S)$, but the power loads in $N(S)$ can receive power from other power sources not in S (see Figure 5a), we can have the following inequality,

$$\sum_{PS_m \in S} O_m \le \sum_{PL_n \in N(S)} I_n \tag{11}$$

On the other side, the total summation of all outgoing power streams from power sources in S can be written as,

$$\sum_{PS_m \in S} O_m = \sum_{PS_i^c \in C(S)} ps_i^c + \sum_{PS_j^f \in F(S)} ps_j^f \ge \sum_{PS_i^c \in C(S)} ps_i^{c-min} + \sum_{PS_j^f \in F(S)} ps_j^f \tag{12}$$

Similarly, the total incoming power flows into $N(S)$ can be represented as,

$$\sum_{PL_n \in N(S)} I_n = \sum_{PL_k^c \in C(N(S))} p\ell_k^c + \sum_{PL_\ell^f \in F(N(S))} p\ell_\ell^f \le \sum_{PL_k^c \in C(N(S))} p\ell_k^{c-max} + \sum_{PL_\ell^f \in F(N(S))} p\ell_\ell^f \tag{13}$$

By combining (11)–(13), we can conclude that Condition *1-1*, is satisfied whenever the system has a feasible solution. The necessity of Condition *1-2* can be presented in a same way, in which the following inequality is a key to show such situation.

$$\forall T \subseteq \mathcal{PL}, \quad \sum_{PS_m \in N(T)} O_m \ge \sum_{PL_n \in T} I_n \tag{14}$$

The above inequality holds since each power load in T is getting power from power sources in $N(T)$, but the power sources in $N(T)$ can give power to loads not in T. $\quad\square$

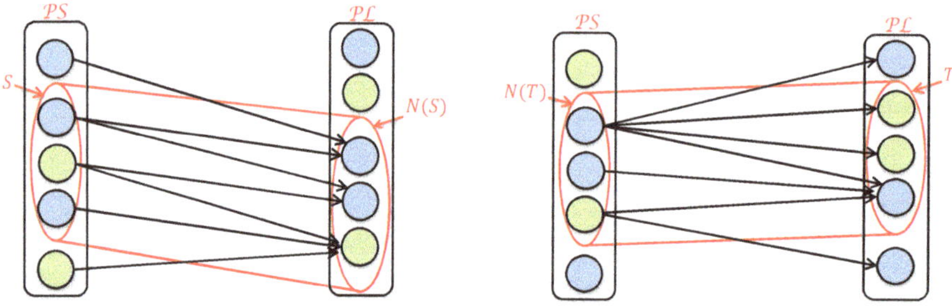

Figure 5. Illustration of subsets of power sources and loads. (**a**) Subset S of power sources and neighbor/connected set $N(S)$; (**b**) Subset T of power loads and its neighbor/connected set $N(T)$.

To provide the sufficiency, we will introduce several definitions and an auxiliary optimization problem generated from our original power flow control problem. The goal of this sufficiency proof is, supposing that Condition 1-1 and Condition 1-2 are satisfied, to show the "existence" of a feasible solution of our original power flow control problem, i.e., a power flow assignment $x : \chi \to R_+$ which satisfies (7)–(10). The following definitions and an auxiliary optimization problem are introduced for this purpose.

Definition 1. *Each power source could have three states: Power-High, Power-Low, and Power-Balanced. Power-High: When $ps_i^{c-max} > O_i^c$ holds for a controllable power source PS_i^c, there is room to increase the outgoing power flow O_i^c. Such power source is called "power-high" node. Similarly, when $ps_j^f > O_j^f$ for a fluctuating power source PS_j^f, PS_j^f is also called "power-high" node.*

Power-Low: When $ps_i^{c-min} < O_i^c$ holds for PS_i^c, there is room to decrease outgoing power flow O_i^c. Such power source is called "power-low" node. Similarly, when $ps_j^f < O_j^f$ for PS_j^f, PS_j^f is also called "power-low" node.

Power-Balanced: When $ps_i^{c-min} \leq O_i^c \leq ps_i^{c-max}$ holds for PS_i^c, we can control the power level of PS_i^c so that the generating power and the outgoing power are balanced. Such PS_i^c is called "power-balanced" node. Similarly, when $ps_j^f = O_j^f$ for PS_j^f, PS_j^f is called "power-balanced" node.

According to the above definition, a controllable power source PS_i^c with $ps_i^{c-min} < O_i^c < ps_i^{c-max}$ is power-high, power-low and power-balanced simultaneously. It means, for such power source, the outgoing power O_i^c can be increased ("power-high"), decreased ("power-low") or kept unchanged ("power-balanced") by controlling the generating power level of PS_i^c within its minimum and maximum power limits. On the other hand, the definitions of power-high, power-low and power-balanced for a fluctuating power source are disjoint, since the generating power level is given and cannot be altered for a fluctuating power source, and power-balanced state needs $ps_j^f = O_j^f$ exactly.

Three states, power-high, power-low and power-balanced, are defined also for a power load as follows. Power-High: When $I_k^c > pl_k^{c-min}$ holds for a controllable power load PL_k^c, or $I_\ell^f > pl_\ell^f$ for fluctuating power load PL_ℓ^f, there is room to decrease the incoming power flow I_k^c or I_ℓ^f, respectively. Such power load is called "power-high" node.

Power-Low: When $I_k^c < pl_k^{c-max}$ holds for PL_k^c, or $I_\ell^f < pl_\ell^f$ for PL_ℓ^f, there is room to increase the incoming power flow I_k^c or I_ℓ^f, respectively. Such power load is called "power-low" node.

Power-Balanced: When $pl_k^{c-min} \leq I_k^c \leq pl_k^{c-max}$ holds for PS_k^c, or $O_\ell^f = pl_\ell^f$ for PL_ℓ^f, such power lord is called "power-balanced" node.

Similar to the definitions for a power source, the definitions of these three states for a controllable power load are overlapped, while they are not overlapped for a fluctuating power load.

Definition 2. *A power path is an alternative sequence of devices/nodes and power flows/connections, where each device in a path is either an initial node followed by a power flow/connection incident to this device, an intermediate device which is incident to the previous and the following power flow/connections or a finishing device which is incident to the previous connection. A power path can contain "forward edges" with similar direction with path direction and "backward edges" with the reverse direction with the power path direction. If every backward edge has increment in the power flow amount, then the power path is called "alternating path". The power flow obligation on each flow/connection of an alternating path is shown in Figure 6.*

Definition 3. *An alternating power path which initialize from "power-high" device/node and ends/terminates on "power-low" device is called an augmenting path (Figure 7).*

Definition 4. *With regard to an augmenting path, the procedure to increase power flow or amount of power on each connection in the path uniformly by $\triangle > 0$ ($+\triangle$ for a forward edge, and $-\triangle$ for a backward edge) is called "power flow augmentation". Please note that by this power flow management/augmentation, the total incoming/outgoing power of each device/node changes only at an initial device and an ending device.*

Please note that the words "alternating path" and "augmenting path" are borrowed from Graph Theory. In addition, our Theorem-1 can be considered to be an extension of Hall's theorem in Bipartite Matching [38].

Our proof of the sufficiency of Theorem-1 begins with the introduction of the following Optimization Problem-1.

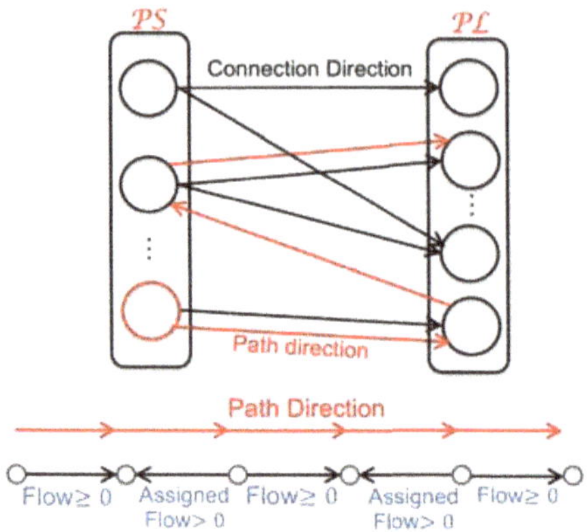

Figure 6. Alternating Path.

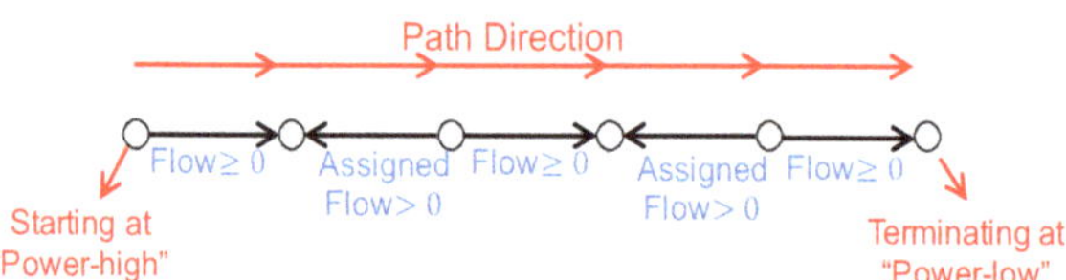

Figure 7. Augmenting Path.

Optimization Problem-1: Find $x : \mathcal{X} \rightarrow R_+$ such that

$$min \left(\sum_{i=1}^{I} \phi \left(ps_i^{c-min} - O_i^c \right) + \sum_{j=1}^{J} \mid ps_j^f - O_j^f \mid + \sum_{k=1}^{K} \phi \left(p\ell_k^{c-min} - I_k^c \right) + \sum_{\ell=1}^{L} \mid p\ell_\ell^f - I_\ell^f \mid \right)$$

where

$$\phi(p) = \left\{ \begin{array}{ll} p & : p > 0 \\ 0 & : p \leq 0 \end{array} \right.$$

with following constraints,

$$O_i^c \leq ps_i^{c-max} \tag{15}$$

$$O_j^f \leq ps_j^f \tag{16}$$

$$I_k^c \leq p\ell_k^{c-max} \tag{17}$$

$$I_\ell^f \leq p\ell_\ell^f \tag{18}$$

Our goal of this proof is to show that if Conditions *1-1* and *1-2* are satisfied, Optimization Problem-1 always has an optimum solution which makes the objective function zero, i.e.,

$$ps_i^{c-min} \leq O_i^c \leq ps_i^{c-max}, \quad \forall PS_i^c \in C(\mathcal{PS}) \tag{19}$$

$$ps_j^f = O_j^f, \quad \forall PS_j^f \in F(\mathcal{PS}) \tag{20}$$

$$p\ell_k^{c-min} \leq I_k^c \leq p\ell_k^{c-max}, \quad \forall PL_k^c \in C(\mathcal{PL}) \tag{21}$$

$$p\ell_\ell^f = I_\ell^f, \quad \forall PL_\ell^f \in F(\mathcal{PL}) \tag{22}$$

It is clear that this type of optimum solution is a feasible solution of our original Power Flow Control Problem.

Please note that since $x(PS_m, PL_n) = 0$ for all $(PS_m, PL_n) \in \mathcal{X}$ is a feasible solution for Optimization Problem-1, there always exists an optimum solution of Optimization Problem-1.

To achieve our goal by contradiction, we assume that the optimum solution $x^* : \mathcal{X} \to R_+$ does not achieve the objective function equal to zero. This shows that there exists PS_m such that $O_m^c < ps_m^{c-min}$ or $O_{m-1}^f < ps_{m-1}^f$ or there exists PL_n such that $I_n^c < p\ell_n^{c-min}$ or $I_{n-K}^f < p\ell_{n-K}^f$.

Case 1. *In case PS_m such that $O_m^c < ps_m^{c-min}$ or $O_{m-1}^f < ps_{m-1}^f$ exists:*
Let A be the group/set of power sources and B be the group/set of power loads which can be extended from PS_m by alternating paths. As an alternating path can be stretched from a power source device/node to a power load device/node without any constraint, $B = N(A)$ holds. Conversely, power loads in B can have connection (that must be a zero-power flow) with power sources not exists in A, i.e., $A \subseteq N(B)$. The power used by power loads in B is provided by power sources in A, since power flows amount on connections from $\mathcal{PS} \setminus A = \{PS \mid PS \in \mathcal{PS} \text{ and } PS \notin A\}$ to B are zero (see Figure 8), which means $\sum_{PS_a \in A} O_a = \sum_{PL_b \in N(A)} I_b$. Now we can consider possibilities as given below.

[Case 1-1]: A includes a "power-low" node or B includes a "power-low": If $PS_t \in A$ ($PL_t \in B$) is a "power-low" node, The alternating path from PS_m to PS_t (PL_t, respectively) is an augmenting path, and the power flow augmentation is applied to get a new power flow assignment which has the difference $\phi(ps_m^{c-min} - O_m^c)$ or $\mid ps_{m-1}^f - O_{m-1}^f \mid$ smaller than x^*, while none of the other differences $ps_i^{c-min} - O_i^c$, $ps_j^f - O_j^f$, $p\ell_k^{c-min} - I_k^c$ and $p\ell_\ell^f - I_\ell^f$ becomes larger than that in x^*. It means that the new power flow assignment is a better solution that x^*, which contradicts the optimality of x^*.

[Case 1-2]: Neither A nor B includes "power-low" node: Every node in A and B is either "power-balanced" or "power-high", which means

$$ps_i^{c-min} \geq O_i^c, \quad \forall PS_i^c \in C(A)$$

$$ps_j^f \geq O_j^f, \quad \forall PS_j^f \in F(A)$$

$$I_k^c \geq pl_k^{c-max}, \quad \forall PL_k \in C(B)$$

$$I_\ell^f \geq pl_\ell^f, \quad \forall PL_\ell \in F(B)$$

Together with the fact that PS_m also exists in A, we have,

$$\sum_{PS_i^c \in C(A)} ps_i^{c-min} + \sum_{PS_j^f \in F(A)} ps_j^f > \sum_{PS_i^c \in C(A)} O_i^c + \sum_{PS_j^f \in F(A)} O_j^f = \sum_{PL_k^c \in C(N(A))} I_k^c + \sum_{PL_\ell^f \in F(N(A))} I_\ell^f$$

$$\geq \sum_{PL_k^c \in C(N(A))} p\ell_k^{c-max} + \sum_{PL_\ell^f \in F(N(A))} p\ell_\ell^f$$

which is the contradiction to Condition 1-1.

Case 2. *In case PL_n such as $I_n^c < p\ell_n^{c-min}$ or $I_{n-K}^f < p\ell_{n-K}^f$ exists:*
Now, let D and E be the sets of power loads and sources, respectively, which can be a starting node of an alternating path terminating at PL_n. Since a starting node of an alternating path can be reached from a load to a source without any restriction, $E = N(D)$ holds. The sources in E can have connection (must have zero-power flow) with power loads outside D, i.e., $D \subseteq N(E)$ (see Figure 9). The power generated by power sources in E is supplied to only power loads in D, which means $\sum_{PS_e \in N(D)} O_e = \sum_{PL_d \in D} I_d$. We can consider the following two possibilities.

[Case 2-1]: D includes "power-high" node or E includes "power-high" node: We can find augmenting path starting from a "power-high" node $PL_s \in D$ or $PS_s \in E$ and terminating at PL_n, and apply the power flow augmentation along this augmenting path to get a new power flow assignment which is better than the assumed optimum solution x^*. It is the contradiction to the assumption.

[Case 2-2]: *Neither* D *nor* E *includes "power-high" node:* *Every node in D and E is either "power-balanced" or "power-low" node, which means,*

$$I_k^c \leq pl_k^{c-min}, \quad \forall PL_k^c \in C(D)$$

$$I_\ell^f \leq pl_\ell^f, \quad \forall PL_\ell^f \in F(D)$$

$$ps_i^{c-max} \leq O_i^c, \quad \forall PS_i^c \in C(E)$$

$$ps_j^f \leq O_j^f, \quad \forall PS_j^f \in F(E)$$

Together with the fact that PL_n also exists in C, we have,

$$\sum_{PS_i^c \in C(N(D))} ps_i^{c-max} + \sum_{PS_j^f \in F(N(D))} ps_j^f \leq \sum_{PS_i^c \in C(N(D))} O_i^c + \sum_{PS_j^f \in F(N(D))} O_j^f = \sum_{PL_k^c \in C(D)} I_k^c$$

$$+ \sum_{PL_\ell^f \in F(D)} I_\ell^f < \sum_{PL_k^c \in C(D)} p\ell_k^{c-min} + \sum_{PL_\ell^f \in F(D)} p\ell_\ell^f,$$

which is the contradiction to Condition 1-2.

As stated above, if we assume that the optimum solution of Optimization Problem-1 does not make the objective function zero, it always incurs a contradiction. Hence, if Conditions *1-1* and *1-2* are satisfied, Optimization Problem-1 always has an optimum solution which makes the objective function zero, and hence our original Power Flow Control Problem has a feasible solution.

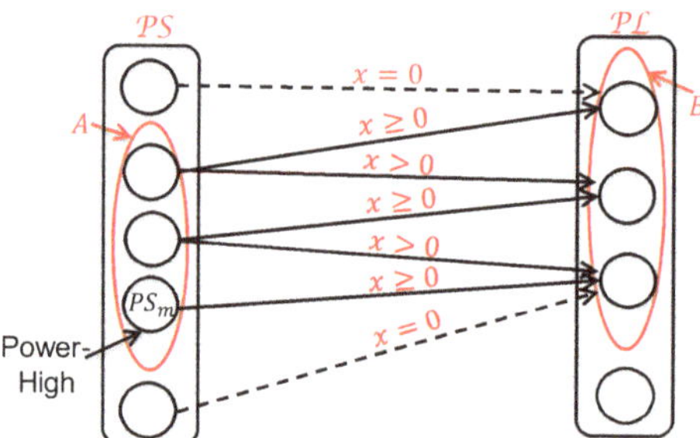

Figure 8. Illustration of sets *A* and *B* of power sources and power loads, respectively, which are reachable from PS_m by alternating paths.

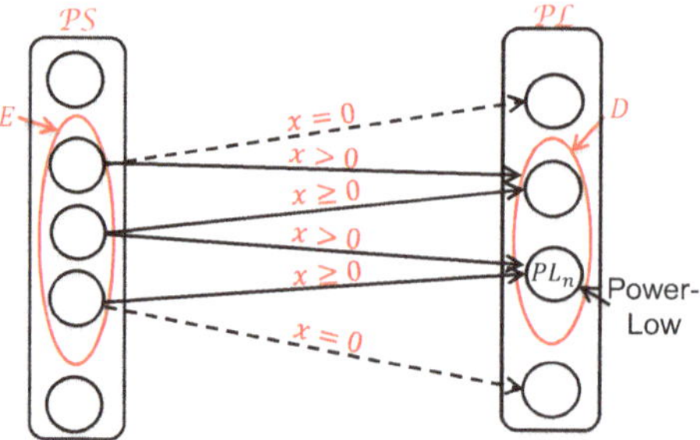

Figure 9. Illustration of a subset *E* of sources and set *D* of loads, which can be starting nodes of alternating paths terminating at PL_n

5. System Condition for the Robustness against Fluctuation

In this section, we assume that each fluctuating power device has an arbitrary power level within the specified minimum and maximum levels. Even though the power level of individual fluctuating device is not specified as a concrete number, we can guarantee the existence of a feasible solution of the power flow control problem if some structural conditions are satisfied. The following Theorem 2 describes these structural conditions.

Theorem 2. *The power flow control problem continuously has a feasible solution if and only if the succeeding two conditions are fulfilled.*
Condition 2-1:

$$\forall S \subseteq PS, \quad \sum_{PS_i^c \in C(S)} ps_i^{c-min} + \sum_{PS_j^f \in F(S)} ps_j^{f-max} \leq \sum_{PL_k^c \in C(N(S))} p\ell_k^{c-max} + \sum_{PL_\ell^f \in F(N(S))} p\ell_\ell^{f-min}$$

Condition 2-2:

$$\forall T \subseteq PL, \quad \sum_{PS_i^c \in C(N(T))} ps_i^{c-max} + \sum_{PS_j^f \in F(N(T))} ps_j^{f-min} \geq \sum_{PL_k^c \in C(T)} p\ell_k^{c-min} + \sum_{PL_\ell^f \in F(T)} p\ell_\ell^{f-max}$$

Proof of Theorem 2. Here, we presented the sufficiency of above system conditions. Let S be any group/subset of power sources and $N(S)$ be the connected/neighbor set of S (see Figure 10), then from Condition 2-1, Condition 1-1 can be held as,

$$\sum_{PS_i^c \in C(S)} ps_i^{c-min} + \sum_{PS_j^f \in F(S)} ps_j^f \leq \sum_{PS_i^c \in C(S)} ps_i^{c-min} + \sum_{PS_j^f \in F(S)} ps_j^{f-max}$$

$$\leq \sum_{PL_k^c \in C(N(S))} p\ell_k^{c-max} + \sum_{PL_\ell^f \in F(N(S))} p\ell_\ell^{f-min} \leq \sum_{PL_k^c \in C(N(S))} p\ell_k^{c-max} + \sum_{PL_\ell^f \in F(N(S))} p\ell_\ell^f,$$

This determines that if the system Condition 2-1 is satisfied, then system Condition 1-1 is always satisfied for any power level generated/consumed of fluctuating power devices. Correspondingly, for any group/subset of power loads T and its connected/neighbor group/subset $N(T)$ of power sources (see Figure 11), Condition 1-2 can be obtained as,

$$\sum_{PS_i^c \in C(N(T))} ps_i^{c-max} + \sum_{PS_j^f \in F(N(T))} ps_j^f \geq \sum_{PS_i^c \in C(N(T))} ps_i^{c-max} + \sum_{PS_j^f \in F(N(T))} ps_j^{f-min}$$

$$\geq \sum_{PL_k^c \in C(T)} p\ell_k^{c-min} + \sum_{PL_\ell^f \in F(T)} p\ell_\ell^{f-max} \geq \sum_{PL_k^c \in C(T)} p\ell_k^{c-min} + \sum_{PL_\ell^f \in F(T)} p\ell_\ell^f$$

This proves that Condition 2-2 is satisfied, then Condition 1-2 is always fulfill for any situation of fluctuating power devices.

A system, which has a feasible solution for any situation of fluctuating devices, must have a feasible solution even for the case $ps_j^f = ps_j^{f-max}$ and $p\ell_\ell^f = p\ell_\ell^{f-min}$ for all fluctuating devices. From the necessity of Condition 1-1 with this specific situation of fluctuating devices, the necessity of Condition 2-1 is shown. Similarly, considering the necessity of Condition 1-2 for a specific case with $ps_j^f = ps_j^{f-min}$ and $p\ell_\ell^f = p\ell_\ell^{f-max}$, the necessity of Condition 2-2 is shown. $\square$

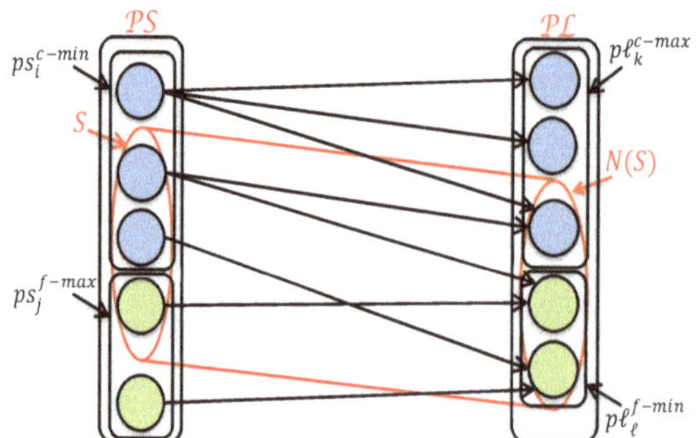

Figure 10. Illustration of a subset S of power sources and set $N(S)$ of power loads.

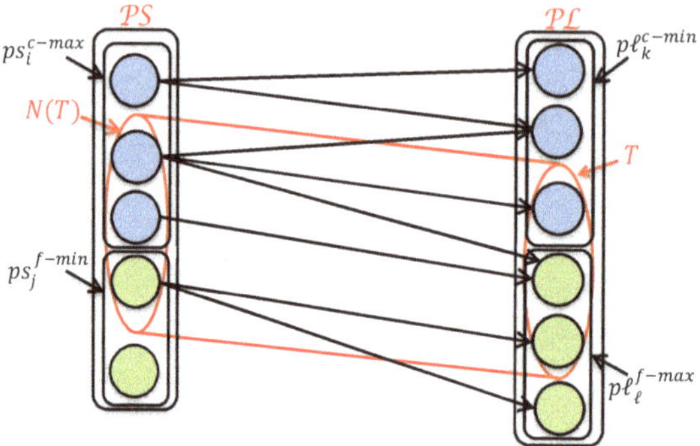

Figure 11. Illustration of a subset T of power loads and set $N(T)$ of power sources.

6. Numerical Results

This subsection signifies the application of the proposed theorem to demonstrate the existence of a feasible solution. We design the system with three power suppling devices and three power consuming devices with connections. One power sources is chosen as controllable (PS_1^c), and the remaining two sources are chosen as fluctuating (PS_1^f and PS_2^f). Likewise, one of the load is chosen as controllable (PL_1^c), and remaining two loads are fluctuating as PL_1^f and PL_2^f.

In Figure 12, given power levels by fluctuating power sources are represented as, $ps_1^f = 7$, and $ps_2^f = 2$. The power request levels by fluctuating loads are indicated as, $p\ell_1^f = 1$, and $p\ell_2^f = 5$, respectively. The controllable power devices are restricted between power limitations i.e., maximum and minimum power limits as, $ps_1^{c-max} = 4$, and , $ps_1^{c-min} = 0$. The power boundaries for controllable load are given as, $p\ell_1^{c-max} = 6$, and $p\ell_1^{c-min} = 2$. At first, we demonstrate that Condition *1-1* is fulfilled for all subsets of power sources.

Table 1 indicates all possible subsets of power sources with connected/neighbor subsets with power generation and consumption calculation according to Condition *1-1*. For each subset, the summation of minimum power levels for controllable and given power levels of fluctuating is less or equal to the sum of maximum power levels for controllable load and given power levels for fluctuating loads. Similarly, we can represent that the Condition *1-2* is also fulfilled and finally we discover that the system fulfilled conditions *1-1* and *1-2*, and we have feasible solution.

Table 1. List of subset S of $\mathcal{PS}$ and $N(S)$

S	$N(S)$	$ps_i^{c-min} + ps_j^f$	$pl_k^{c-max} + p\ell_\ell^f$
$\{PS_1^c\}$	$\{PL_2^f\}$	0	5
$\{PS_1^f\}$	$\{PL_1^c, PL_1^f\}$	7	7
$\{PS_2^f\}$	$\{PL_1^f, PL_2^f\}$	2	6
$\{PS_1^c, PS_1^f\}$	$\{PL_1^c, PL_1^f, PL_2^f\}$	7	12
$\{PS_1^c, PS_2^f\}$	$\{PL_1^f, PL_2^f\}$	2	6
$\{PS_1^f, PS_2^f\}$	$\{PL_1^c, PL_1^f, PL_2^f\}$	9	12
$\{PS_1^c, PS_1^f, PS_2^f\}$	$\{PL_1^c, PL_1^f, PL_2^f\}$	9	12

As for the power flow allocation for each connection, first it is considered to be "zero". Since all sources are "power-high", the system will try to find an augmenting path to increase power by choosing a source randomly. For example, the augmenting path commenced with PS_2^f and ended at "power-low" node PL_1^f is chosen and the power is increased on this connection by "1" to satisfy the power demand (Figure 12(1)). This makes this power load a "power-balanced" node. Also, we select a path from PS_2^f to PL_2^f and augmented power flow by "1". The next augmenting path is chosen from PS_1^f to PL_2^f through PL_1^f and PS_2^f and power flow along this path is augmented by "1" so that the power flow on each connection does not become negative as displayed in the Figure 12(2). Since PS_2^f and PL_1^f became "power-balanced" nodes, the next augmenting path is chosen from PS_1^f to PL_1^c for power increase by "6" (Figure 12(3)). Now all power devices are "power-balanced" except PS_1^c and PL_2^f, so system chosen augmenting path starting from PS_1^c and ending at PL_2^f to increase power by "3" (Figure 12(4)). Here, we select generated power by fluctuating power devices as much as possible to keep power supply of controllable power sources.

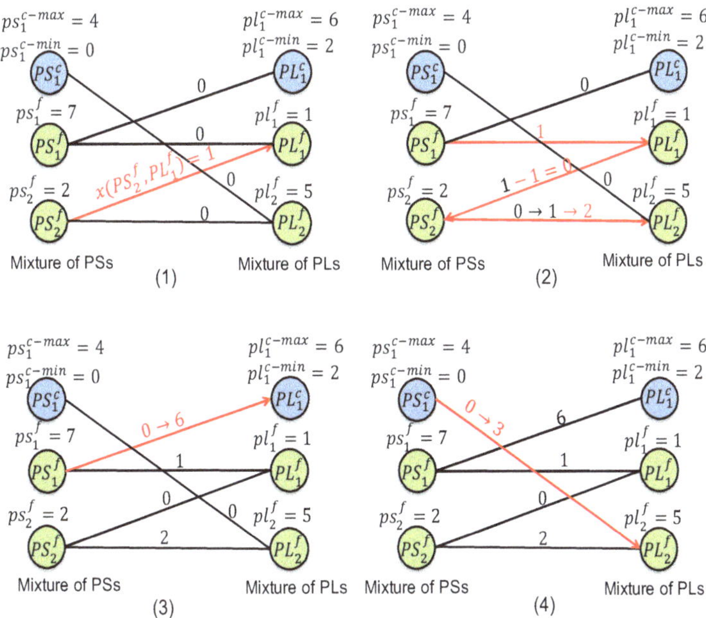

Figure 12. Demonstration example for feasible solution case.

In Figure 13, non-feasible case is discussed. The power supply and consumption levels for fluctuating power devices are same with the previous example but the maximum power levels for controllable sources and loads are different. We observed that by changing the maximum power limits, the conditions *1-1*, and *1-2* are not satisfied for subsets S, and T and their neighboring devices shown in Figure 13. For this case, we cannot find the feasible solution for the given system.

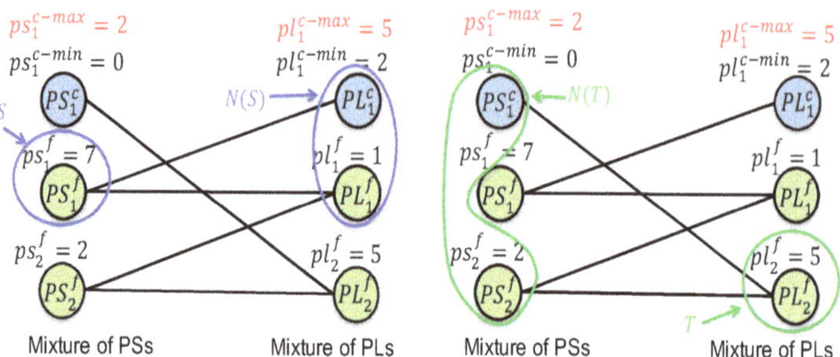

Figure 13. Demonstration example for non-feasible solution case.

7. Concluding Remarks

Renewable energy sources such as wind power and photovoltaic power generation system play very important roles because of low impact against the environment. However, the generated power from the renewable energy sources varies greatly, resulting in a risk of fluctuations which is uncontrollable. The increased penetration of renewable energy sources has an effect on the power system's stability and quality of power. In addition, the power fluctuations caused by power loads also have the same consequences on power system. From such point of view, in which power supply and demand changes dynamically, a power flow control mechanism is introduced which assigns power levels for controllable power devices and connections between power devices to absorb the power fluctuations caused by fluctuating devices.

In order for a power system to continue a safe operation under the presence of fluctuating power levels of fluctuating devices, the power flow control algorithm should be designed properly and, at the same time, a power system itself should be designed properly. This paper proposed structural system conditions for a power system to possess the robustness against fluctuation, i.e., the condition for a system to have always a feasible solution of the power flow control problem with any power levels of fluctuating devices (Theorem 2). These conditions are described in terms of the connectivity of connections between power sources and loads and the minimum and the maximum power levels of individual power devices. Most important application area of our result might be a power system design which includes the allocation of power devices, allocation of connections between power devices, and specification of the maximum and the minimum power levels of individual power devices.

As a first attempt to investigate structural conditions for a power flow system to possess the robustness against fluctuation, we have discussed the power flow control problem and structural conditions based on a relatively simple system model. For example, the capacity of individual connections is assumed to be large enough so that it does not affect the existence/inexistence of a feasible solution of the power flow control problem. The improved structural conditions regarding more sophisticated system model, e.g., consideration of the capacity of individual connections, and the limitation of speed of change in power levels of power devices remain as future problems.

Author Contributions: Conceptualization, S.J., M.K., and Y.T.; writing—original draft preparation, S.J.; writing—review and editing, S.J., and M.K.; supervision, M.K., and Y.T. All authors have read and agreed to the published version of the manuscript.

Funding: This research received no external funding.

Conflicts of Interest: The authors declare no conflict of interest.

References

1. Maegaard, P. Balancing fluctuating power sources. In Proceedings of the World-Non-Grid-Connected Wind Power and Energy Conference, Kunming, China, 13–17 October 2010.
2. Soroudi, A.; Ehsan, M.; Caire, R.; Hadjsaid, N. Possibilistic evaluation of distributed generations impacts on distribution networks. *IEEE Trans. Power Syst.* **2011**, *26*, 2293–2301. [CrossRef]
3. Umer, S.; Tan, Y.; Lim, A.O. Stability analysis for smart homes energy management system with delay consideration. *J. Clean Energy Technol.* **2014**, *2*, 332–338. [CrossRef]
4. Umer, S.; Tan, Y.; Lim, A.O. Priority based power sharing scheme for power consumption control in smart homes. *Int. J. Smart Grid Clean Energy* **2014**, *3*, 340–346. [CrossRef]
5. Umer, S.; Kaneko, M.; Tan, Y.; Lim, A.O. System design and analysis for maximum consuming power control in smart house. *J. Autom. Control Eng.* **2014**, *2*, 43–48. [CrossRef]
6. Lumbreras, S.; Ramos, A.; Banez-Chicharro, F. Optimal transmission network expansion planning in real-sized power systems with high renewable penetration. *Electr. Power Syst. Res.* **2017**, *149*, 76–88. [CrossRef]
7. Han, J.; Choi, C.; Park, W.; Lee, I.; Kim, S. Smart home energy management system including renewable energy based on ZigBee and PLC. *IEEE Trans. Consum. Electron.* **2014**, *60*, 198–202. [CrossRef]
8. Han, J.; Choi, C.; Park, W.; Lee, I.; Kim, S. PLC-based photovoltaic system management for smart home energy management system. *IEEE Trans. Consum. Electron.* **2014**, *60*, 184–189. [CrossRef]
9. Hong, I.; Kang, B.; Park, S. Design and implementation of intelligent energy distribution management with photovoltaic system. *IEEE Trans. Consum. Electron.* **2012**, *58*, 340–346. [CrossRef]
10. Rashidi, Y.; Moallem, M.; Vojdani, S. Wireless Zigbee system for performance monitoring of photovoltaic panels. In Proceedings of the 2011 37th IEEE Photovoltaic Specialists Conference, Seattle, WA, USA, 19–24 June 2011; pp. 3205–3207.
11. Xiaoli, X.; Daoe, Q. Remote monitoring and control of photovoltaic system using wireless sensor network. In Proceedings of the 2011 International Conference on Electric Information and Control Engineering, Wuhan, China, 15–17 April 2011; pp. 633–638.
12. Biabani, M.; Golkar, M.A.; Johar, A.; Johar, M. Propose a home demand -side management algorithm for smart nano-grid. In Proceedings of the 4th Annual International Power Electronics, Drive Systems and Technologies Conference, Tehran, Iran, 13–14 February 2013; pp. 487–494.
13. Asare-Bediako, B.; Kling, W.L.; Robeiro, P.F. Home energy management systems: Evolution, trends and frameworks. In Proceedings of the 2012 47th International Universities Power Engineering Conference (UPEC), London, UK, 4–7 September 2012; pp. 1–5.
14. Shwehdi, M.H.; RajaMuhammad, S. Proposed smart DC Nano-Grid for green buildings a reflective view. In Proceedings of the 2014 International Conference on Renewable Energy Research and Application (ICRERA), Milwaukee, WI, USA, 19–22 October 2014; pp. 765–769.
15. Kinn, M.C. Proposed components for the design of a smart Nano-Grid for a domestic electrical system that operates at below 50V DC. In Proceedings of the IEEE PES International Conference and Exhibition on Innovative Smart Grid Technologies (ISGT Europe), Manchester, UK, 5–7 December 2011; pp. 1–7.
16. Schonberger, J.; Duke, R.; Round, S.D. DC-Bus signaling: A distributed control strategy for a hybrid renewable Nano-grid. *IEEE Trans. Ind. Electron.* **2006**, *53*, 1453–1460. [CrossRef]
17. Latha, S.H.; Mohan, S.C. Centralized power control strategy for 25 kW Nano-Grid for rustic electrification. In Proceedings of the 2012 International Conference on Emerging Trends in Science, Engineering and Technology (INCOSET), Tiruchirappalli, India, 13–14 December 2012; pp. 456–461.
18. Kondoh, J.; Higuchi, N.; Sekine, S.; Yamaguchi, H.; Ishii, I. Distribution system research with an analog simulator. In Proceedings of the First Industrial Conference on Power Electronics for Distributed and Co-generation, Irvine, CA, USA, 22–24 March 2004.
19. Kondoh, J.; Aki, H.; Yamaguchi, H.; Murata, A.; Ishii, I. Consumed power control of time deferrable loads for frequency regulation. In Proceedings of the IEEE PES Power Systems Conference and Exposition, New York, NY, USA, 10–13 October 2004.

20. Okabe, Y.; Sakai, K. *QoEn (Quality of Energy) Routing toward Energy on Demand Service in the Future Internet*; ICE-IT, Academic Center for Computing and Media Studies, Kyoto University: Kyoto, Japan, 2009.
21. Abe, R.; Taoka, H.; McQuilkin, D. Digital Grid: Communicative electrical grids of the future. *IEEE Trans. Smart Grid* **2011**, *2*, 399–410. [CrossRef]
22. Takuno, T.; Kitamori, Y.; Takahashi, R.; Hikihara, T. AC power routing system in home based on demand and supply utilising distributed power sources. *Energies* **2011**, *4*, 717–726. [CrossRef]
23. Javaid, S.; Kurose, Y.; Kato, T.; Matsuyama, T. Cooperative distributed control implementation of the power flow coloring over a Nano-grid with fluctuating power loads. *IEEE Trans. Smart Grid* **2017**, *8*, 342–352. [CrossRef]
24. Javaid, S.; Kato, T.; Matsuyama, T. Power flow coloring system over a Nano-grid with fluctuating power sources and loads. *IEEE Trans. Ind. Inform.* **2017**, *13*, 3174–3184. [CrossRef]
25. Yamaguchi, H.; Kondoh, J.; Aki, H.; Murata, A.; Ishi, I. Power fluctuation analysis of distribution network introduced a large number of photovoltaic generation system. In Proceedings of the 18th International Conference and Exhibition on Electricity Distribution, Turin, Italy, 6–9 June 2005.
26. Riffonneau, Y.; Bacha, S.; Ploix, S. Optimal power flow management for grid connected PV systems with batteries. *IEEE Trans. Sustain. Energy* **2011**, *2*, 309–320. [CrossRef]
27. Denholma, P.; Margolis, R.M. Evaluating the limits of solar pho-tovoltaics (PV) in electric power systems utilizing energy storage and other enabling technologies. *Energy Policy* **2007**, *35*, 4424–4433. [CrossRef]
28. Perrin, M.; Saint-Drenan, Y.M.; Mattera, F.; Malbranche, P. Lead- acid batteries in stationary applications: Competitors and new markets for large penetration of renewable energies. *J. Power Sources* **2005**, *144*, 402–410. [CrossRef]
29. Cesena, E.A.M.; Capuder, T.; Mancarella, P. Flexible distributed multienergy generation system expansion planning under uncertainty. *IEEE Trans. Smart Grid* **2016**, *7*, 348–357. [CrossRef]
30. Wei, J.; Zhang, Y.; Wang, J.; Cao, X.; Khan, M.A. Multi-period planning of multi-energy microgrid with multi-type uncertainties using chance constrained information gap decision method. *Appl. Energy* **2020**, *260*, 1–19. [CrossRef]
31. Garcez, C.G. Distributed electricity generation in Brazil: An analysis of policy context, design and impact. *Util. Policy* **2017**, *49*, 104–115. [CrossRef]
32. Sharma, P.; Tandon, A. Techniques for optimal placement of DG in radial distribution system: A review. In Proceedings of the Communication, Control and Intelligent Systems (CCIS), Mathura, India, 7–8 November 2015; pp. 453–458.
33. Kazemi, A.; Bayat, P.M. Determination and allocation of reactive power loss in deregulated power systems in the presence of multilateral multiple contracts. In Proceedings of the International Conference on Power Engineering, Energy and Electrical Drives, Setubal, Portugal, 12–14 April 2007; pp. 454–459.
34. Caampued, C.P.C.C.; Aguirre, R.A. Determination of penetration limit of wind distributed generation considering multiple bus integration. In Proceedings of the IEEE Asia-Pacific Power and Energy Engineering Conference (APPEEC), Kota Kinabalu, Malaysia, 7–10 October 2018; pp. 370–375.
35. Ebrahim, A.F.; Youssef, T.A.; Mohammed, O.A. Power quality improvements for integration of hybrid AC/DC nanogrids to power systems. In Proceedings of the IEEE Green Technologies Conference, Denver, CO, USA, 29–31 March 2017; pp. 171–176.
36. Egghtedarpour, N.; Farjah, E. Power control and management in a hybrid AC/ DC microgrid. *IEEE Trans. Smart Grid* **2014**, *5*, 1494–1505. [CrossRef]
37. Nejabatkhah, F.; Li, Y.W. Overview of power management strategies of hybrid AC/DC microgrid. *IEEE Trans. Power Electron.* **2015**, *30*, 7072–7089. [CrossRef]
38. Axler, S.; Ribet, K.A. *Graduate Texts in Mathematics*; Springer: Berlin, Germany, 2008.

Article

Qualitative and Quantitative Transient Stability Assessment of Stand-Alone Hybrid Microgrids in a Cluster Environment

Kishan Veerashekar *, Halil Askan and Matthias Luther

Institute of Electrical Energy Systems, Friedrich-Alexander-Universität Erlangen-Nürnberg, Cauerstr. 4 (House No. 1), 91058 Erlangen, Germany; halil.askan@fau.de (H.A.); matthias.luther@fau.de (M.L.)
* Correspondence: kishan.veerashekar@fau.de

Received: 29 January 2020; Accepted: 8 March 2020; Published: 10 March 2020

Abstract: Neighboring stand-alone hybrid microgrids with diesel generators (DGs) as well as grid-feeding photovoltaics (PV) and grid-forming battery storage systems (BSS) can be coupled to reduce fuel costs and emissions as well as to enhance the security of supply. In contrast to the research in control and small-signal rotor angle stability of microgrids, there is a significant lack of knowledge regarding the transient stability of off-grid hybrid microgrids in a cluster environment. Therefore, the large-signal rotor angle stability of pooled microgrids was assessed qualitatively and also quantitatively in this research work. Quantitative transient stability assessment (TSA) was carried out with the help of the—recently developed and validated—micro-hybrid method by combining time-domain simulations and transient energy function analyses. For this purpose, three realistic dynamic microgrids were modelled regarding three operating modes (island, interconnection, and cluster) as well as the conventional scenario "classical" and four hybrid scenarios ("storage", "sun", "sun & storage", and "night") regarding different instants of time on a tropical partly sunny day. It can be inferred that, coupling hybrid microgrids is feasible from the voltage, frequency, and also transient stability point of view. However, the risk of large-signal rotor angle instability in pooled microgrids is relatively higher than in islanded microgrids. Along with critical clearing times, new stability-related indicators such as system stability degree and corrected critical clearing times should be taken into account in the planning phase and in the operation of microgrids. In principle, a general conclusion concerning the best operating mode and scenario of the investigated microgrids cannot be drawn. TSA of pooled hybrid microgrids should be performed—on a regular basis especially in the grid operation—for different loading conditions, tie-line power flows, topologies, operating modes, and scenarios.

Keywords: critical clearing times; diesel generators; grid-feeding photovoltaics; grid-forming battery storage systems; stand-alone hybrid microgrids; system stability degree; the micro-hybrid method; transient stability assessment

---

1. Introduction

Autonomous hybrid microgrids with diesel and/or biogas engine-driven synchronous generators as well as grid-feeding photovoltaics (PV) and grid-forming battery storage systems (BSS) are being widely installed in developing and underdeveloped countries [1–4]. By coupling spatially close stand-alone hybrid microgrids, it is possible to curtail fuel costs and emissions, and to enhance the security of supply [5,6].

Such clustered microgrids should be investigated from the system stability point of view. Voltage stability analysis has been carried out in an interconnected AC-DC microgrid under fault conditions [7]. In [8], clustering two microgrids comprising a synchronous generator and two inverter-based systems was studied with respect to blackouts. Small-signal stability of AC and DC microgrids in a cluster environment was analyzed in [9–13]. Ref [9] deals with the implementation of a hierarchical control scheme for DC microgrid clusters. Eigenvalue analysis was performed from the control point of view in [10] considering two interconnected DC microgrids. On the other hand, [11] dealt with the small-signal stability assessment of clustered AC microgrids comprising inverter-based distribution generation units. Identification of critical microgrid clusters was carried out using a stability indicator, i.e., stability margin based on the active power droop of the inverters. In [12], multiple inverter-based microgrid clusters were analyzed from the small-signal stability analysis and the dynamic behavior point of view. In addition to the eigenvalue analysis of a cluster of four identical microgrids, optimal parametrization of controllers was performed—using the particle swarm optimization—to improve the system stability [13]. However, there is a significant lack of research in the qualitative and especially in the quantitative transient stability assessment (TSA) of clustered microgrids.

Large-signal rotor angle stability of clustered microgrids with synchronous generators should be analyzed both in the short-term and long-term planning phase as well as in the operation [14,15]. A detailed dynamic microgrid modelling and knowledge of system response in case of three-phase faults—especially critical faults—are essential for an effective qualitative TSA, which is performed with the help of time-domain simulations (TDS) [14,16]. The qualitative TSA of microgrids comprising diesel generators (DGs) and PV, acting as current sources as well as static loads (SL) was investigated taking different microgrid models in [17] and [18].

On the other hand, online and/or offline quantitative TSA using hybrid methods—combining TDS and transient energy function (TEF) analyses—allows determining the dynamic (operational) limits in microgrids [15]. In [19], the new hybrid method was proposed and applied in an one-machine infinite bus microgrid and two machine microgrid taking classical and detailed models of DGs with and without controllers into account. Furthermore, the other five hybrid methods have been briefly discussed in [19].

Due to a discrepancy in the profiles of kinetic energy (KE) and potential energy (PE), the new hybrid technique was improved by adjusting the TEF for microgrids [20]. In [20] the micro-hybrid method (the improved new hybrid method) was verified and applied in a stand-alone microgrid with three DGs. Furthermore, quantitative TSA was performed in off-grid DG-based microgrids—without inverter-based systems such as PV and BSS as well as dynamic loads—for different coupling degrees by considering a short-circuit location [20].

Regarding microgrids in a cluster environment, there is a need to analyze the transient stability of off-grid hybrid microgrids, comprising synchronous generators and inverter-based PV and BSS, both qualitatively and quantitatively. The dynamic performance of stand-alone hybrid microgrids with DGs, grid-feeding PV, grid-forming BSS, and static and dynamic loads, under three-phase short-circuit conditions has not been investigated in-depth so far. This calls for a detailed dynamic system modelling. From the feasibility and system stability point of view, the influence of different degrees of clustering should be investigated. The quantitative TSA performed using the (recently proposed and verified) new hybrid method [19] and the micro-hybrid method [20], in microgrids with only DGs, can be now applied in hybrid microgrids. Further, no research has yet been performed in off-grid microgrids, taking system's and critical machine's critical energy (KE/PE) into account. Hence, in the framework of this research, the following tasks were executed with the help of software DIgSILENT® (Gomaringen, Germany) PowerFactory™ (v2017.0.2), and MathWorks® (Natick, USA) MATLAB™ (vR2016b):

- Detailed dynamic modelling of classical and hybrid microgrids—representing four hybrid scenarios with respect to different points in time on a summer day—comprising DGs, grid-feeding PV, grid-forming BSS, as well as static and dynamic loads in a cluster environment.

- Comparison of critical clearing time (CCT) profiles of the microgrids with respect to the four hybrid scenarios and the classical scenario (with DGs only) as well as three operating modes, namely island, interconnection, and cluster mode.

- Qualitative stability assessment by comparing the behavior of the four hybrid microgrids (i.e., hybrid scenarios in cluster mode) with that of the classical microgrid for a three-phase short-circuit location.
- Qualitative investigation of the impact of different degrees of coupling (i.e., operating modes) of microgrids on the system stability.
- Quantitative TSA using the micro-hybrid method in the scenarios and operating modes by investigating the system stability degree with respect to different clearing times for the same fault location.
- Quantitative analysis of the critical energy and of the effect of the fault location on the critical energy in the cluster mode.

In this research paper fundamentals will not be given in a separate chapter, since the relevant basics have already been presented in recently published journal articles [19,20]. Hence, the focus will be laid on methodology and results. The rest of the paper is structured as follows: In Chapter 2, the methodological approach will be explained in detail. The modelled microgrids, the scenarios and the operating modes together with the dimensioning of the controllers in PV and BSS will be described. The last subchapter deals with the approach with respect to the qualitative and quantitative TSA. Chapter 3 describes the simulation results and the corresponding discussion. The first part of this chapter focuses on the results of the qualitative TSA, whereas the second part deals with the quantitative TSA. A detailed summary of the research work along with the key findings, as well as an outlook, will be given in Chapter 4.

2. Methodology

2.1. System Modelling

The model of the three coupled or pooled rural microgrids comprising DGs and static loads has been presented in previous papers [17,20,21]. The topology of the stand-alone microgrids was slightly extended in this research work (see Figure 1) in order to connect grid-feeding PV, grid-forming BSS, as well as dynamic loads. The installed capacity of the clustered microgrid is ca. 950 kW. The ratio of the static to dynamic loads was exemplarily selected to be 0.8:0.2. The load was assumed to be high, such that the actual load is equal to 80% of the nominal power demand. The modelling and RMS-simulations of these microgrids were performed in software DIgSILENT® PowerFactory™.

The dimensioning of the DGs, PV, and BSS as well as the powers was performed according to following criterion:

- High mechanical (engine) loading of DGs (70%–80%)
- High PV feed-in (75%–85% of the nominal real power)
- High consumption/feed-in of BSS (60%–90% of the nominal real power)
- Allowable maximum and minimum voltage and frequency values according to the international norm ISO 8528-5 in the steady-state and sudden load change.

The details of the 5th order modelling of diesel engine-driven synchronous generators [22] from Stamford® [23] along with speed governor and voltage controller have been given in [17,20,21]. The nominal real power of DGs is between 24 and 112 kW (power factor of 0.8)—apparent power between 30 and 140 kVA. In total 12 DGs (4 DGs per microgrid) were installed in the cluster mode, such that 6 DGs were assumed to be reserve DGs. They were connected or disconnected depending on the point in time (scenario), whereas the other 6 DGs were operated continuously.

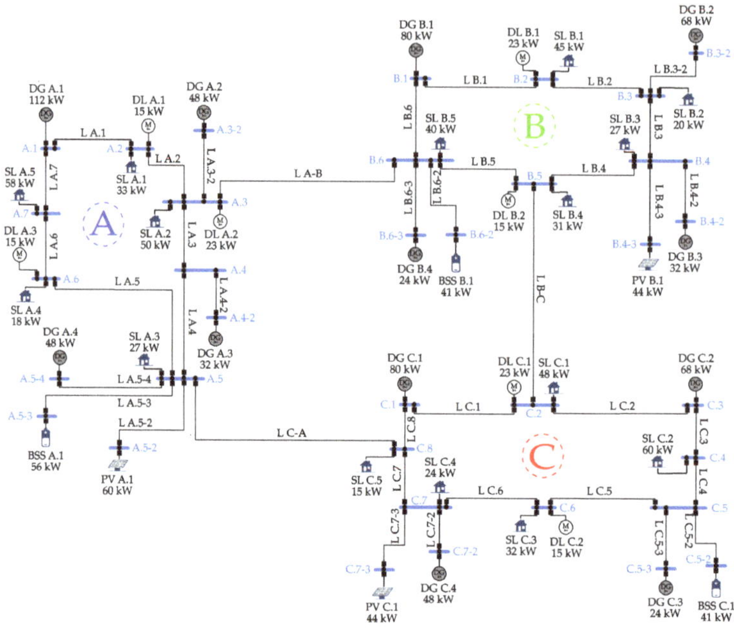

Figure 1. Topology of the investigated microgrids in the cluster mode with diesel generators (DGs), photovoltaics (PV), battery storage systems (BSS), and static and dynamic loads.

PV systems were modelled considering the DC-side. The dimensioning of the main components of PV systems—PV generator, DC-link capacitor, and L filter—has been explained in [17]. In this research work, each microgrid comprises one PV system. The nominal active power of PV A.1 is 60 kW and PV B.1 and C.1 is 44 kW, with a constant power factor of 0.95 (cap.). Each PV generator is made up of 28 modules in series and 6–9 modules in parallel—belonging to Sharp® NQ-R258H [24]. The employed inverters are Sunny TP 60 and Sunny TP Core1 [25,26]. The investigated day was assumed to be a partly sunny day in a tropical country, where the solar irradiance and the PV module temperature is equal to 1000 W/m² and 60 °C, respectively. On the other hand, the DC-side of BSS was not considered in the framework of this research work. The inverter of BSS was connected to an ideal DC voltage source. The nominal power of BSS A.1 and BSS B.1 and C.1 is 56kW and 41 kW, respectively, with a power factor of 0.95 (cap.). The dimensioning of the controllers used in the grid-feeding PV systems and grid-forming BSS will be described in Section 2.3.

In contrast to transmission systems, modelling microgrids' loads exclusively as SL does not lead to convincing results. [27] Off-grid microgrids are usually relatively smaller in size and lack historical measurement data. Hence, dynamic loads were modelled directly in the studied microgrids instead of considering composite dynamic load models [28,29]. In this research work—the most widely employed dynamic loads—double-cage induction motors were taken into account for small-scale agricultural activities, namely water pumps and sugarcane crushers [30,31]. On the other hand, the majority (80%) of the loads were modelled as static and constant impedance loads, which are voltage dependent and frequency independent. According to ISO 8528-5, the modelled static loads belong to class G2 [32,33]. Each microgrid has a total nominal power of loads of roughly 200 kW.

2.2. Overview of Scenarios and Operating Modes

In the framework of this research, five scenarios and three operating modes were analyzed. Depending on the point in time during the partly sunny day, four scenarios were considered as hybrid. The classical scenario was also analyzed to compare the hybrid microgrids with the traditional microgrids comprising DGs only. Table 1 lists the investigated scenarios, where the BSS act as loads and generating units in scenario "storage" and "sun & storage", respectively. The load demand remains the same in all the scenarios, i.e., 80% of the nominal power. Further, the solar irradiance and the PV module temperature were assumed to be constant in the corresponding hybrid scenarios. BSS were disconnected in scenario "sun", whereas PV systems were considered to be inactive in scenario "night".

Table 1. Overview of the five investigated scenarios.

Scenario	Name (Time of Day)	Short Form	Active Equipment
Classical	Classical (-)	Cl	DGs
Hybrid	Storage (12 pm)	St	DGs + PV + BSS (load)
	Sun (2 pm)	Su	DGs + PV
	Sun & Storage (3 pm)	S&S	DGs + PV + BSS (generation)
	Night (8 pm)	Ni	DGs + BSS (generation)

The total installed capacity of DGs, PV, and BSS in each hybrid microgrid is listed in Table 2. The nominal active power of the clustered microgrid lies around 950 kW, whereas the total load demand is about 650 kW. In the classical scenario with DGs only, the installed capacity of the DGs remains unchanged—i.e., equal to 665 kW, which is slightly higher than that of loads. The ratio of the total nominal active power of DGs, PV, and BSS lies around 70:15:15, independent of the operating mode.

Table 2. Nominal active power (in kW) of DGs, PV, BSS, and loads in the hybrid microgrids.

	Generation				Load	DG:PV:BSS
	DG	PV	BSS	Total		
Microgrid A	240	60	56	355	239	67:17:16
Microgrid B	204	44	41	289	201	71:15:14
Microgrid C	220	44	41	305	217	72:15:13
Cluster	**664**	**148**	**138**	**949**	**656**	**70:16:14**

Depending on the time of day (scenario) and degree of coupling (operating mode), reserve DGs were activated or deactivated, such that the engine loading lies between 70% and 80%—see Section 2.1. According to Table 3, the minimum number of active DGs is equal to 2 ("sun & storage" in the island mode), while the maximum number of operating DGs is 12 ("classical" and "storage" in the cluster mode). In this research work, only microgrid C is studied in the islanded (ISD) mode, whereas the coupling of microgrid A and C is investigated in the interconnected (INT) mode. Microgrid A, B, and C are connected with each other in the form of a ring in the clustered (CLS) mode.

Table 3. Overview of the number of active DGs according to scenarios and operating modes.

	Classical	Storage	Sun	Sun & Storage	Night
Island	4	4	3	2	3
Interconnection	8	8	6	4	6
Cluster	12	12	8	6	8

The percentage share of the actual active power of DGs, PV, and BSS in each operating mode is illustrated in Figure 2. Since BSS act as loads in "storage" unlike in "sun & storage", BSS are not shown in the graphs. The highest amount of inverter-based systems is observed in "sun & storage". It should be noted that, the power exchange between the microgrids is negligible in both the interconnected and clustered mode. However, the effect of pooling microgrids with significant power exchange on the system stability can be analyzed in the future research work.

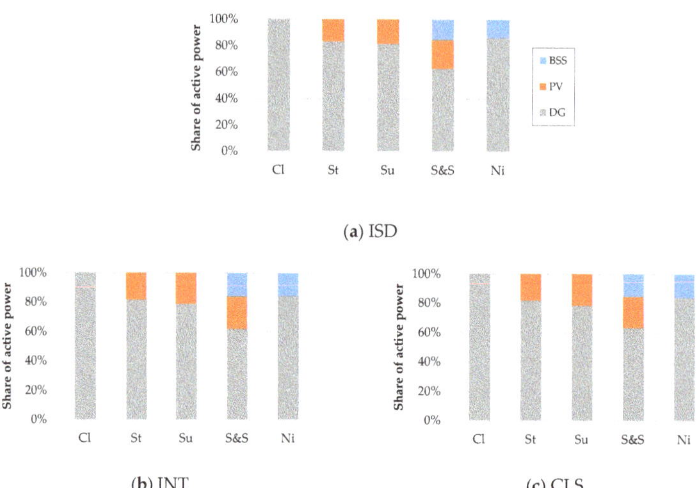

(a) ISD

(b) INT

(c) CLS

Figure 2. Percentage of actual real power of DGs, PV, and BSS in the scenarios: island (**a**), interconnection (**b**), and cluster (**c**) mode.

2.3. Dimensioning of Controllers in Photovoltaics and Battery Storage Systems

2.3.1. Photovoltaics

Figure 3 depicts the block diagram of grid-feeding PV systems with respective controllers. The inner and outer control loop corresponds to the current and DC voltage controller, respectively. The fundamentals of this block diagram have been given in [17,34,35]. In this section, the dimensioning of the PI controller in the current and DC voltage control block will be discussed.

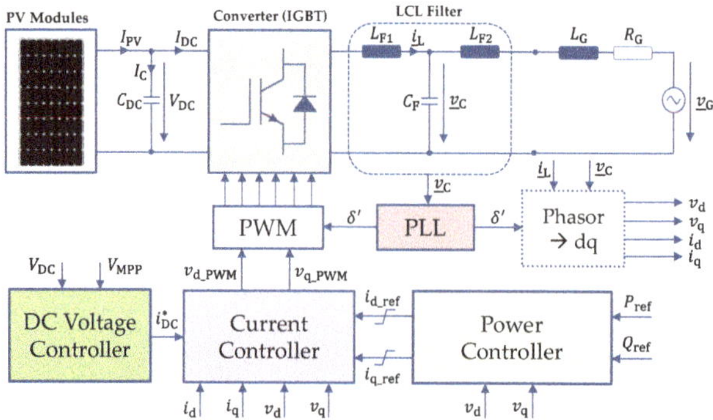

Figure 3. Block diagram of grid-feeding photovoltaics with corresponding controllers, according to [35,36].

Current Controller

The block diagram of the current control loop comprises a PI controller and a controlled plant, which is made up of three first-order delay blocks [35]—cf. Figure 4. In order to reduce the complexity in representing the transfer function of the plant, an L filter was taken into consideration [35]. In principle, modulus optimum can be employed to dimension the parameters of the PI controller (K_{pc} and T_{ic}), since there is no integrator block in the controlled plant [37].

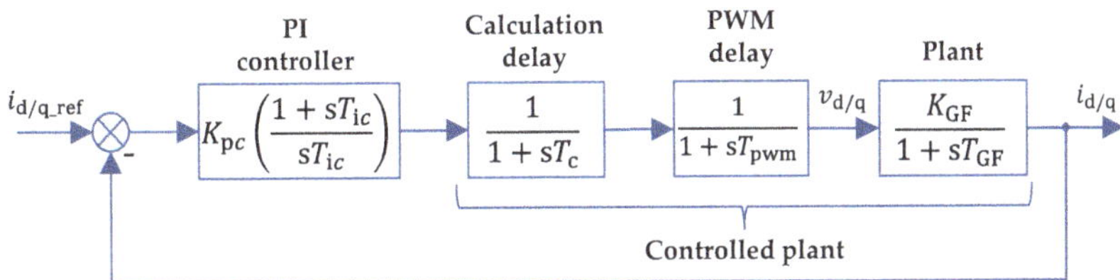

Figure 4. Block diagram of the inner control loop in PV, according to [35].

However, due to the fact that $T_{GF} > 4T_\alpha$, symmetrical optimum was used for dimensioning K_{pc} and T_{ic} [37]. $T_{GF} = L_{GF}/R_{GF}$, where the numerator and denominator are the sum of equivalent short-circuit inductance and resistance at the point of common coupling as well as relatively smaller inductance and resistance of the L filter, respectively. T_α describes the summation of the calculation delay T_c and PWM delay T_{pwm} (set integration time-step in the RMS-Simulations) [35].

The corresponding formula of K_{pc} and T_{ic} is shown in Equation (1), where a represents the variation factor [37]. According to the Bode magnitude and phase plot based on the open-loop transfer function as well as the step response of the closed-loop transfer function for a equal to 5, the control system was assessed to be stable.

$$K_{pc} = \frac{T_{GF}R_{GF}}{aT_\alpha}$$
$$T_{ic} = a^2 T_\alpha \tag{1}$$

Table 4 lists the PI controller parameter values of the three PV systems in the respective microgrids with regard to the three operating modes in scenario "storage". The values corresponding to the other scenarios will not be shown. However, the calculation methodology is identical with that of scenario "storage".

Table 4. Values of K_{pc} and T_{ic} in the current controller of PV of scenario "storage" with respect to the operating modes.

	PV A.1		PV B.1		PV C.1	
	K_{pc}	T_{ic}	K_{pc}	T_{ic}	K_{pc}	T_{ic}
Island	–	–	–	–	0.21	0.004
Interconnection	0.14	0.004	–	–	0.15	0.004
Cluster	0.12	0.004	0.14	0.004	0.14	0.004

DC Voltage Controller

Figure 5 illustrates the block diagram of the outer control loop, which comprises a PI controller and a controlled plant. The latter consists of a calculation delay, inner (current) control loop, gain, and plant. The plant is an integrator block due to the DC-link capacitance. Hence, symmetrical optimum was employed for dimensioning the parameters of the PI controller $K_{pv,DC}$ and $T_{iv,DC}$ [37,38].

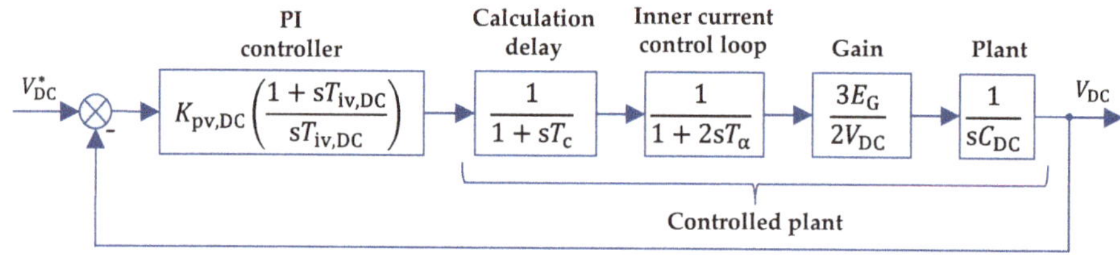

Figure 5. Block diagram of the outer control loop in PV, according to [38].

The corresponding formula of $K_{pv,DC}$ and $T_{iv,DC}$ is shown in Equation (2). C_{DC} and V_{DC} are the DC-link capacitance and its voltage, respectively. E_G represents the peak value of the line-ground voltage, whereas T_β is the summation of the time delays. [37,38]

$$K_{pv,DC} = \frac{2C_{DC}V_{DC}}{3aE_G T_\beta}$$
$$T_{iv,DC} = a^2 T_\beta \tag{2}$$

For a equal to 2.4 the control system was analyzed to be stable based on the Bode magnitude and phase plot as well as the step response. However, the resulting values of $K_{pv,DC}$ and $T_{iv,DC}$ lead to unacceptable oscillations in the investigated microgrids. The calculated values of $K_{pv,DC}$ and $T_{iv,DC}$ were therefore exemplarily divided and multiplied by a factor of 15 and 1000, respectively. The resulting grid simulations (without any fault or disturbance) were acceptable and plausible. The values of $K_{pv,DC}$ and $T_{iv,DC}$ of the PV systems for scenario "storage" with respect to the different operating modes are listed in Table 5.

Table 5. Values of $K_{pv,DC}$ and $T_{iv,DC}$ in the DC voltage controller of PV of scenario "storage" with respect to the operating modes.

	PV A.1		PV B.1		PV C.1	
	$K_{pv,DC}$	$T_{iv,DC}$	$K_{pv,DC}$	$T_{iv,DC}$	$K_{pv,DC}$	$T_{iv,DC}$
Island	-	-	-	-	1.14	2.0
Interconnection	1.27	2.0	-	-	1.14	2.0
Cluster	1.27	2.0	1.14	2.0	1.14	2.0

Phase Locked Loop (PLL)

Grid-feeding PV systems are synchronized with the rest of the microgrid using PLL [35,39]. The block diagram of the PLL comprising a PI controller and a controlled plant is illustrated in Figure 6. In the steady-state, the voltage signal v_q is equal to zero.

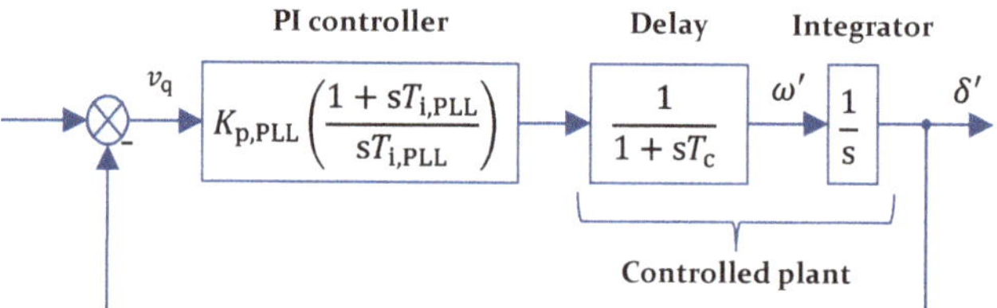

Figure 6. Block diagram of the phase locked loop in PV, according to [39].

As a result of the integrator block in the controlled plant, symmetrical optimum was employed for dimensioning the controller parameters: $K_{p,PLL}$ and $T_{i,PLL}$. The corresponding formulae are shown in Equation (3). The variation factor is given by $b = 1/\omega_{co}T_{pwm}$, where ω_{co} is the crossover frequency. V_n is the nominal grid voltage in pu, i.e., per unit. [39]

$$K_{p,PLL} = \frac{1}{bV_nT_{pwm}}$$
$$T_{i,PLL} = b^2T_{pwm} \tag{3}$$

Table 6 lists the calculated controller parameters of the PLL in PV systems. The values are identical in each PV, since the parameters are independent of the grid connection point.

Table 6. Values of $K_{p,PLL}$ and $T_{i,PLL}$ in the phase locked loop of PV.

	$K_{p,PLL}$	$T_{i,PLL}$
PV A.1		
PV B.1	314	0.0101
PV C.1		

2.3.2. Battery Storage Systems

The block diagram of BSS with grid-forming inverter, which can be operated in parallel mode with other generators, is shown in Figure 7. Hence, the PLL is absent in the control loops. Neglecting the DC-side dynamics of BSS, the control structure comprises of three controllers: current, voltage (AC) and droop controller. The current and voltage controller represents the inner and outer control loop, respectively. [36,40,41] In the droop controller, there exists a correlation between active and reactive power with angle δ' and the input signal v'_{ref} [42]. The details of the grid-forming inverters will not be dealt with in this paper.

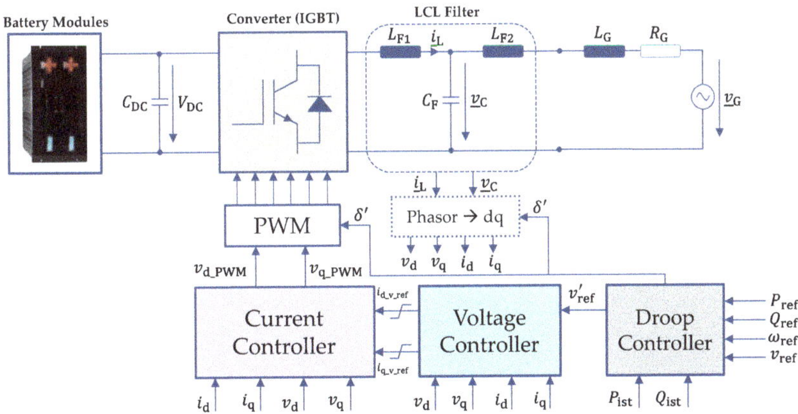

Figure 7. Block diagram of grid-forming BSS with corresponding controllers, according to [36,42].

In the steady-state operating conditions, the grid-forming inverters act as voltage sources. However, the inverters behave as controlled current sources in case of short-circuits. [41] This is due to the fact that, the input signals of the current controller are limited to a maximum value of 1 pu with the help of a current limitation block. Hence, the output current of BSS will be not greater than the nominal value. [41,43] It should be noted that, virtual impedance strategy, described in [36,41], leads to frequent spikes in the output current, which is unfavorable in microgrids in case of short-circuits.

Current Controller

The block diagram of the inner control loop is identical to that of PV (see Figure 4). The Bode plots and the step response of the transfer functions were acceptable. The parameters in the current controller of BSS were calculated using Equation (1). The corresponding values are listed in Table 7.

Table 7. Values of K_{pc} and T_{ic} in the current controller of BSS of scenario "storage" with respect to the operating modes.

	BSS A.1		BSS B.1		BSS C.1	
	K_{pc}	T_{ic}	K_{pc}	T_{ic}	K_{pc}	T_{ic}
Island	-	-	-	-	0.60	0.0012
Interconnection	0.76	0.0012	-	-	0.55	0.0012
Cluster	0.68	0.0012	0.50	0.0012	0.50	0.0012

AC Voltage Controller

Figure 8 shows the block diagram of the voltage controller in grid-forming BSS. The controlled plant comprises, among others, the integrator block due to the filter capacitance. Hence, symmetrical optimum was applied to determine the control parameters: K_{pv} and T_{iv}. [37,41,44] The Bode plots and the step response of the transfer functions for a equal to 4 were acceptable and the control system was accessed to be stable.

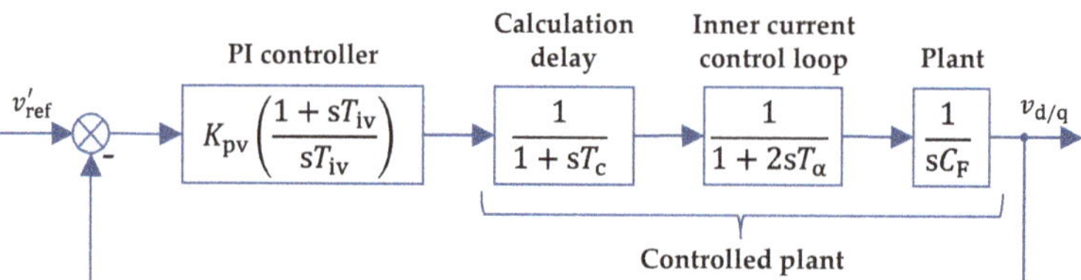

Figure 8. Block diagram of the outer control loop in BSS, according to [44].

The calculation of the control parameters was performed with the help of Equation (4), where C_F and T_γ represents the filter capacitance and the summation of time delays, respectively. [41,44]

$$K_{pv} = \frac{C_F}{aT_\gamma}$$
$$T_{iv} = a^2 T_\gamma \tag{4}$$

Long-lasting oscillations were observed in grid simulations for the calculated values of K_{pv} and T_{iv}—like in PV systems. Hence, K_{pv} and T_{iv} were exemplarily multiplied by a factor of 1.5 and 3000, respectively. Consequently, no oscillations were noticed. Table 8 lists the values of the parameters in the voltage controller of BSS in scenario "storage" regarding the three different operating modes.

Table 8. Values of K_{pv} and T_{iv} in the voltage controller of BSS of scenario "storage" with respect to the operating modes.

	BSS A.1		BSS B.1		BSS C.1	
	K_{pv}	T_{iv}	K_{pv}	T_{iv}	K_{pv}	T_{iv}
Island	-	-	-	-	0.63	8.4
Interconnection	0.91	8.4	-	-	0.63	8.4
Cluster	0.91	8.4	0.63	8.4	0.63	8.4

2.4. Fault Behaviour of Microgrids in the Island Mode

Before performing dynamic short-circuit analysis, microgrid C, along with the other two microgrids, in the islanded mode was verified according to the ISO norm 8528-5 [32,33] with respect to the five scenarios—both in the steady-state and with sudden step-load change. The latter is performed by increasing as well as decreasing the actual power consumption of the static loads—at the studied operating point—up to 40% and 90%, respectively. According to [17], a significant reduction in the solar irradiance of ca. 55% within 2.5 s does not have negative consequences in the analyzed clustered microgrid model comprising DG and PV, which is similar to the microgrid model in scenario "sun".

The three-phase fault location in microgrid C in scenario "storage" (representing all the installed grid equipment) is shown in Figure 9. Since the short-circuit location is very close to a DG and a grid-forming BSS, the fault is assumed to be critical. The fault clearance is achieved by tripping the affected line only, i.e., L C.5, with the help of differential protection. The clearing time is identical in each scenario and corresponds to the minimum value of the CCT among the scenarios. It should be noted that, in this research work the resistance of the fault was assumed to be 0.02 Ω.

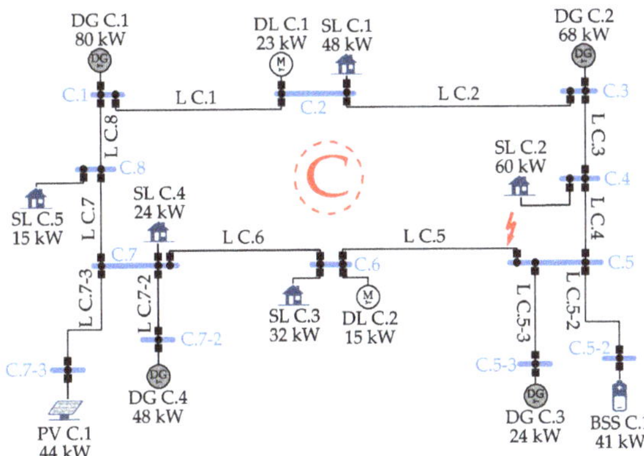

Figure 9. Microgrid C in the islanded mode (scenario "storage") with the investigated fault location.

If the bus voltage drops below 0.15 pu, inverter-based systems can be disconnected from the system according to [45]. Further, DGs can be tripped in case of a voltage drop of 0.3 pu or less [45]. However, DGs, PV, and BSS were purposely not disconnected in this research work—irrespective of the bus voltage. On the other hand, induction motors, i.e., dynamic loads, were tripped 40 ms after the fault occurrence using undervoltage protection [46]. Soon after the fault incident, induction motors will provide subtransient short-circuit currents (similar to DGs), which are beneficial also in microgrids. However, the reactive power demand of induction motors during the fault-on and the post-fault time period can be very critical in microgrids and also can lead to voltage or system collapse.

The short-circuit behavior of the classical and the hybrid microgrids in the islanded mode was studied—before performing qualitative and quantitative TSA—not only at the system level, but also at the equipment level. First and foremost, the measured voltage and frequency on the buses were compared regarding the scenarios. The considered variables of DGs are relative rotor angle (i.e., rotor angle in the center of inertia frame, COI), actual rotor angular frequency, and relative rotor angular frequency deviation. Furthermore, electrical and mechanical torque of DGs were also analyzed taking "storage" and "sun & storage" (comprising DGs, PV, and BSS) into account.

The active and reactive power as well as the output current of DGs, PV, and BSS were also investigated. The profiles of the static and dynamic loads will not be shown in this paper. By analyzing the system and the equipment behavior under short-circuit conditions, the modelling of the classical and hybrid microgrids were verified and also compared with each other.

2.5. Effect of Pooling Microgrids on the System Stability

The influence of coupling microgrids on the system stability should be analyzed not only qualitatively, but also quantitatively. It has been already mentioned that only microgrid C was studied in the islanded mode. The interconnected mode corresponds to microgrid A and C, whereas the ring topology of microgrid A, B, and C was considered in the clustered mode. It should be noted that the power exchange between the microgrids in the coupled operating modes was negligible. These three operating modes along with five scenarios lead to 15 different grid models, in other words 15 different cases. Henceforth, other combinations of the microgrids were not considered in this research work.

Firstly, the *CCT* of microgrids with respect to the 15 cases were calculated by performing RMS-simulations (fault on lines) with a fault resistance of 0.02 Ω. This allows to compare the *CCT* values with respect to the scenarios and operating modes, so that the sensitivity of the microgrids can be assessed in terms of transient stability. The methodology of the calculation of the *CCT* has been given in [20]. The *CCT* values were categorized according to Table 9. The minimum feasible clearing time (differential protection) in the studied microgrids is approximately 60 ms [47–50].

Table 9. Classification of the *CCT* in the studied microgrids.

CCT Range	Risk Level
less than 60 ms	extreme
60–100 ms	very high
101–150 ms	high
151–200 ms	medium
201–300 ms	low
greater than 300 ms	very low

Secondly, the behavior of microgrids in the case of the three-phase fault in microgrid C (illustrated in Figure 9) was investigated in detail. In contrast to the analysis in the islanded mode, the individual variables and parameters of the grid equipment will not be shown. In this research, work voltage and frequency stability will be analyzed along with rotor angle stability.

2.6. Quantitative Transient Stability Assessment

In the 1990s, five hybrid methods were proposed for quantitative TSA in transmission systems combining the advantages of TDS and TEF [51–55]. These techniques have been discussed in [19], which cannot be employed in microgrids comprising engine-driven synchronous generators with relatively fast reacting speed governor. Hence, a new hybrid method has been proposed in [19], which has been improved in [20]. This hybrid technique valid for microgrids is called "the micro-hybrid method". Figure 10 lists the hybrid methods to quantitatively access the transient stability of electrical energy systems.

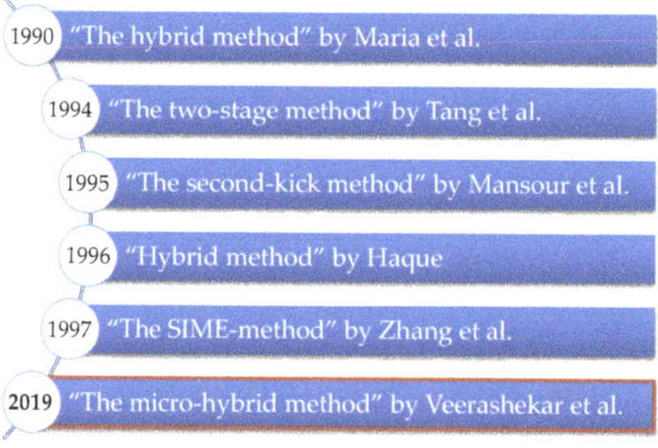

Figure 10. Overview of the hybrid methods for quantitative TSA.

Using the micro-hybrid method, three stability related terms can be determined, namely system stability degree (SSD) as well as stability reserve degree (SRD) and participation factor (PF) of individual synchronous generators. The corresponding results of the TDS can be analyzed further with the help of the TEF given in Equation (5). E, E_{kin}, and E_{pot} represents the total energy, kinetic energy (KE), and potential energy (PE), respectively. The detailed derivation, based on [22,56], can be found in [19,20].

$$E = \sum_{i=1}^{n} H_i \omega_n \Delta\omega_{iCOI,pu}^2 - \sum_{i=1}^{n} \int_{\delta_{iCOI_0}}^{\delta_{iCOI_1}} \tau_{iCOI\Sigma,pu} d\delta_{iCOI}$$

$$E_{kin} = \sum_{i=1}^{n} H_i \omega_n \Delta\omega_{iCOI,pu}^2 \quad \text{and} \quad E_{pot} = \sum_{i=1}^{n} \int_{\delta_{iCOI_0}}^{\delta_{iCOI_1}} \tau_{iCOI\Sigma,pu} d\delta_{iCOI}$$

(5)

H_i, ω_n, and $\Delta\omega_{iCOI,pu}$ are the inertia constant of the i-th synchronous generator, nominal rotor angular velocity, and change in the rotor angular velocity in the COI frame, respectively. On the other hand, δ_{iCOI_0} and δ_{iCOI_1} represents the pre-fault and post-fault rotor angle of the i-th synchronous generator, respectively. Further, the rotor angle of each generator in the COI frame is given by δ_{iCOI}. $\tau_{iCOI\Sigma,pu}$ describes the equivalent torque of the individual generators in the COI frame (see Equation (6)).

$$2H_i \frac{d\Delta\omega_{iCOI,pu}}{dt} = \tau_{mi,pu} - \tau_{eli,pu} - \frac{H_i}{H_\Sigma} \tau_{COI,pu} = \tau_{iCOI\Sigma,pu}$$

(6)

$\tau_{mi,pu}$ and $\tau_{ei,pu}$ represent the mechanical and electrical torque of the i-th synchronous generator, respectively. The summation of the inertia constant of each generator is given by H_Σ. Equation (7) describes the equivalent torque of the COI reference machine $\tau_{COI,pu}$, where $\Delta\omega_{COI,pu}$ is the change in the rotor angular velocity of the COI reference machine.

$$\tau_{COI,pu} = 2H_\Sigma \frac{d\Delta\omega_{COI,pu}}{dt} = \sum_i \tau_{mi,pu} - \tau_{eli,pu}$$

(7)

The methodology of the calculation of the KE and PE has been explained in detail in [19,20]. Based on the energy values of all the generators until the point of time corresponding to the end of the forward swing of the critical generator, the critical energy E_{cr} and clearing energy E_{cl} can be calculated [19,20]. E_{cr} and E_{cl} correspond to the CCT and (stable) fault clearing time, respectively. Due to a large number of investigated cases (15 in total) regarding operating modes and scenarios, only the SSD has been taken into account in this research work, which is given by Equation (8). SSD defines the percentage margin of a microgrid towards instability boundary for a (stable) fault clearing time, e.g., SSD of 75% indicates that the microgrid has a reserve of 75% with respect to the rotor angle stability. In other words, the microgrid has lost 25% of its stability reserve.

$$SSD = \frac{E_{cr} - E_{cl}}{E_{cr}} \cdot 100 \%$$

(8)

In [20], the threshold value of the SSD to determine the corrected critical clearing time (CCCT) was discussed. Since a microgrid cannot be modelled without any inaccuracies and assumptions, a SSD-threshold of 10% can be assumed. As a result, the new CCT, i.e., $CCCT$, can be considered as the maximum allowable time to clear the fault.

Furthermore, the critical energy of the system E_{cr} and the critical machine E_{cr_cm} were analyzed taking the CCT into consideration. In a one-machine infinite bus system, the CCT is inversely proportional to the critical energy. The correlation between these critical energies and the CCT was studied regarding the operating modes and scenarios. Similarly, the effect of the fault location on the critical energy was analyzed in the clustered mode with respect to the three-phase fault on both ends of L C.5, i.e., very close to busbar (BB) C.5 and C.6.

3. Results and Discussion

The results of the performed simulations will be presented and discussed in this section. Firstly, the *CCT* profile of the different operating modes and scenarios will be described in Section 3.1. The next section (Section 3.2) deals with the system behavior under fault conditions in island mode. Consequently, the influence of coupling microgrids on the voltage, the frequency, and the rotor angle stability will be qualitatively analyzed in Section 3.3. Furthermore, the profiles of the SSD versus clearing times—based on the quantitative TSA using the micro-hybrid method—will be discussed in Section 3.4 with respect to the operating modes and scenarios. The last two sections (Sections 3.5 and 3.6) deal with the critical energy of the system and critical machine.

3.1. Critical Clearing Times

3.1.1. Comparison of Scenarios

In this section, the percentage distribution of *CCT* values of the various scenarios with respect to the islanded mode (microgrid C), interconnected mode (microgrid A and C), and clustered mode (microgrid A, B, and C) will be discussed—see Figure 11. It has been mentioned in Section 2.3 that faults were simulated on lines within each microgrid except tie-lines. The categorization and comparison of the *CCT* values is based on Table 9.

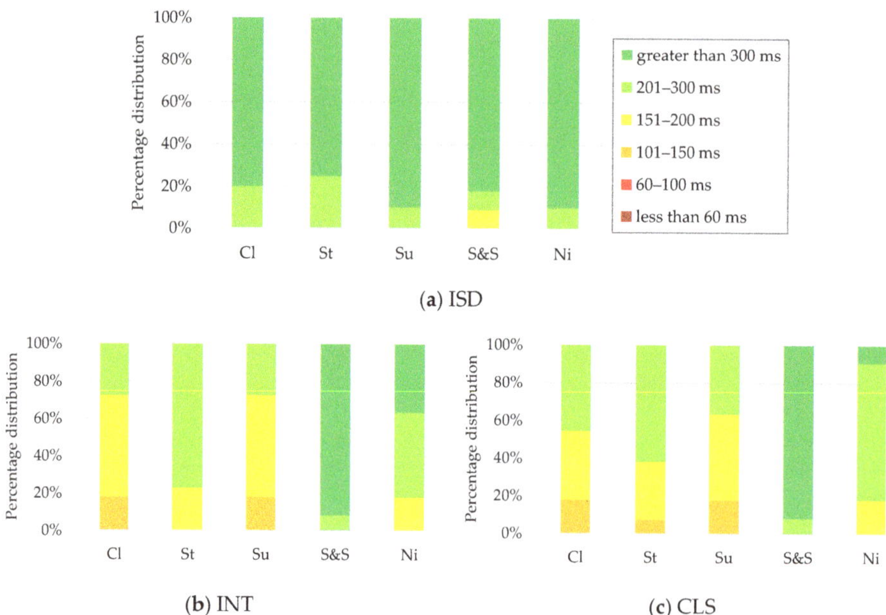

Figure 11. Percentage distribution of *CCT* with respect to different scenarios in: island (**a**), interconnection (**b**), and cluster (**c**) mode.

Island

Figure 11a corresponds to the islanded mode (i.e., microgrid C). The risk level of the majority —approximately 80%—of the *CCT* values in each scenario is very low (*CCT* > 300 ms). In scenario "sun & storage" with DGs, PV, and BSS (acting as generating units), several *CCT* values correspond to the risk level belonging to medium, which is considered to not be critical. In the following subsections, the *CCT* profile of the interconnected and clustered mode will be discussed.

Interconnection

According to Figure 11b, the risk level corresponding to extreme ($CCT < 60$ ms) and very high (CCT between 60–100 ms) has not been observed in any of the scenarios in the interconnected mode. However, expect in scenario "sun & storage" the risk level varies between high and very low. The risk of one of the DGs losing synchronism with the system is higher in the interconnected mode.

Cluster

Figure 11c shows that the CCT profile of each scenario is almost identical to that of the interconnected mode—cf. Figure 11b. Scenario "sun & storage" can be highly recommended in the interconnected and clustered mode, whereas any scenario can be preferred in the islanded mode.

Considering a three-phase fault on line L C.5 (see Figure 9), an in-depth analysis of the effect of coupling off-grid microgrids on the system stability will be performed in Section 3.3. In the next section, a quantitative comparison of the CCT profiles will be performed taking only microgrid C into account.

3.1.2. Comparison of Scenarios with Respect to Operating Modes

Interconnection vs. Island

With the help of Equation (9), the CCT of 11 overhead lines in microgrid C has been quantitatively compared regarding the interconnected and islanded mode. ΔCCT, CCT_{INT}, and CCT_{ISD} correspond to the relative CCT difference in percentage, absolute CCT value (ms) in the interconnected mode and islanded mode, respectively.

$$\Delta CCT = \frac{CCT_{INT} - CCT_{ISD}}{CCT_{ISD}} \cdot 100\,\% \tag{9}$$

It can be observed in Figure 12 that except scenario "sun & storage" a negative difference is characterized in each scenario. The maximum positive and negative difference lies around +40% and −80%, respectively. The fault on line L C.5—close to a DG and BSS, and with a CCT change of roughly +10% and −70%—will be investigated in the further sections.

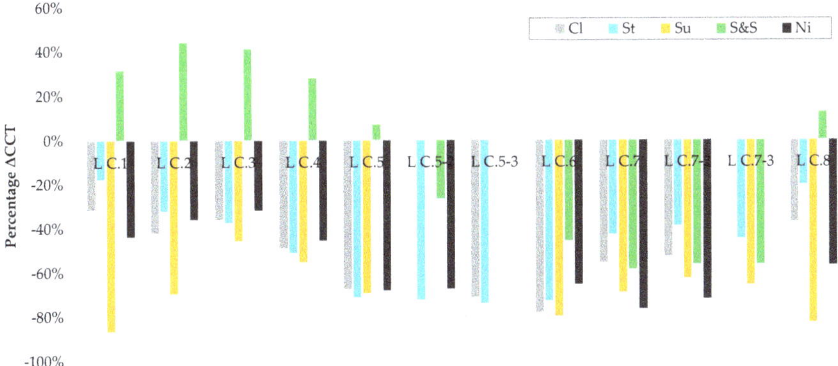

Figure 12. Relative percentage difference of CCT of microgrid C in interconnection and island mode.

Cluster vs. Island

Since the *CCT* profiles are nearly similar in the coupled microgrids, the comparison has been performed—between the clustered and islanded mode—using Equation (10). According to Figure 13, the ΔCCT values of microgrid C remain almost identical to the values illustrated in Figure 12.

$$\Delta CCT = \frac{CCT_{CLS} - CCT_{ISD}}{CCT_{ISD}} \cdot 100\,\% \tag{10}$$

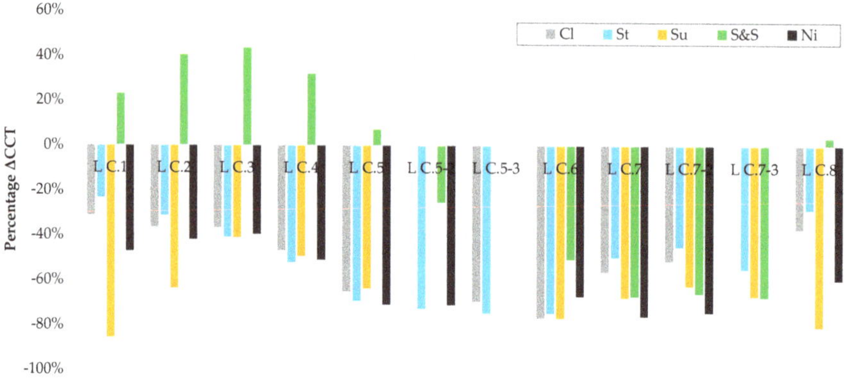

Figure 13. Relative percentage difference of CCT of microgrid C in cluster and island mode.

3.1.3. Comparison of Operating Modes

For the TSA in the grid planning phase and/or grid operation, not only different points of time (i.e., scenarios) are essential, but also different operating modes. Hence, the *CCT* profiles will be presented according to the scenarios—taking solely microgrid C into consideration—so that different operating modes can be compared directly.

Classical

Figure 14a represents the percentage distribution of *CCT* in the classical scenario of microgrid C regarding the islanded, interconnected, and clustered mode. The other profiles in Figure 14 correspond to the hybrid scenarios.

In case of microgrids comprising DGs only, the choice of the optimal operating mode can be easily made based on Figure 14a: island over pooling. However, coupling the microgrid with neighboring microgrids is not characterized by unacceptable *CCT* values.

Hybrid

In scenarios "storage" and "sun", i.e., Figure 14b,c, island operation can be preferred to coupling modes for microgrid C. However, cluster mode can be favored in scenario "sun & storage" (cf. Figure 14d). The choice can be made between the islanded and interconnected mode in scenario "night" (cf. Figure 14e) from the microgrid C's point of view.

In general, scenario "sun & storage" can be preferred to other scenarios in the interconnected and clustered mode. On the other hand, the selection of scenarios is not critical in the islanded mode. Furthermore, coupling classical or hybrid microgrids does not pose serious problems regarding transient stability, whereas the stability risk in the pooled modes is higher as against the islanded mode. However, the effect of extended coupling of microgrids (with different topologies) on the CCT values needs further investigations.

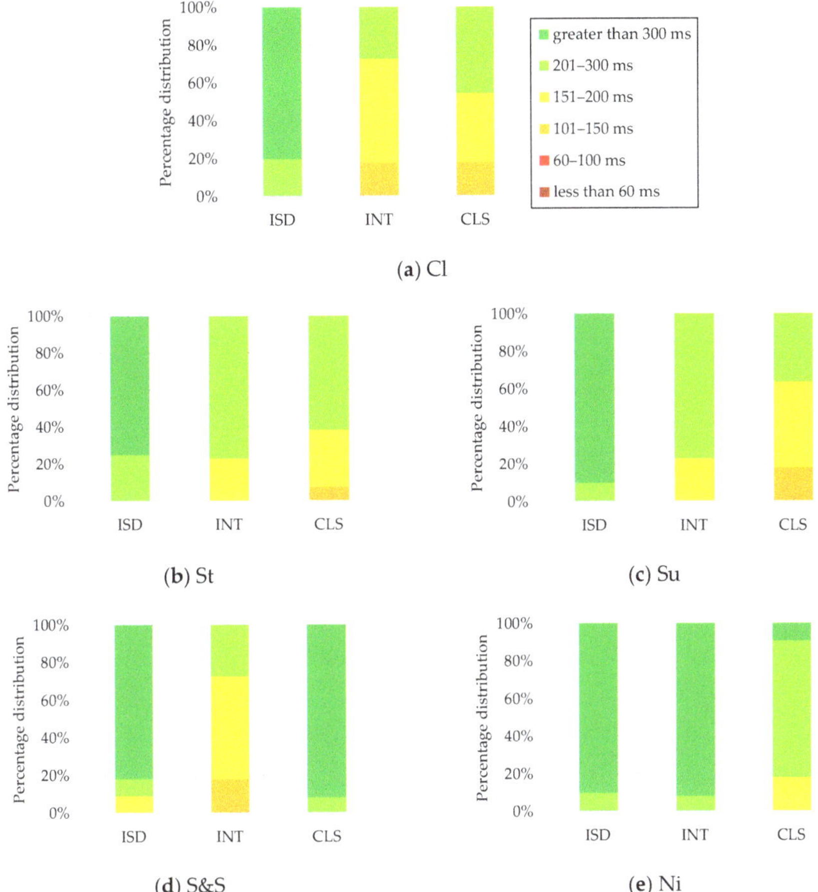

Figure 14. Percentage distribution of *CCT* in microgrid C with respect to different operating modes in: classical (**a**), storage (**b**), sun (**c**), sun and storage (**d**), and night (**e**) scenario.

3.2. Fault Behaviour of Microgrid C in Island Mode

Before investigating the fault behavior of the coupled classical and hybrid microgrids, the response of microgrid C at both system and equipment level, will be studied. Further, the microgrid modelling can be verified by analyzing the system and equipment profiles in the pre-fault, fault-on, and post-fault period.

The fault clearing time was chosen such that it corresponds to the minimum *CCT* of the scenarios. The *CCT*s of the scenarios in the ascending order is as follows: sun and storage (531 ms), classical (588 ms), sun (667 ms), storage (793 ms), and night (1058 ms). Hence, the fault on line L C.5 (see Figure 9) is cleared after 531 ms.

3.2.1. Bus Voltage and Frequency

Figure 15 shows the voltage profiles of the busbars (BB) in microgrid C for different scenarios. Due to the relatively smaller dimension of microgrid C, voltage at all the busbars drops significantly soon after the fault incident. It should be noted that induction motors were disconnected 40 ms after the fault occurrence. The resulting voltage fluctuation (marginal increase) is negligibly small. During the fault-on period, i.e., until the fault clearance, the corresponding bus voltages are almost similar in each scenario. Further, the post-fault voltage recovery is relatively quick—within about 1 s—in both the classical and hybrid scenarios. The influence of the employment of PV and BSS (hybrid scenarios) is not significant in the islanded mode of the studied microgrid C.

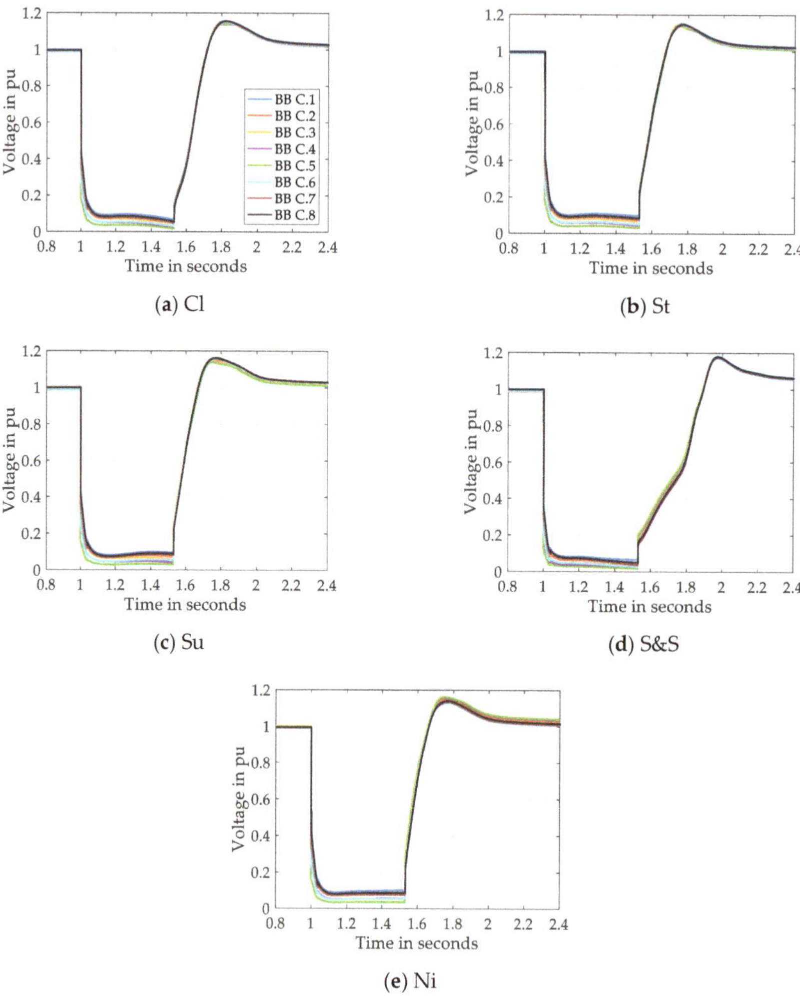

Figure 15. Voltage of busbars of microgrid C (island mode) in: classical (**a**), storage (**b**), sun (**c**), sun and storage (**d**), and night (**e**) scenario.

The observed frequency at the busbars of microgrid C of each scenario is illustrated in Figure 16. The minimum and maximum frequency in every scenario, except "sun & storage" is 45 Hz and 55 Hz, respectively—i.e., ±10% of the nominal value. The frequency range in "sun & storage" with 2 DGs is between 43 Hz and 57 Hz (±14%). The frequency drops significantly in the subtransient phase due to the provision of very high short-circuit current/power by the DGs. The frequency fluctuation due to the disconnection of induction motors 40 ms after the fault incident is noticeable, however not critical. During fault-on period, the frequency values lie in the overfrequency range (greater than 50 Hz) as a result of the very fast reaction of the speed governor of the DGs. This will be discussed more in the following subsections.

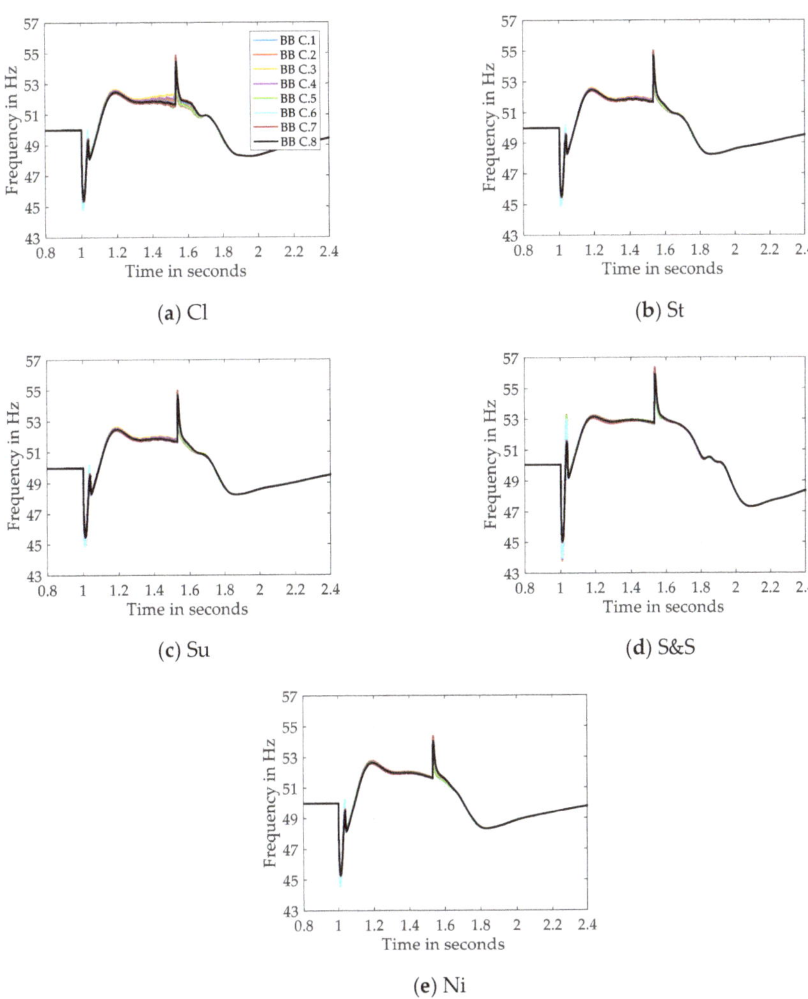

Figure 16. Frequency of busbars of microgrid C (island mode) in: classical (**a**), storage (**b**), sun (**c**), sun and storage (**d**) and night (**e**) scenario.

A sharp frequency increase is noticed in each scenario soon after the fault clearance, which is as a result of the positive difference between the total generation and load in the microgrid. The frequency recovery in the post-fault period lasts only about 1 s. Due to the disconnection of the induction motors and absence of secondary frequency control (in principle not necessary for stability analyses) in the studied microgrids, the steady-state frequency in the post-fault period is approximately equal to 50 Hz and not exactly 50 Hz.

3.2.2. Relative Rotor Angle, Actual Rotor Angular Frequency and Relative Rotor Angular Frequency Deviation of DGs

Relative Rotor Angle

The rotor angle of the DGs in microgrid C of each scenario in the COI reference frame is depicted in Figure 17. The list of the active DGs in every scenario is shown in Table 3. The critical machine is DG C.2 in all the scenarios except in scenario "sun & storage", where DG C.1 loses synchronism. Even though scenario "sun & storage" has the minimum *CCT* (corresponding to the clearing time), the observed value of the rotor angle of the critical machine is slightly higher than 80°. In general, the end of the forward and backswing of the DGs in each scenario occurs almost at the same point of time.

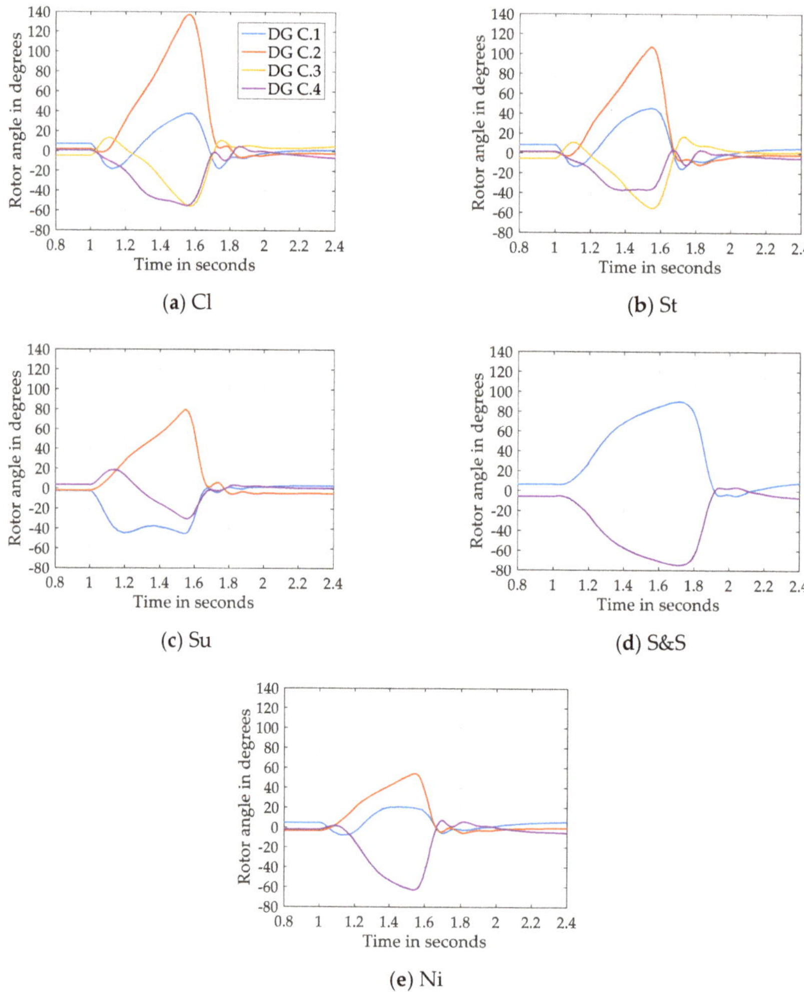

Figure 17. Relative rotor angle of DGs of microgrid C (island mode) in: classical (**a**), storage (**b**), sun (**c**), sun and storage (**d**) and night (**e**) scenario.

Actual Rotor Angular Frequency

Figure 18 illustrates the actual rotor angular frequency of the DGs in microgrid C of each scenario. Soon after the fault incident, the speed of the rotors drops due to the provision of (subtransient) short-circuit currents by DGs. As a result of the decrease in the magnitude of the short-circuit currents and increase in the mechanical moment of the speed governor during the fault-on period, the rotor angular frequency of the DGs increases. The DGs regain synchronism within roughly 500 ms after the fault clearance. In scenarios with the maximum number of DGs, i.e., classical and storage, relatively larger oscillations can be noticed.

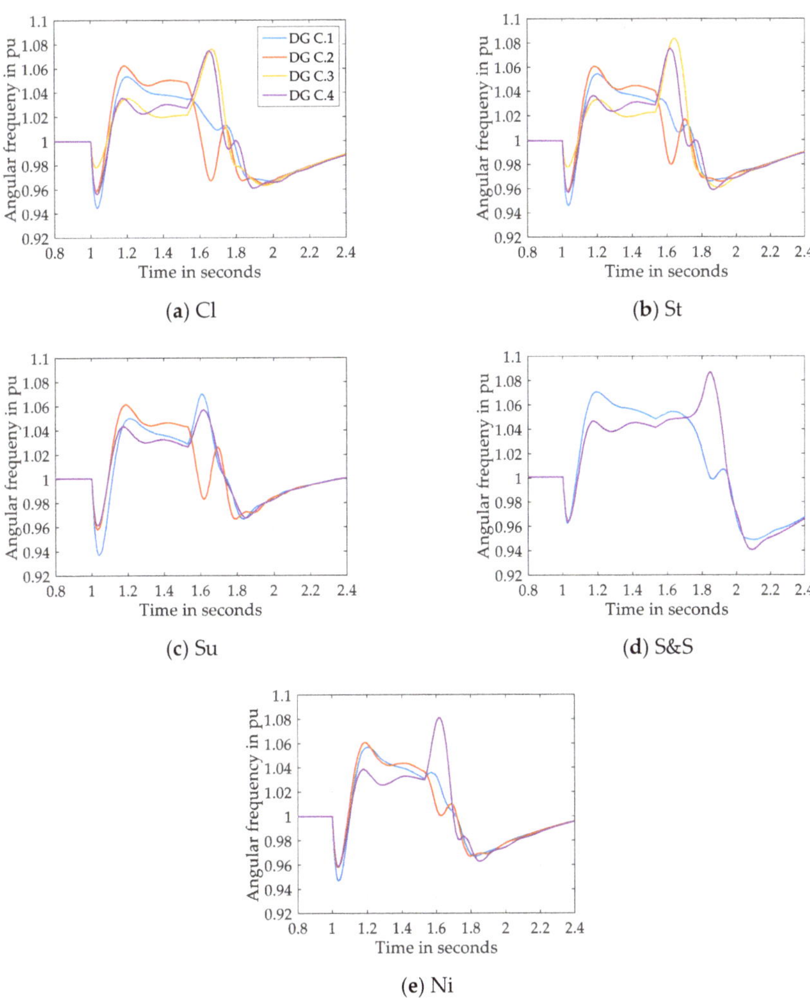

Figure 18. Actual rotor angular frequency of DGs of microgrid C (island mode) in: classical (**a**), storage (**b**), sun (**c**), sun and storage (**d**), and night (**e**) scenario.

Relative Rotor Angular Frequency Deviation

The relative rotor angle of the DGs is directly related to their relative rotor angular frequency deviation (see Figure 19), which is also represented in the COI reference frame. It can be observed that the DGs with the rotor acceleration during the fault-on period exhibit deceleration soon after the fault clearance, and vice-versa. The maximum relative rotor angle (see Figure 17) is noticed just after clearing the fault, where the rotor angle deviation is predominantly due to the rotor acceleration. The rotor angle increases further—only slightly—after tripping the faulty line as a result of the rotor deceleration. The negative relative angular frequency of the critical machine corresponds to the sharp decrease in the rotor angle.

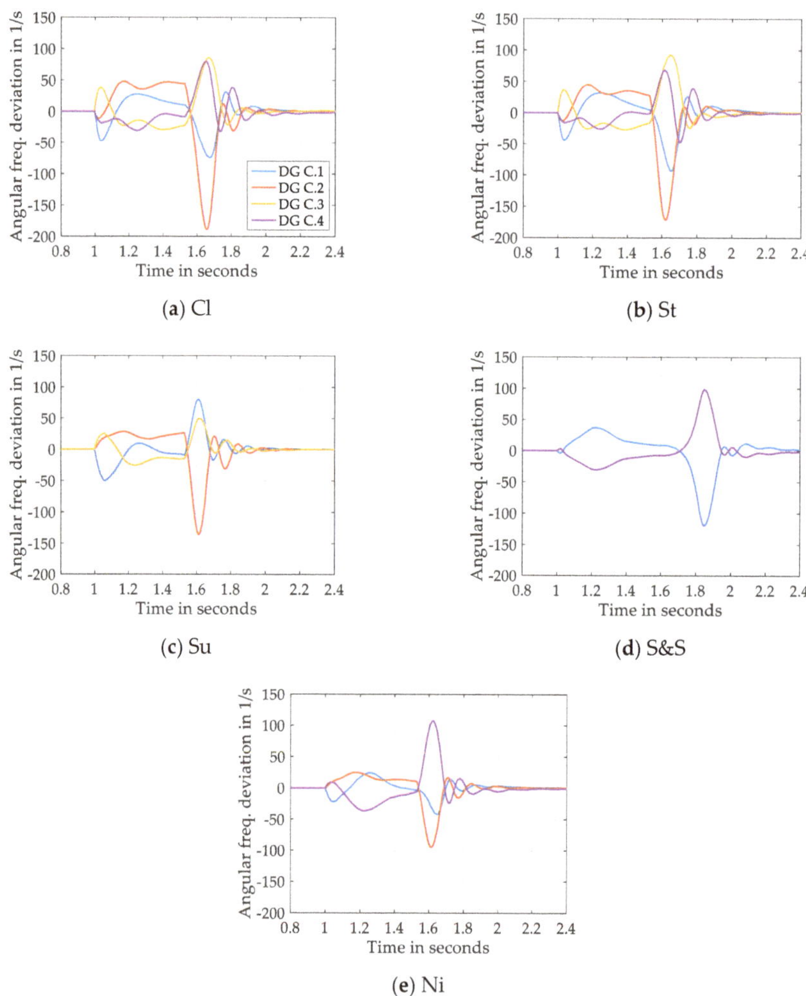

Figure 19. Relative rotor angular frequency deviation of DGs of microgrid C (island mode) in: classical (**a**), storage (**b**), sun (**c**), sun and storage (**d**), and night (**e**) scenario.

- Electrical and Mechanical Torque of DGs

In the following subsections, the scenarios "storage" and "sun & storage" will be considered, where the BSS acts as a load and a generation unit, respectively. Further, these two scenarios—representing DGs, PV, and BSS being active—correspond to the maximum and minimum number of the DGs. Figure 20 illustrates the electrical (in principle, electromagnetic) torque of the DGs in microgrid C. The sharp increase of the torque—soon after the fault incident—corresponds to the subtransient short-circuit current, which will be shown at the end of this section. Since the voltage drop in scenario "sun & storage" is slightly higher than in storage, the corresponding maximum torque of DG C.1 in scenario "sun & storage" is 3.5 pu, whereas the value is 2.6 pu in scenario "storage". As a result of the (sustained) fault until the line tripping, a relatively smaller short-circuit current leads to an electrical torque of less magnitude.

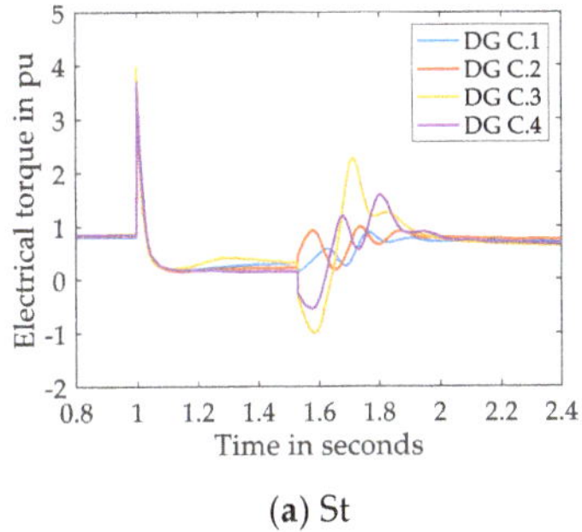
(a) St

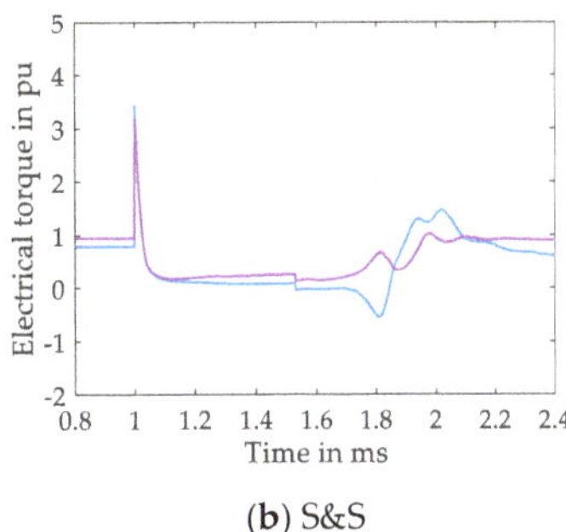
(b) S&S

Figure 20. Electrical torque of DGs of microgrid C (island mode) in: storage (a) and sun and storage (b) scenario.

Several tens of milliseconds after the fault occurrence, the mechanical torque of the DGs will be increased by the speed governor DEGOV1 of the DGs due to the drop in the frequency (cf. Figure 21). In case of transmission systems, the speed governor in synchronous generators during fault-on period does not change the mechanical torque due to relatively large time constants (few seconds) [22]. However, the set point of torque/power can be changed very quickly in engine-driven generators. The maximum and the minimum limit of torque in DEGOV1 corresponding to 1 pu and 0 pu, respectively, can be observed in the profiles. Once the short-circuit current (electrical torque) of the DGs decreases, the difference between the mechanical and electromagnetic torque becomes positive, which causes an increase in the rotor speed. Consequently, the mechanical torque is reduced by DEGOV1 from 1 pu to almost 0 pu. After the fault clearance, the torque will be increased such that the speed deviation becomes zero.

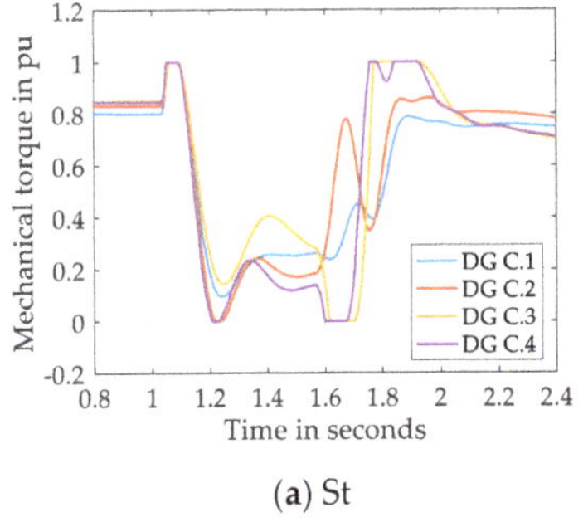
(a) St

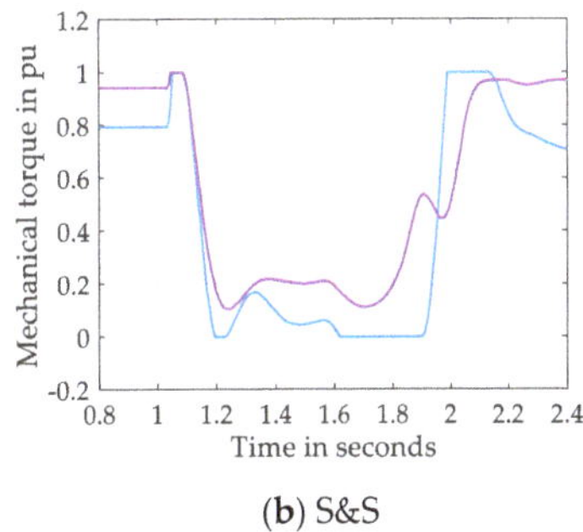
(b) S&S

Figure 21. Mechanical torque of DGs of microgrid C (island mode) in: storage (a) and sun and storage (b) scenario.

- Active and Reactive Power as well as Output Current of DGs, PV, and BSS

  i.   Active Power

Since the voltage is nearly 0.1 pu during fault-on period, the electrical power is close to 0 pu. However, due to the residual voltage and very high short-circuit current in the subtransient phase the active power is significantly high—cf. Figure 22.

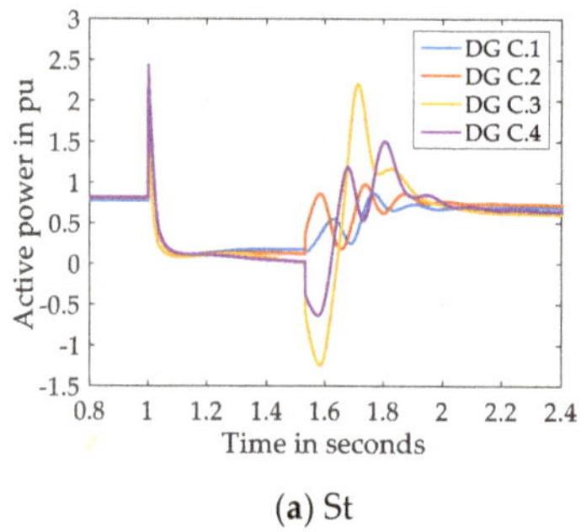

(a) St

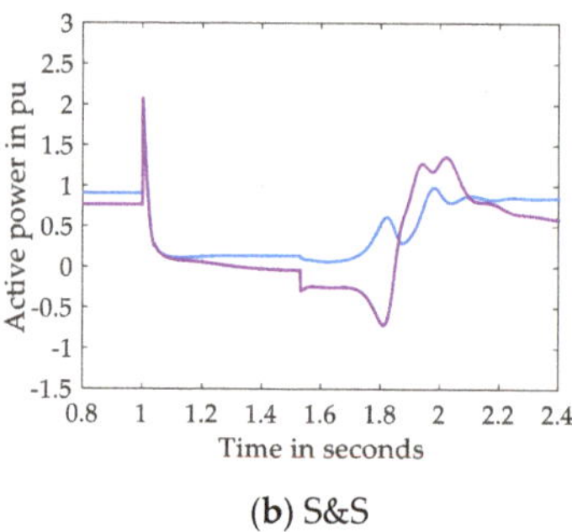

(b) S&S

Figure 22. Real power of DGs of microgrid C (island mode) in: storage (a) and sun and storage (b) scenario.

The active power of the grid-feeding PV in scenario "storage" and "sun & storage" is shown in Figure 23. Unlike DGs the provision of the short-circuit current in PV is limited. Hence, the profile of the active power is significantly dependent on the terminal voltage. Due to the relatively slower (post-fault) voltage recovery in sun and storage, it takes slightly longer to reach the pre-fault value of the active power.

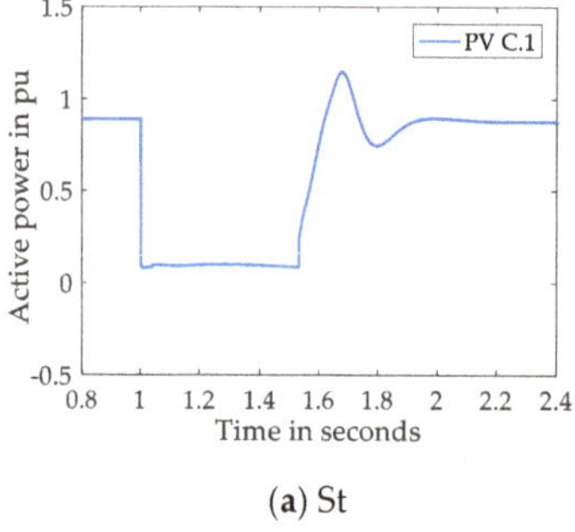

(a) St

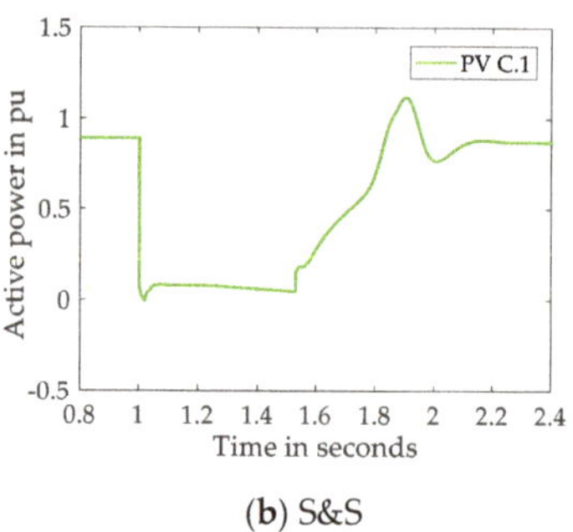

(b) S&S

Figure 23. Real power of PV of microgrid C (island mode) in: storage (a) and sun and storage (b) scenario.

Figure 24 illustrates the active power of the grid-forming BSS, where it operates as a load and generating unit in storage and sun and storage, respectively. Even though BSS acts as a voltage source, the behavior during the fault-on period is similar to that of grid-feeding inverters, i.e., current sources. According to Section 2.3.2, the short-circuit current of BSS, like in PV, is restricted to 1 pu. Henceforth, the active power profiles are similar to the corresponding (terminal) voltage profiles.

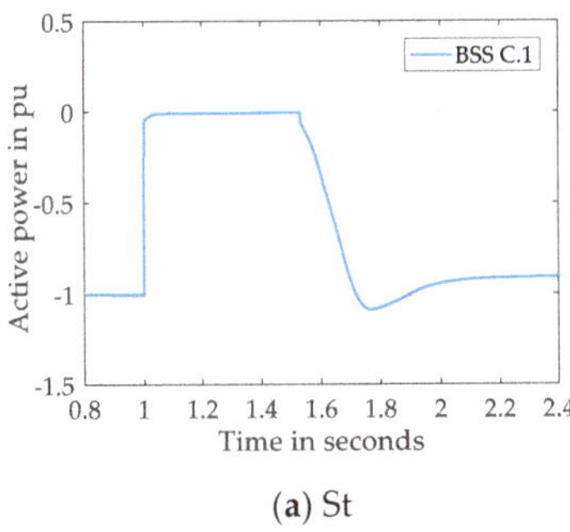
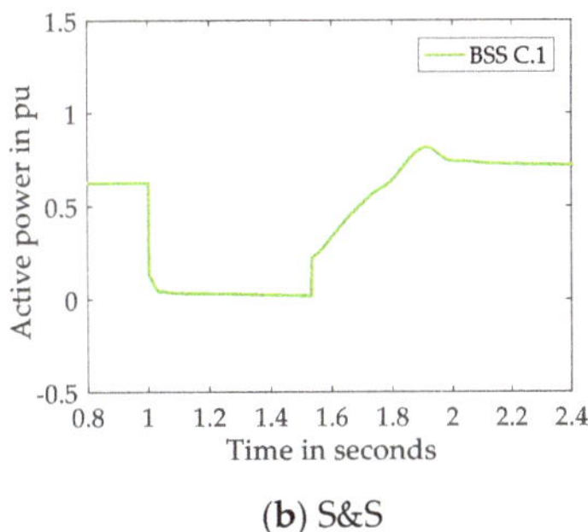

(a) St   (b) S&S

Figure 24. Real power of BSS of microgrid C (island mode) in: storage (**a**) and sun and storage (**b**) scenario.

ii.   Reactive Power

In Figure 25 the reactive power of the DGs in microgrid C is depicted. Due to the resistive-inductive (static and dynamic) loads, the DGs provide capacitive reactive power. Similar to the active power, the reactive power depends on the terminal voltage and output current of the DGs. Thus, the reactive power is relatively less during the fault-on period. After the fault clearance, the DGs close to the fault location act as inductive loads, since the corresponding terminal voltages are relatively less than the terminal voltages of the DGs that are relatively far away from the short-circuit location. Due to the disconnection of inductive motors, the power set-points of the DGs should be adjusted once the post-fault steady-state has been reached.

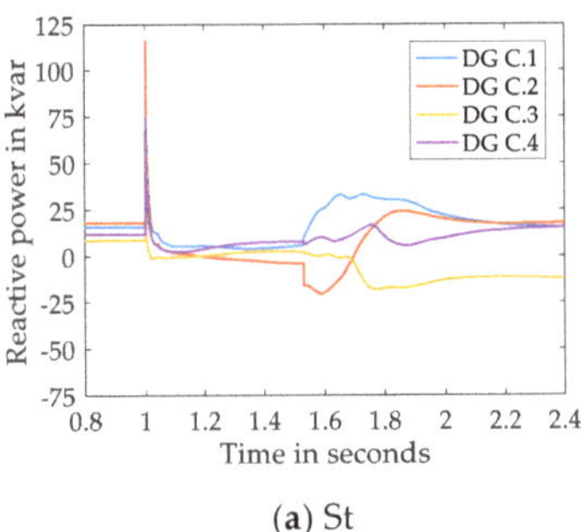
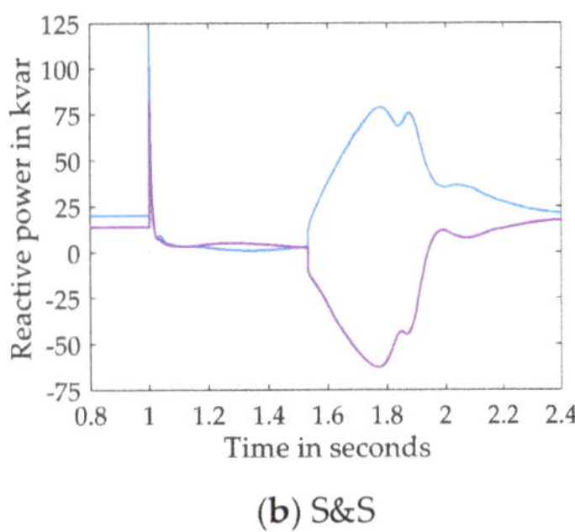

(a) St   (b) S&S

Figure 25. Reactive power of DGs of microgrid C (island mode) in: storage (**a**) and sun and storage (**b**) scenario.

The grid-feeding PV system—along with DGs and BSS—provides capacitive reactive power in the pre-fault period—see Figure 26. In contrast to the DGs, the reactive power of PV can be controlled during fault-on period. The output signals of the power control block (cf. Figure 3) along with the current limiter block are i_{d_ref} and i_{q_ref}. Due to the disconnection of induction motors 40 ms after the fault incident, the (inductive) reactive power demand in the microgrid is relatively not high during fault-on and especially post-fault period. Further, by giving 100% priority to active power (i.e., no reactive power provision) under short-circuit conditions, a positive effect on the transient stability has been noticed in [17]. Thus, the PV system is made to reduce its reactive power completely to zero.

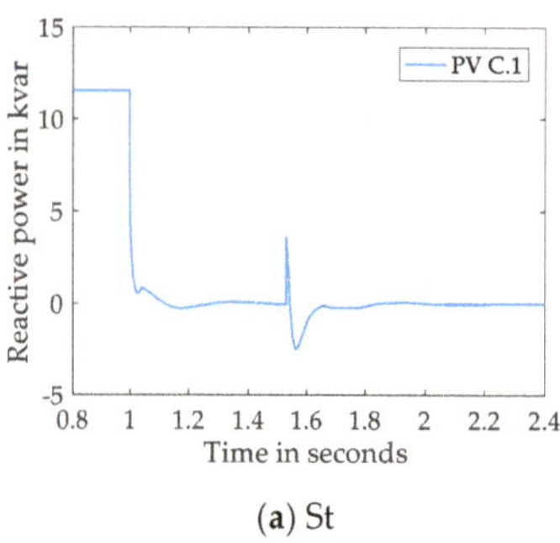

(a) St

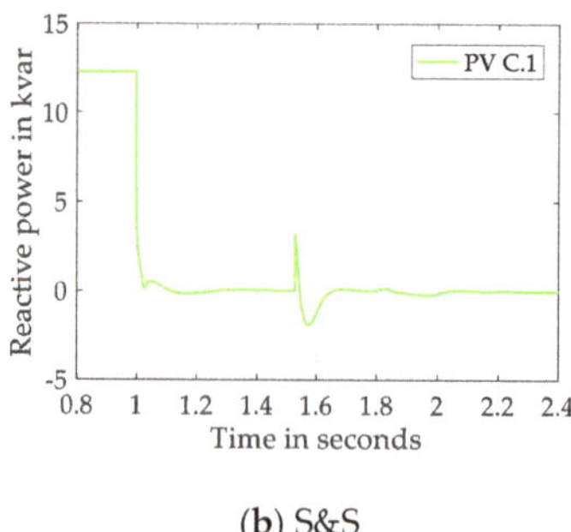

(b) S&S

Figure 26. Reactive power of PV of microgrid C (island mode) in: storage (a) and sun and storage (b) scenario.

Similarly, the main input signals of the current control block in grid-forming BSS are (cf. Figure 7) $i_{d\_v\_ref}$ and $i_{q\_v\_ref}$. In case of a short-circuit, modified very high reference current signals, i.e., the mentioned signals, will be fed into the current control block, such that $i_{q\_v\_ref}$ is equal to zero. Hence, the grid-forming BSS acting now as a current source is forced not to provide any capacitive reactive power directly after the fault incident—see Figure 27. Since the BSS acts as a load in scenario "storage", the capacitive reactive power is shown with a minus sign.

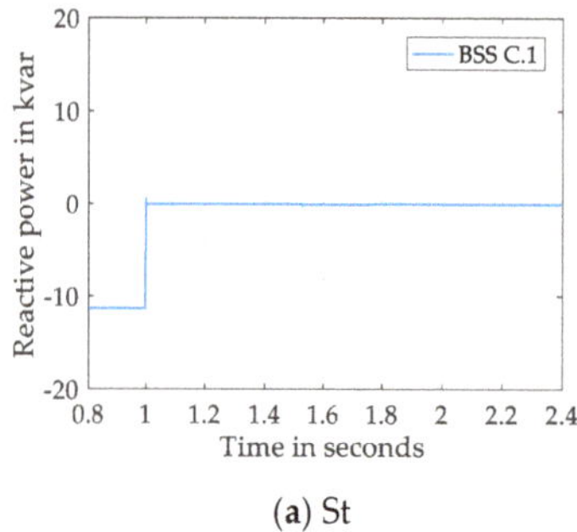

(a) St

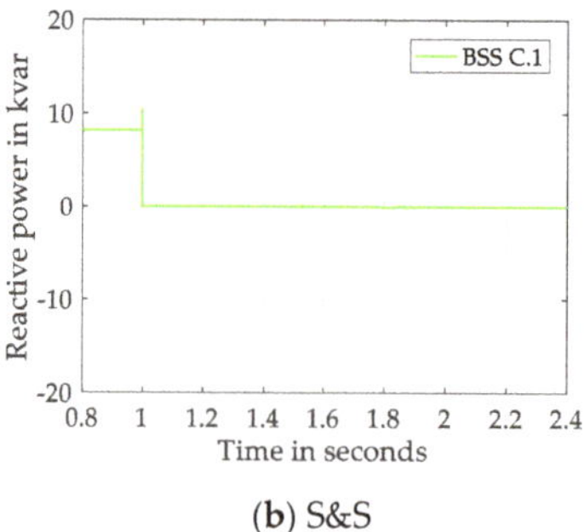

(b) S&S

Figure 27. Reactive power of BSS of microgrid C (island mode) in: storage (a) and sun and storage (b) scenario.

iii.   Output Current

The initial and the sustained short-circuit currents of the DGs (cf. Figure 28) are within the acceptable limits [23]. DG C.3 in scenario "storage" is characterized with a relatively higher output current due to the closeness to the fault location.

In contrast to DGs, the output current of PV and BSS can be controlled by varying the input current signals of the current controller block. Otherwise, the power electronic components will be overloaded and/or damaged due to very high short-circuit currents. [35,43] The output current profile of the grid-feeding PV system is shown in Figure 29.

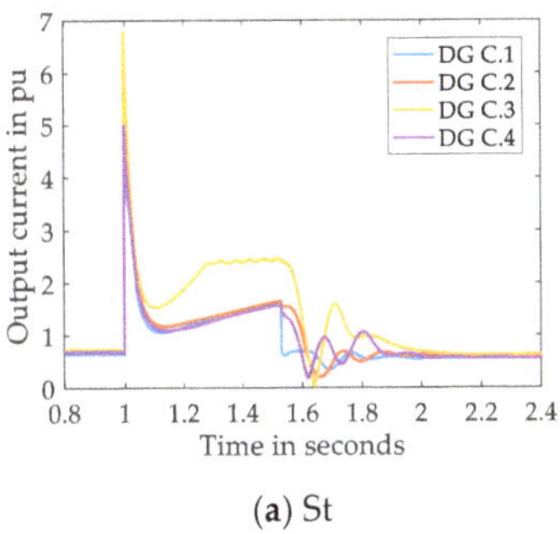
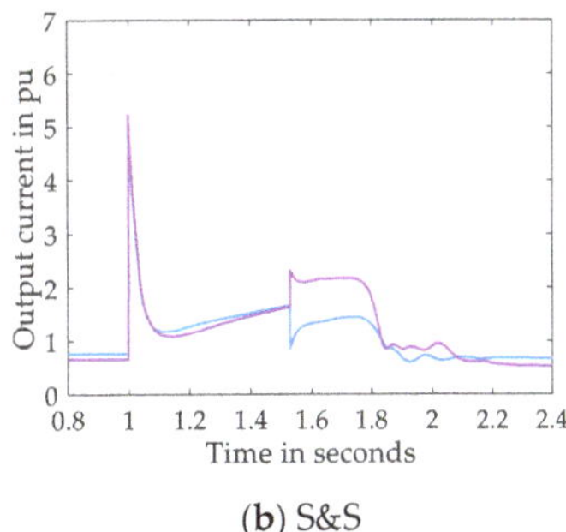

(a) St

(b) S&S

Figure 28. Current of DGs of microgrid C (island mode) in: storage (**a**) and sun and storage (**b**) scenario.

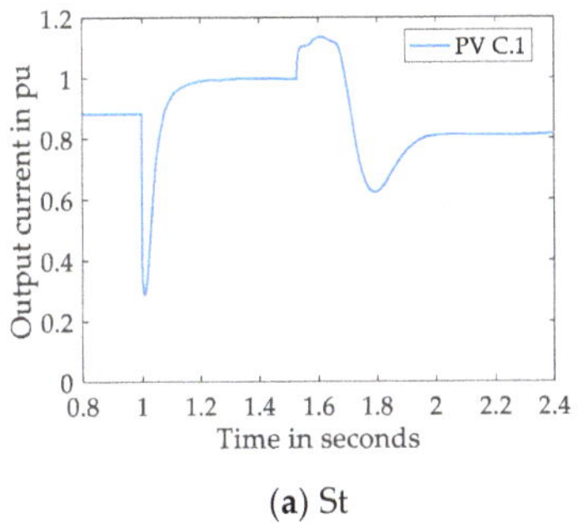
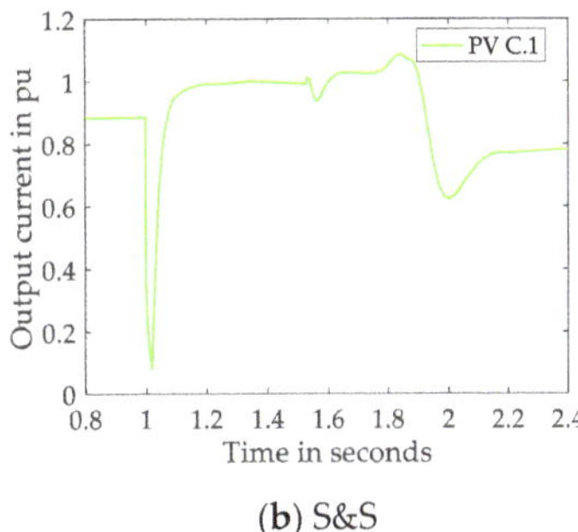

(a) St

(b) S&S

Figure 29. Current of PV of microgrid C (island mode) in: storage (**a**) and sun and storage (**b**) scenario.

Soon after the sudden voltage drop due to the fault, the difference between the DC currents I_{DC} and I_{PV} in PV (see Figure 3) becomes positive. As a result of the increase of I_C, the DC-link voltage V_{DC} rises. With the help of the DC voltage controller, the output current of PV is indirectly increased by increasing the equivalent (d-component) input signal of the current controller block. [17] At the same time, i_{q_ref} is set to zero by giving entire priority to i_{d_ref}. Due to the corresponding time constants in the control loops, it takes several milliseconds until the output current reaches 1 pu. Due to the reduction in the reactive power soon after the short-circuit occurrence, the post-fault (steady-state) current is different from the pre-fault current. However, the active power output remains unaltered (cf. Figure 23), since the solar irradiance and the module temperature were assumed to be constant in the transient stability analyses.

The BSS in microgrid C acts as a generating unit in scenario "sun & storage". Since the DC side of the storage system was modelled by a constant DC voltage source in this research work, a sudden increase in the output current is noticed in the BSS (see Figure 30). Further, the post-fault current is strongly dependent on the terminal voltage. On the other hand, the output current of the BSS acting as a load in scenario "storage" is reduced, since the operating mode (charging/storing) of the BSS remains unchanged.

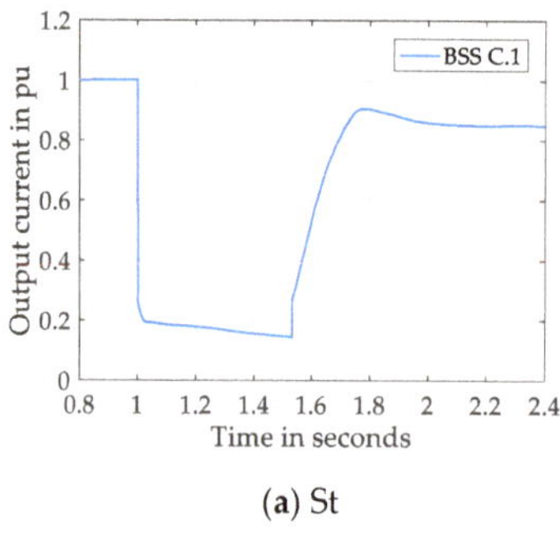
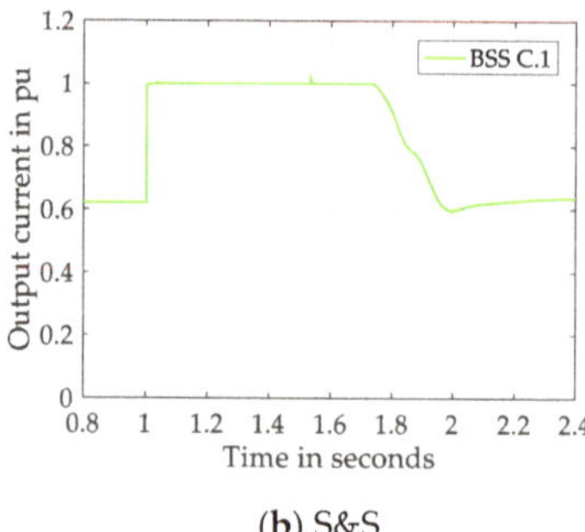

(**a**) St   (**b**) S&S

Figure 30. Current of BSS of microgrid C (island mode) in: storage (**a**) and sun and storage (**b**) scenario.

3.3. Effect of Pooling Microgrids on the System Stability

In this section, the simulation results of the three-phase short-circuit analysis on line L C.5 in microgrid C corresponding to the three operating modes (island, interconnection, and cluster) will be presented with respect to voltage, frequency and rotor angle stability. The studied fault was cleared after 192 ms, which corresponds to the minimum *CCT* among the scenarios and operating modes—classical scenario and interconnection mode.

3.3.1. Voltage Stability

In Figure 31, the measured voltage on busbar C.5 in the different operating modes with respect to the five scenarios is illustrated. A positive effect on the minimum voltage during fault-on is noticed in the interconnected and clustered mode as against the islanded mode—independent of the scenarios. Except in scenario "night", the clustered mode is characterized by a better voltage profile during fault-on and post-fault period. Even though the improvement in the voltage profiles of the scenarios between the fault incident and clearance is subtle, the influence on the frequency and rotor angle stability is significant in microgrids.

3.3.2. Frequency Stability

Due to space constraints, the frequency stability will be discussed taking the reference scenario, i.e., classical, and the hybrid scenario (sun and storage) with the least number of DGs. Figure 32 shows the actual rotor angular frequency of the DGs in the three different operating modes. Similar to the voltage stability, the coupling of microgrids leads to an improvement in the frequency stability. Further, a positive impact is noticed also in the other scenarios, which are not shown in this research paper. Due to the disconnection of the induction motors after the fault occurrence, the post-fault steady-state value of the angular frequency of the DGs is not exactly equal to 1 pu. Further, secondary frequency control is not implemented in the microgrids, since the focus of this research work lies on the short-term stability analysis.

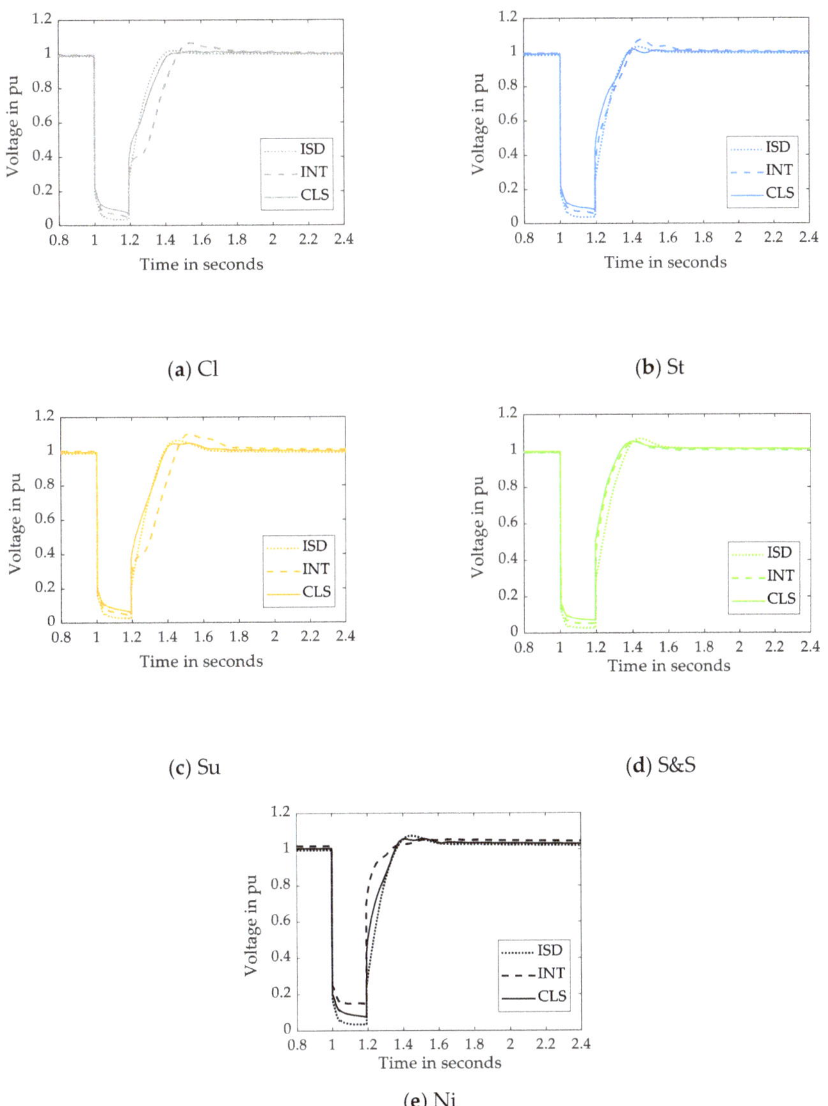

Figure 31. Voltage on busbar C.5 in the islanded, interconnected and clustered mode in: classical (**a**), storage (**b**), sun (**c**), sun and storage (**d**) and night (**e**) scenario.

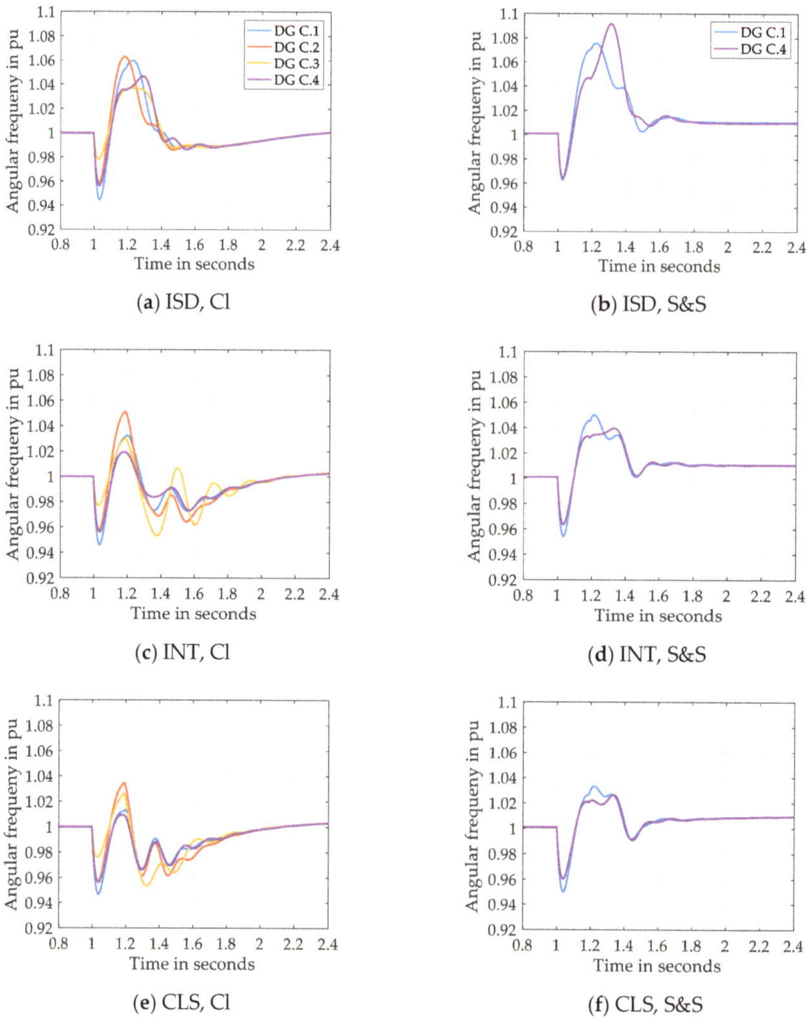

Figure 32. Actual rotor angular frequency of the DGs in the islanded, interconnected and clustered mode in: classical (**a**,**c**,**e**) and sun and storage (**b**,**d**,**f**) scenario.

3.3.3. Rotor Angle Stability

The rotor angle (in the COI frame) of the DGs belonging to microgrid C in the three operating modes and scenario "classical" and "sun & storage" is presented in Figure 33. Regarding the classical scenario, the rotor angle oscillations (excursions) increase in coupled microgrids as against the islanded operating mode, thus a negative impact on the rotor angle stability. However, a marginal positive effect is observed in scenario "sun & storage", which can be also noticed in the positive ΔCCT—see Section 3.1.2. According to Figures 12 and 13 as well as the profiles of the other hybrid scenarios, rotor angle stability is deteriorated by coupling microgrids, like in the classical scenario.

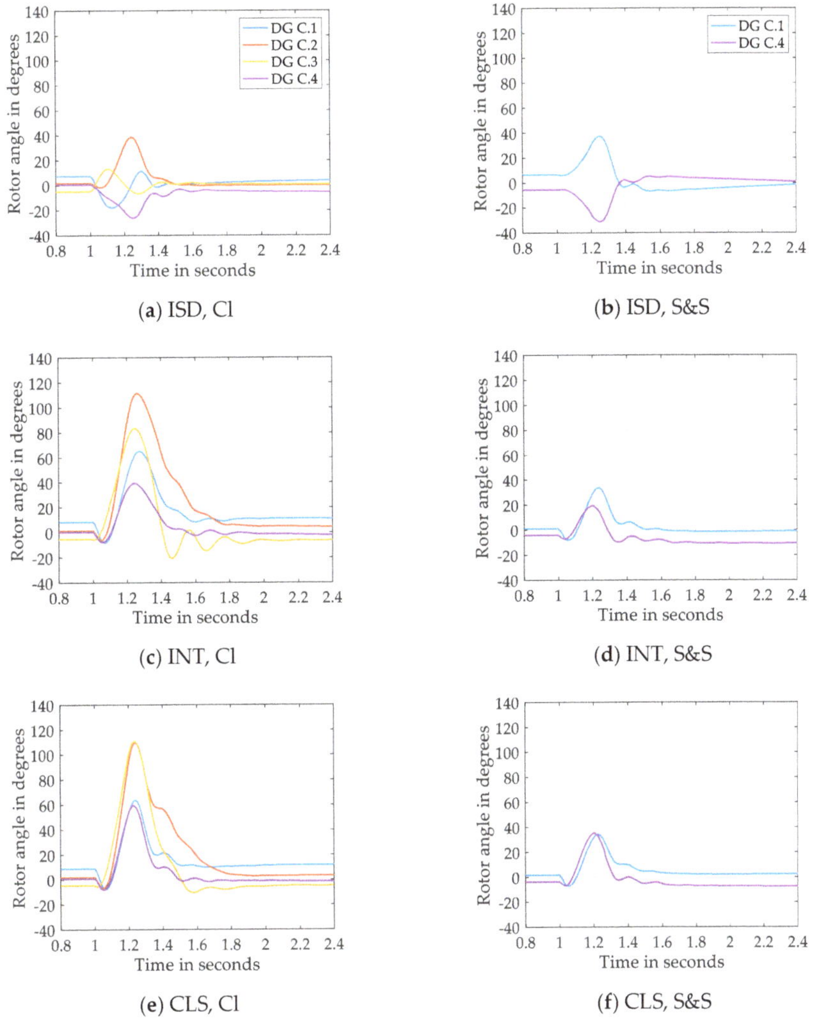

Figure 33. Relative rotor angle of the DGs in the islanded, interconnected and clustered mode in: classical (**a,c,e**), and sun and storage (**b,d,f**) scenario.

Based on the three types of system stability—for the fault on L C.5—optimal scenarios and operating modes can be selected (see Table 10) in the grid planning and/or during near real-time grid operation. The qualitative assessment with respect to the voltage stability was performed by analyzing the voltage drop soon after the fault incident as well as the post-fault voltage recovery. The frequency and the rotor angle stability were assessed qualitatively by studying the excursions and oscillations during fault-on and post-fault period. As against the hybrid scenarios, the scenario "classical" does not outperform in the investigated cases.

Table 10. Overview of the optimal scenarios in the different operating modes.

		Cl	St (12 pm)	Su (2 pm)	S&S (3 pm)	Ni (8 pm)
Island	V		✓	✓		✓
	f			✓		✓
	δ		✓	✓		✓
Interconnection	V		✓		✓	✓
	f			✓	✓	✓
	δ		✓		✓	✓
Cluster	V		✓		✓	✓
	f			✓	✓	✓
	δ				✓	✓

According to the qualitative rotor angle stability analysis (highlighted in Table 10), it can be concluded that:

- Storage, sun and night scenario can be chosen in the **islanded** operation of microgrid C.
- Furthermore, storage, sun and storage, and night scenario outperform the other scenarios in the **interconnected** operating mode.
- In the **clustered** operating mode, sun and storage, and night scenario are beneficial.

Scenario "night" is the most optimal scenario in each operating mode with respect to the system stability, whereas scenario "classical" is not advantageous. In practice, it is significantly tedious to draw conclusions taking every fault location and every microgrid topology into account. Hence, the selection of the optimal scenarios and operating modes should be performed by considering only the critical fault(s). Nevertheless, the decision-making process based on the qualitative stability assessment leads to general conclusions with respect to numerous scenarios and operating modes as well as different topologies in a relatively larger cluster environment.

3.4. System Stability Degree in Operating Modes and Scenarios

3.4.1. Operating Modes

The values of the SSD for the corresponding fault clearing time—calculated using the micro-hybrid method according to Equation (8)—have been plotted for the different cases (13 out of 15) in Figure 34. The minimum fault clearing time of 60 ms has been highlighted by a vertical dashed line. It has been discussed in Section 2.6 that an SSD of 10% was chosen as the threshold value for determining *CCCT*, which has been also shown in the form of a horizontal dashed line. It should be noted that the stability reserve degree and the participation factor of individual synchronous generators will be analyzed in the upcoming research work. However, the corresponding analysis taking the classical scenario into account can be found in [20].

If the fault is cleared 1 ms (impossible in practice) after the fault occurrence, the SSD lies around 100%. If the fault clearance is increased, the system stability reserve drops. The system can be characterized as uncritical (or the fault can be considered as severe) in case of a low value of the gradient of the SSD profile. Profiles with a very high slope are classified to be critical, since the SSD value gets reduced significantly. Any delay in clearing the fault can lead to instability. It can be inferred from Figure 34a that the profile in the islanded mode is less critical than that in the interconnected and clustered mode. Furthermore, the difference between the *CCT* and *CCCT* values lies between 35% and 45%.

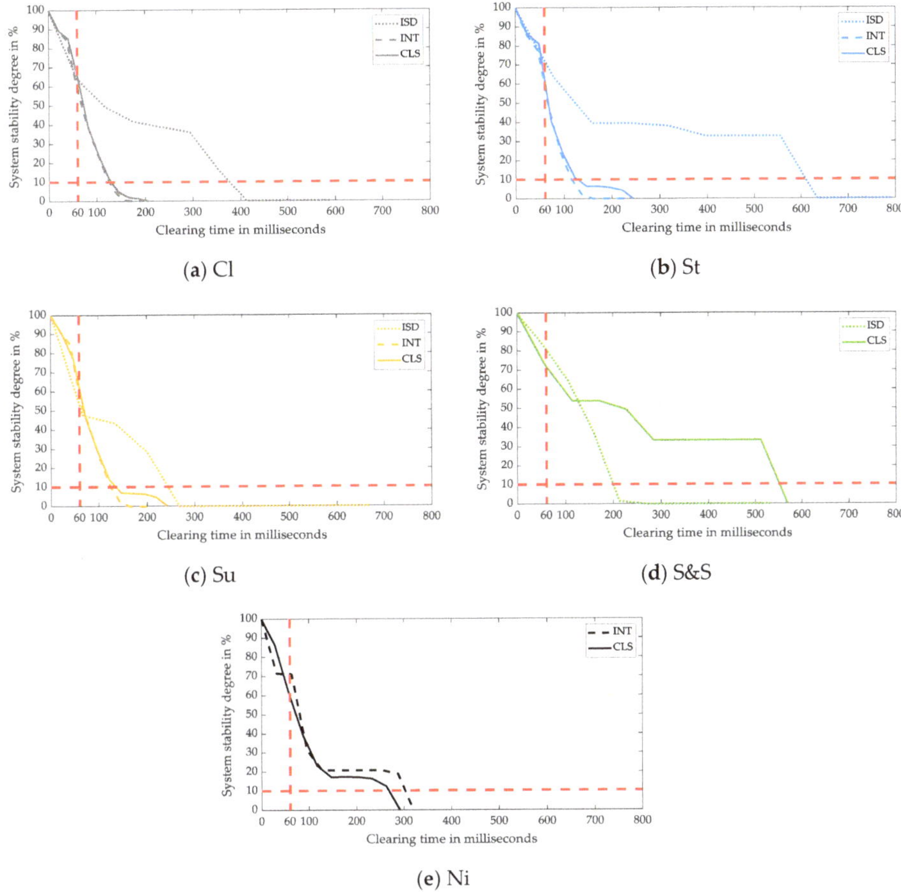

Figure 34. System stability degree versus fault clearing times regarding the different operating modes in: classical (**a**), storage (**b**), sun (**c**), sun and storage (**d**), and night (**e**) scenario.

The SSD profile of the hybrid scenarios is illustrated in Figure 34b–e. The profiles in the coupled modes are similar as against the islanded mode. The interconnected and islanded mode in scenarios "sun & storage" and "night" have been purposefully not shown in the figure, since the SSD—in these two cases—based on the performed TDS were not plausible. A detailed analysis of these two profiles was not performed in this research work. Further, the SSD of several cases (5 out of 13), e.g., scenario "sun" in the islanded mode, lies around 0% for a clearing time less than the *CCT*. The corresponding clearing times up to the *CCT* were not plausible. Hence, the values of the SSD were assumed to be zero. These profiles should be further investigated.

In general, the islanded mode, independent of the scenarios, is characterized by the best SSD profile—in terms of the *CCT*, the *CCCT* and the slope of SSD profiles. The profiles in interconnection and in cluster mode are acceptable. However, the sensitivity of the SSD for various clearing times is relatively high. The trend (slope) of the SSD profiles should be analyzed in case of scenarios and operating modes with higher *CCT* values.

The corrected critical clearing time (*CCCT*) can be determined based on the set threshold value in the SSD profiles, i.e., 10% SSD. Table 11 lists the *CCT*, *CCCT*, and their difference Δ*CCT* in the investigated cases. *CCCT* values less than 60 ms are not noticed in any of the cases. It is recommended to adjust the settings of protection systems based on the *CCT*, instead of the *CCT*, so that the risk of system losing stability can be reduced further.

Table 11. Overview of the *CCT*, *CCCT*, and ΔCCT in the three operating modes with respect to the five scenarios.

		CCT in ms	*CCCT* in ms	ΔCCT
Classical	ISD	588	377	36%
	INT	192	126	34%
	CLS	206	131	36%
Storage	ISD	793	610	23%
	INT	230	116	50%
	CLS	244	129	47%
Sun	ISD	667	243	64%
	INT	205	124	40%
	CLS	242	135	44%
Sun & Storage	ISD	531	200	62%
	INT	568	-	-
	CLS	571	552	3%
Night	ISD	1058	-	-
	INT	321	303	6%
	CLS	292	269	8%

3.4.2. Scenarios

In case of analyzing the SSD regarding the scenarios in the different operating modes, the profiles depicted in Figure 35 can be taken into consideration. Similar to the previous section, the sensitivity or the slope of the SSD profiles should be taken into account while selecting the optimal scenario in each operating mode. The scenario with the highest value of *CCT* in each operating mode is characterized by the best SSD profile. The magnitude of the slope of the profiles corresponds to the respective *CCT* value. If the (differential) protection system is supposed to clear a fault in cluster mode after, e.g., 100 ms, scenario "sun & storage" can be preferred as the optimal scenario. On the other hand, the *CCCT* values determined based on the SSD's threshold value are listed in Table 12.

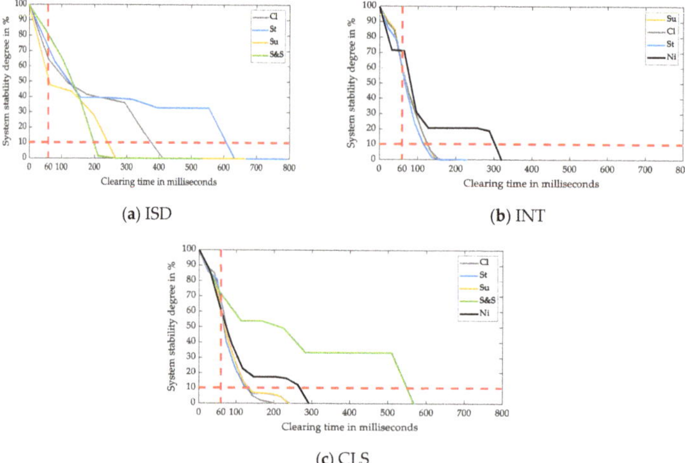

Figure 35. System stability degree versus fault clearing times regarding the different scenarios in: island (**a**), interconnection (**b**), and cluster (**c**) mode.

Table 12. Overview of the *CCT*, *CCCT*, and Δ*CCT* in the five scenarios with respect to the three operating modes.

		CCT in ms	*CCCT* in ms	Δ*CCT*
Island	Cl	588	377	36%
	St	793	610	23%
	Su	667	243	64%
	S&S	531	200	62%
	Ni	1058	-	-
Interconnection	Cl	192	126	34%
	St	230	116	50%
	Su	205	124	40%
	S&S	568	-	-
	Ni	321	303	6%
Cluster	Cl	206	131	36%
	St	244	129	47%
	Su	242	135	44%
	S&S	571	552	3%
	Ni	292	269	8%

3.5. Comparison of Critical Energy in Scenarios and Operating Modes

In this section, the critical energy of the system and of the critical machine will be compared with each other—considering the islanded and clustered operating modes—in order to analyze the ratio of the critical energies. The calculation of the SSD is based on the system critical energy. According to the classical scenario—cf. Figure 17a—and other hybrid scenarios corresponding to the respective *CCT*, the critical DG can be generally easily noticed. Unlike in large transmission systems, a distinct formation of a critical group of DGs, approaching the stability limit, has been not observed in the microgrids, which has been also given in [19,20].

3.5.1. Island

Figure 36 shows the absolute value of E_{cr} and E_{cr_cm} as well as their ratio in each scenario for the fault on L C.5. Further, the *CCT* of the scenarios has been illustrated. The system critical energy decreases with a reduction in the number of the operating DGs (cf. Table 3). The trend of the profile of E_{cr_cm} in the scenarios cannot be compared with that of E_{cr}. The ratio of E_{cr_cm} and E_{cr} is between ca. 20% and 60%, where the scenarios with fewer numbers of DGs are characterized by a higher ratio. A direct correlation between the critical energies and *CCT* values cannot be found in this research work, which will be analyzed in Section 3.6.

3.5.2. Cluster

The critical energies and the *CCT* in the clustered mode are depicted in Figure 37. Similar to the islanded mode, the share of E_{cr_cm} in E_{cr} in the clustered (and also interconnected) operating mode lies between about 20% and 50%. Further, a direct proportionality between E_{cr} and the number of DGs can be seen. However, the absolute values of E_{cr} and E_{cr_cm} differ significantly as against the values in the islanded mode.

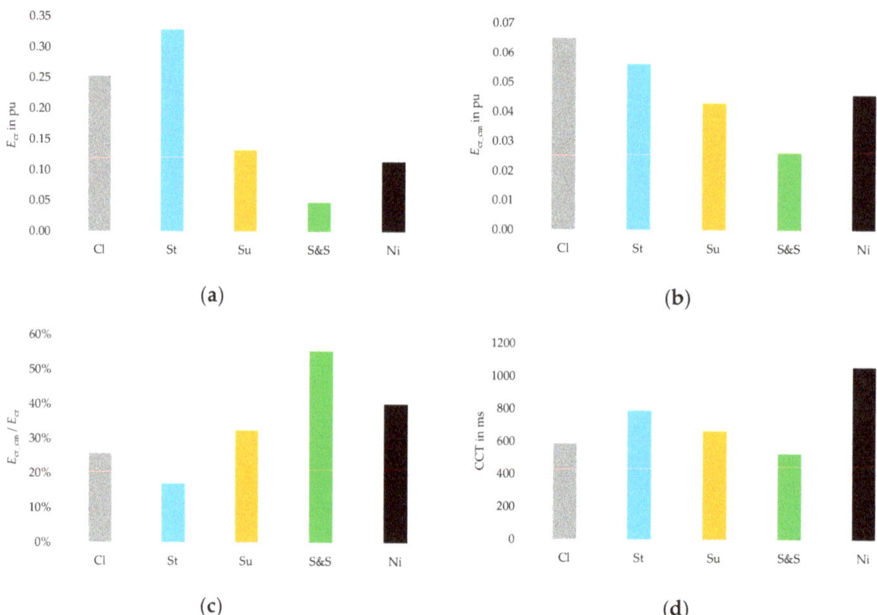

Figure 36. Analysis of the critical energies in the islanded mode: system critical energy (**a**), critical energy of the critical machine (**b**), ratio of the critical energies (**c**), and *CCT* (**d**).

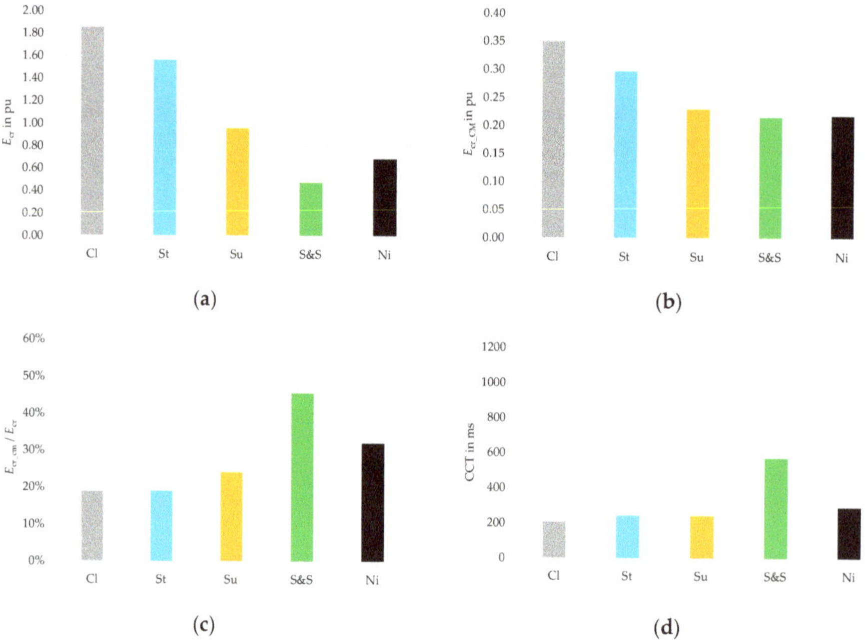

Figure 37. Analysis of the critical energies in the clustered mode: system critical energy (**a**), critical energy of the critical machine (**b**), ratio of the critical energies (**c**), and *CCT* (**d**).

3.6. Influence of Fault Location on Critical Energy in Cluster Mode

The three-phase fault on both ends of L C.5, i.e., very close to BB C.5 and BB C.6, was studied to determine a correlation between the *CCT* and the critical energy of the system and of the critical machine. The critical generator in each scenario in the clustered operating mode is DG C.2, whereas DG C.1 represents the critical machine in scenario "sun & storage". It should be noted that the scenarios will not be compared with each other in this section. However, the *CCT*, E_{cr}, and E_{cr_cm} will be analysed in each scenario—cf. Figure 38.

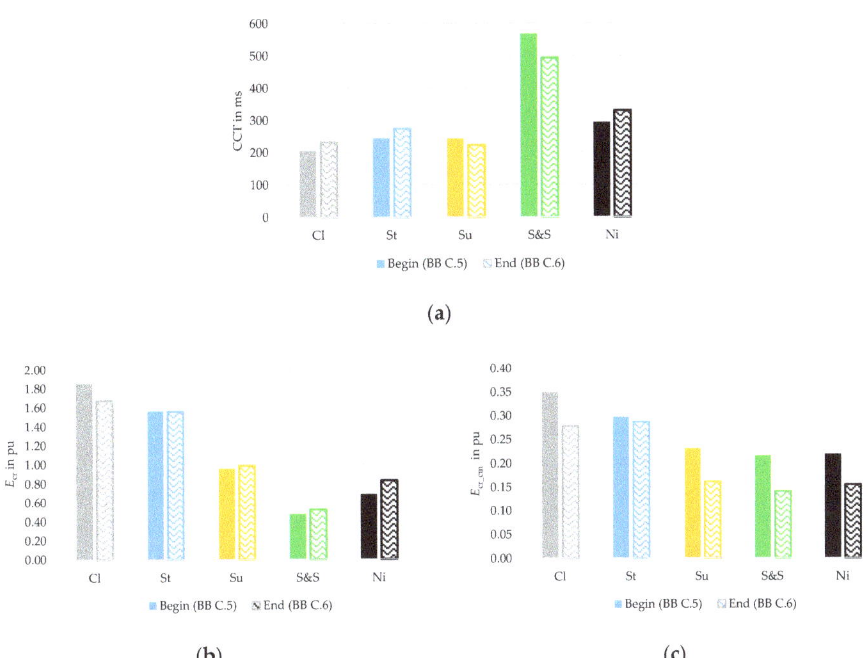

Figure 38. Minimum allowable clearing time and critical energy in each scenario: *CCT* (**a**), system critical energy (**b**) and critical energy of the critical machine (**c**).

In a one-machine infinite bus system, an inverse proportionality exists between the *CCT* and critical energy of the machine, modelled as a classical 2nd order generator. In case of a three-phase fault very close to the synchronous generator, the terminal voltage is almost zero during fault-on. The resulting change in the KE and corresponding PE is relatively higher than that of a case (with the identical pre-fault operating point) with an insignificant drop in the terminal voltage—like in a single-phase fault. The *CCT* of the critical fault is less as against the *CCT* of the fault with a residual voltage during fault-on period.

As against the short-circuit close to BB C.5, the fault next to BB C.6 corresponds to a relatively higher residual terminal voltage of the critical machine DG C.2 (see Figure 9, however in the clustered mode) in every scenario, except in scenario "sun & storage". According to Figure 38, it can be inferred that the *CCT* values and critical energies are both directly and inversely proportional. In scenarios "classical", "storage", and "night", an increase in the *CCT* is observed. This corresponds to an indirect proportionality with E_{cr_cm}, like in the above mentioned one-machine infinite bus system. However, a negative difference of the *CCT* is related to a negative difference of E_{cr_cm} in scenarios "sun" and "sun & storage" (critical machine being DG C.1).

In principle, finding a direct correlation between the *CCT* and the system critical energy with respect to different fault locations, especially in the clustered operating mode, can be misleading. In the future research work, the correlation between the *CCT* and energies should be investigated in detail. For example, by analyzing the *CCT* (of critical faults) and the critical energy of every DG or group of "critical" DGs in the microgrids, the steady-state operating point of DGs can be adjusted.

4. Summary and Outlook

4.1. Summary

Pooling nearby off-grid hybrid microgrids—comprising diesel engine-driven synchronous generators as well as grid-feeding photovoltaics (PV) and grid-forming battery storage systems (BSS)—leads to a reduction in the fuel costs and greenhouse gas emissions as well as to an increase in the security of supply. In the planning phase and in near real-time operation, coupling of microgrids should be investigated not only in the steady-state, but also from the transient stability point of view: both qualitatively and quantitatively. This calls for a detailed dynamic microgrid modelling and three-phase short-circuit analysis. Using the recently developed micro-hybrid method—combining time-domain simulations and transient energy function analyses—quantitative transient stability assessment can be performed.

Hence, three spatially close realistic off-grid microgrids (A, B, and C) were modelled and analyzed in the framework of this research regarding three operating modes (island "C", interconnection "A and B", and cluster "A, B, and C") as well as five scenarios with respect to different instants of time on a partly sunny day. Scenario "classical" represents microgrids with diesel generators (DGs) only, whereas hybrid scenarios "storage", "sun", "sun & storage", and "night" correspond to microgrids with DGs, PV, and BSS (either as loads or as generation units).

Firstly, critical clearing time (CCT) profiles of the microgrids were compared considering 15 cases, by categorizing three operating modes and five scenarios. Secondly, qualitative stability assessment was performed in microgrid C (island mode) for a (critical) short-circuit location very close to a DG and a BSS. The fault behavior of the hybrid microgrids was verified at the system and the equipment level, and was also compared with the response of the classical microgrid. Thirdly, the effect of different operating modes of microgrids was studied among the scenarios from the system stability—i.e., voltage, frequency, and rotor angle stability—point of view. In the last two sections, transient stability was accessed quantitatively with the help of the micro-hybrid method. Regarding the studied fault location in microgrid C, a system stability degree (SSD) was calculated for each fault clearing time in the operating modes and scenarios. Further, the critical energy of the system and the critical machine as well as the effect of the fault location on the critical energy were investigated among scenarios and operating modes.

The key findings of this research work can be summed up as follows:

- **Critical clearing times:** According to the *CCT* profiles of the microgrids (based on the risk level) with respect to the operating modes and scenarios, it can be concluded that, any scenario can be preferred in the islanded mode. However, scenario "sun & storage" is characterized by better *CCT* profiles in the interconnected and clustered mode. In general, coupling classical or hybrid microgrids is not critical with respect to the *CCT* values for the considered high load operating point on a partly sunny day. However, interconnection and cluster mode are characterized by a relatively higher risk of transient stability as against the island mode.
- **Fault behavior of microgrid C in island mode:** The response of microgrid C—in the pre-fault, fault-on and post-fault period—at both system and equipment level for the simulated short-circuit can be assessed as plausible and thereby the dynamic system modelling can be validated.
- **Effect of pooling microgrids on the system stability:** Scenarios "storage", "sun", and "night" of microgrid C perform better in the islanded operation regarding the critical short-circuit location. On the other hand, scenarios "storage", "sun & storage", and "night" outperform in the

interconnected mode. Further, scenarios "sun & storage" and "night" have a positive impact on the system stability in the clustered mode. In this studied case, scenario "night" can be categorized to be the most optimal in each operating mode, whereas scenario "classical" does not perform better among the scenarios. Analyses regarding the selection of the optimal operating modes and scenarios should be executed on a regular basis—by considering only the critical fault(s) with respect to either location or *CCT* values—in the near real-time grid operation.

- **System stability degree in operating modes and scenarios:** The SSD was determined for each corresponding (stable) fault clearing time in various cases with the help of the micro-hybrid method. The SSD profile of the scenarios in interconnection and cluster mode is, in terms of the severity of the fault and sensitivity of the system, more critical than that in island mode, which is characterized by the slope of the profiles. The relative difference between the *CCT* and the corrected critical clearing time (*CCCT*) of the cases, calculated based on the threshold value of the SSD, lies between 3% and 64%. It is recommended to take *CCCT* into account while setting the protection systems in microgrids, so that the risk of system losing stability can be reduced even further. In addition, the gradient of the SSD profiles should be considered while selecting the optimal scenario in the operating modes and scenarios.

- **Comparison of critical energy in scenarios and operating modes:** According to the analyses of the absolute value of the critical energy of the system and of the critical machine as well as the *CCT* (of the studied fault in different scenarios and operating modes), it can be inferred that, the ratio of the critical energy of the critical machine to that of the system lies between 20% and 60%. The ratio is relatively higher in the scenarios with a smaller number of active DGs.

- **Influence of fault location on critical energy in cluster mode:** In case of a one-machine infinite bus system, an inverse proportionality between the *CCT* and critical energy of the machine (simplified 2nd order model) can be observed. Further, the ratio of the *CCT* of the critical fault to the *CCT* of the fault with a residual voltage during fault-on period is relatively high. It can be concluded from the simulation results (of two nearby fault locations) that the *CCT* and critical energy values of the critical machine are both directly and inversely proportional. Finding a direct correlation between the *CCT* and the critical energy of the system with respect to different fault locations, especially in cluster mode, can be misleading. However, investigation of the *CCT* of critical faults as well as the critical energy of every DG or a group of "critical" DGs can help to set optimal steady-state operating points of DGs—in the grid planning phase or in the near real-time operation.

All in all, the following conclusions can be drawn:

- Coupling classical or hybrid microgrids is, in principle, technically feasible regarding *CCT*. However, interconnection and cluster mode exhibit a relatively higher risk of transient stability as against the island mode.

- The hybrid scenarios have a more positive effect on the system stability than scenario "classical". Furthermore, scenario "night" (DGs and BSS) can be ranked as the most optimal in each operating mode, while scenario "classical" (DGs only) has an overall negative effect compared to the other scenarios.

- According to the quantitative TSA, the profiles of SSD versus clearing times of the scenarios in island mode are better than the SSD profiles in interconnection and cluster mode—with respect to the *CCT*, the *CCCT*, and the slope of SSD profiles. In spite of the relatively higher gradient, the SSD profiles in the coupled modes are also acceptable.

- Further, according to the results regarding the effect of the fault location on the critical energy, there exists a direct and indirect proportionality between the *CCT* and critical energy values of the critical machine.

4.2. Outlook

In the framework of this research, three off-grid microgrids were studied taking five scenarios and three operating modes into account. In case of a cluster of more than three microgrids, apart from ring topology, different coupling possibilities should be analyzed—from system stability point of view, especially large-signal rotor angle stability—in order to choose optimal operating modes and scenarios.

With respect to the modelling of microgrids in, e.g., agriculture-dominated rural areas, the share of DGs can be reduced not only by further increasing the installed capacity of PV and BSS, but also by installing eco-friendly biogas generators and micro hydropower plants. Due to different injection delay times—of the mentioned synchronous generators—in case of disturbances, and consequently different courses of the forward and the backswing of the generators, random rotor swings can be observed. This calls for a detailed qualitative and especially quantitative TSA of such hybrid microgrids in different operating modes.

In hybrid microgrids with a significant share of inverter-based systems, PV and BSS should be modelled using EMT models instead of RMS models. Due to the recent developments in grid modelling and simulation tools [57], co-simulation (RMS and EMT) of microgrids can be performed.

As a result of the extension of power transmission networks—in developing and under-developed countries—with long-term projects connecting remote areas, alongside decentralization, future interconnected stand-alone hybrid microgrids can be operated not only in the islanded mode, but also by connecting to the main transmission and distribution grids. Hence, the dynamic behavior as well as the qualitative and quantitative operational limits of such coupled—off-grid and grid-connected—microgrids with different topologies should be determined in the planning phase and also in near real-time grid operation. Further, switching operations and measures to improve the dynamic stability and security of hybrid microgrids can be investigated in detail.

Author Contributions: Conceptualization, K.V.; methodology, K.V. and H.A.; software, H.A.; validation, K.V. and H.A.; formal analysis, H.A. and K.V.; investigation, H.A.; data curation, K.V.; writing—original draft preparation, K.V.; writing—review and editing, K.V., H.A., and M.L.; visualization, H.A. and K.V.; supervision, K.V. All authors have read and agreed to the published version of the manuscript.

Funding: This research received no external funding.

Conflicts of Interest: The authors declare no conflicts of interest.

References

1. International Renewable Energy Agency (IRENA). Off-Grid Renewable Energy Systems:Status and Methodological Issues. Available online: https://www.irena.org/-/media/Files/IRENA/Agency/Publication/2015/IRENA_Off-grid_Renewable_Systems_WP_2015.pdf (accessed on 13 January 2020).
2. Carnegie Mellon University (CMU). Available online: https://www.cmu.edu/ceic/assets/docs/publications/reports/2014/micro-grids-rural-electrification-critical-rev-best-practice.pdf (accessed on 13 January 2020).
3. Motoren- und Turbinen-Union (MTU). Available online: https://www.mtu-solutions.com/eu/en/stories/power-generation/gas-generator-sets/microgrid-cooperation-with-qinous.html (accessed on 13 January 2020).
4. Caterpillar. Available online: https://www.cat.com/en_US/by-industry/electric-power-generation/Articles/White-papers/white-paper-hybrid-microgrids-the-time-is-now.html (accessed on 13 January 2020).
5. United Nations (UN). Available online: https://sustainabledevelopment.un.org/content/documents/interconnections.pdf (accessed on 13 January 2020).
6. Microgrid Knowledge. Available online: https://microgridknowledge.com/bronzeville-microgrid-cluster/ (accessed on 13 January 2020).
7. Ejal, A.A.; Yazdavar, A.H.; El-Saadany, E.F.; Ponnambalam, K. On the Loadability and Voltage Stability of Islanded AC–DC Hybrid Microgrids During Contingencies. *IEEE Syst. J.* **2019**, *13*, 4248–4259. [CrossRef]
8. Saleh, M.S.; Althaibani, A.; Esa, Y.; Mhandi, Y.; Mohamed, A. Impact of Clustering Microgrids on Their Stability and Resilience during Blackouts. In Proceedings of the International Conference on Smart Grid and Clean Energy Technologies, Offenburg, Germany, 20–23 October 2015; pp. 195–200.

9. Shafiee, Q.; Dragicevic, T.; Vasquez, J.C.; Guerrero, J.M. Hierarchical Control for Multiple DC-Microgrids Clusters. *IEEE Trans. Energy Convers.* **2014**, *29*, 922–933. [CrossRef]

10. Shafiee, Q.; Dragicevic, T.; Vasquez, J.C.; Guerrero, J.M. Modeling, Stability Analysis and Active Stabilization of Multiple DC-Microgrid Clusters. In Proceedings of the ENERGYCON, Dubrovnik, Croatia, 13–16 May 2014; pp. 1284–1290.

11. Nikolakakos, I.P.; Zeineldin, H.H.; El-Moursi, M.S.; Hatziargyriou, N.D. Stability Evaluation of Interconnected Multi-Inverter Microgrids through Critical Clusters. *IEEE Trans. Power Syst.* **2016**, *31*, 3060–3072. [CrossRef]

12. Zhao, Z.; Yang, P.; Wang, Y.; Xu, Z.; Guerrero, J.M. Dynamic Characteristics Analysis and Stabilization of PV-Based Multiple Microgrid Clusters. *IEEE Trans. Smart Grid* **2017**, *10*, 805–818. [CrossRef]

13. He, J.; Wu, X.; Wu, X.; Xu, Y.; Guerrero, J.M. Small-Signal Stability Analysis and Optimal Parameters Design of Microgrid Clusters. *IEEE Access* **2019**, *7*, 36896–36909. [CrossRef]

14. Majumder, R. Some aspects of stability in microgrids. *IEEE Trans. Power Syst.* **2013**, *28*, 3243–3252. [CrossRef]

15. Pavella, M.; Ernst, D.; Ruiz-Vega, D. *Transient Stability of Power Systems: A Unified Approach to Assessment and Control*; Kulwer: Norwell, MA, USA, 2000.

16. CIGRE. Available online: https://e-cigre.org/publication/325-review-of-on-line-dynamic-security-assessment-tools-and-techniques (accessed on 13 January 2020).

17. Veerashekar, K.; Bichlmaier, A.; Luther, M. Transient stability of hybrid stand-alone microgrids considering the DC-side of photovoltaics. In Proceedings of the 4th International Hybrid Power Systems Workshop, Crete, Greece, 22–23 May 2019; pp. 1–10.

18. Veerashekar, K.; Eichner, S.; Luther, M. Modelling and transient stability analysis of interconnected autonomous hybrid microgrids. In Proceedings of the 13th IEEE PES PowerTech Conference, Milan, Italy, 23–27 June 2019; pp. 1–6.

19. Veerashekar, K.; Schuehlein, P.; Luther, M. Quantitative transient stability assessment in microgrids combining both time-domain simulations and energy function analysis. *Int. J. Electr. Power Energy Syst.* **2020**, *115*, 1–12. [CrossRef]

20. Veerashekar, K.; Flick, M.; Luther, M. The micro-hybrid method to assess transient stability quantitatively in pooled off-grid microgrids. *Int. J. Electr. Power Energy Syst.* **2020**, *117*, 1–12. [CrossRef]

21. Veerashekar, K.; Flick, M.; Luther, M. Small-signal stability of interconnected autonomous microgrids (German: Kleinsignalstabilität von vernetzten autonomen Mikronetzen). In Proceedings of the International ETG Congress, Esslingen am Neckar, Germany, 8–9 May 2019; pp. 387–392.

22. Machowski, J.; Bialek, J.W.; Bumby, J.R. *Power System Dynamics: Stability and Control*, 2nd ed.; Wiley: Chichester, UK, 2012.

23. German Generator GmbH. Available online: https://germangenerator.com/products/industrial-power-generators/ (accessed on 15 January 2020).

24. Sharp. Available online: https://www.sharp.de/cps/rde/xchg/de/hs.xsl/-/html/product-details-solar-modules.htm?product=NQR258H (accessed on 15 January 2020).

25. SMA Solar Technology. Available online: https://www.sma.de/en/products/solarinverters/sunny-tripower-60.html (accessed on 16 January 2020).

26. SMA Solar Technology. Available online: https://www.sma.de/en/products/solarinverters/sunny-tripower-core1.html (accessed on 16 January 2020).

27. IEEE PES. Available online: https://resourcecenter.ieee-pes.org/technical-publications/technical-reports/PES_TR0066_062018.html (accessed on 16 January 2020).

28. Stenzel, D.; Hewes, D.; Viernstein, L.; Würl, T.; Witzmann, R. Investigation of demand trends and their impacts on stationary and dynamic aggregated load behaviour. In Proceedings of the International ETG Congress, Esslingen am Neckar, Germany, 8–9 May 2019; pp. 171–176.

29. DIgSILENT/PowerFactory. *Technical Reference Documentation—Asynchronous Machines*; DIgSILENT/PowerFactory: Gomaringen, Germany, 2017; pp. 1–36.

30. Engineers Edge. Available online: https://www.engineersedge.com/motors/classification_electric_motors_13882.htm (accessed on 17 January 2020).

31. University of Oviedo. Available online: http://ocw.uniovi.es/pluginfile.php/5424/mod_resource/content/1/Guia%20de%20General%20Electric%20motores%20I.pdf (accessed on 17 January 2020).

32. FG Wilson. Available online: http://www.fgwilson.ie/files/generator-set-iso8528-5-2005-operating-limits.pdf (accessed on 17 January 2020).
33. International Organization for Standardization (ISO). *ISO 8528-5:2013—Reciprocating Internal Combustion Engine Driven Alternating Current Generating Sets—Part 5: Generating Sets*; ISO: Geneva, Switzerland, 2013.
34. Kroutikova, N.; Hernandez-Aramburo, C.A.; Green, T.C. State-space model of grid-connected inverters under current control mode. *IET Electr. Power Appl.* **2007**, *1*, 329–338. [CrossRef]
35. Teodorescu, R.; Liserre, M.; Rodríguez, P. *Grid Converters for Photovoltaic and Wind Power Systems*; Wiley: Chichester, UK, 2011.
36. Rocabert, J.; Luna, A.; Blaabjerg, F.; Rodriguez, P. Control of power converters in AC microgrids. *IEEE Trans. Power Electron.* **2012**, *27*, 4734–4749. [CrossRef]
37. Schroeder, D. *Electrical Drives—Control of Drive Systems (German: Elektrische Antriebe—Regelung Von Antriebssystemen)*, 4th ed.; Springer: Berlin, Germany, 2015.
38. Tripathi, S.M.; Tiwari, A.M.; Singh, D. Optimum design of proportional-integral controllers in grid integrated PMSG-based wind energy conversion system. *Int. Trans. Electr. Energy Syst.* **2016**, *26*, 1006–1031. [CrossRef]
39. Kaura, V.; Blasko, V. Operation of a phase locked loop system under distorted utility conditions. *IEEE Trans. Ind. Appl.* **1997**, *33*, 58–63. [CrossRef]
40. Vinayagam, A.; Swarna, K.S.V.; Khoo, S.Y.; Oo, A.T.; Stojcevski, A. PV based microgrid with grid-support grid-forming inverter control (simulation and analysis). *Smart Grid Renew. Energy* **2017**, *8*, 1–30. [CrossRef]
41. Gkountaras, A. Modeling Techniques and Control Strategies for Inverter Dominated Microgrids. Ph.D. Thesis, Technical University of Berlin, Berlin, Germany, 2017.
42. De Brabandere, K. Voltage and Frequency Droop Control in Low Voltage Grids by Distributed Generators with Inverter Front-End. Ph.D. Thesis, University of Leuven, Leuven, Belgium, 2006.
43. SMA Solar Technology. Available online: https://files.sma.de/dl/7418/Iscpv-TI-en-17.pdf (accessed on 18 January 2020).
44. Low, H. Control of Grid Connected Active Converter. Master's Thesis, Norwegian University of Science and Technology, Trondheim, Norway, 2013.
45. *Verband der Elektrotechnik Elektronik Informationstechnik (VDE), VDE-AR-N 4105: Power Generating Plants in the Low Voltage Grid (German: Erzeugungsanlagen am Niederspannungsnetz—Technische Mindestanforderungen für Anschluss und Parallelbetrieb von Erzeugungsanlagen am Niederspannungsnetz)*; VDE Verlag GmbH: Berlin, Germany, 2018.
46. Blaschke, H. *Motor Protection (German: Motorschutz)*; VEB Verlag Technik: Berlin, Germany, 1954.
47. Ziegler, G. *Numerical Differential Protection: Principles and Applications*, 2nd ed.; Publicis: Erlangen, Germany, 2011.
48. Larsen & Turbo. Available online: http://corpwebstorage.blob.core.windows.net/media/37784/omega-acb-catalogue.pdf (accessed on 20 January 2020).
49. Eaton. Available online: https://www.eaton.com/ecm/groups/public/@pub/@eaton/@holec/documents/content/ct_255776.pdf (accessed on 20 January 2020).
50. Meshcheryakov, V.P.; Sibatov, R.T.; Samoilov, V.V.; Topchii, A.S. Calculation of an interruption arc current and time of arc quenching in low-voltage current breakers. *Russ. Electr. Eng.* **2008**, *79*, 92–98. [CrossRef]
51. Maria, G.A.; Tang, C.; Kim, J. Hybrid transient stability analysis. *IEEE Trans. Power Syst.* **1990**, *9*, 384–391. [CrossRef]
52. Tang, C.K.; Graham, C.E.; El-Kady, M. Transient stability index from conventional time domain simulation. *IEEE Trans. Power Syst.* **1994**, *9*, 1524–1530. [CrossRef]
53. Mansour, Y.; Vaahedi, E.; Chang, A.Y.; Corns, B.R.; Garrett, B.W.; Demaree, K.; Athay, T.; Cheung, K. Hydro's On-line transient stability assessment (TSA) model development, analysis and post-processing. *IEEE Trans. Power Syst.* **1995**, *10*, 241–253. [CrossRef]
54. Haque, M.H. Hybrid method of determining the transient stability margin of a power system. *IEE Proc. Gener. Transm. Distrib.* **1996**, *143*, 27–32. [CrossRef]
55. Zhang, Y.; Wehenkel, L.; Rousseaux, P.; Pavella, M. SIME: A hybrid approach to fast transient stability assessment and contingency selection. *Int. J. Electr. Power Energy Syst.* **1997**, *19*, 195–208. [CrossRef]

56.   Kundur, P. *Power System Stability and Control*; McGraw-Hill: New York, NY, USA, 1994.
57.   DIgSILENT/PowerFactory. Available online: https://www.digsilent.de/en/stability-analysis.html (accessed on 22 January 2020).

Article

A Model-Based Design Approach for Stability Assessment, Control Tuning and Verification in Off-Grid Hybrid Power Plants

Lennart Petersen [1,2,*], Florin Iov [1] and German Claudio Tarnowski [2]

[1]   Department of Energy Technology, Aalborg University, 9220 Aalborg, Denmark; fi@et.aau.dk
[2]   Vestas Wind Systems, 8200 Aarhus N, Denmark; getar@vestas.com
*   Correspondence: lepte@vestas.com; Tel.: +45-5221-1266

Received: 1 December 2019; Accepted: 17 December 2019; Published: 20 December 2019

Abstract: This paper proposes detailed and practical guidance on applying model-based design (MBD) for voltage and frequency stability assessments, control tuning and verification of off-grid hybrid power plants (HPPs) comprising both grid-forming and grid-feeding inverter units and synchronous generation. First, the requirement specifications are defined by means of system, functional and model requirements. Secondly, a modular approach for state-space modelling of the distributed energy resources (DERs) is presented. Flexible merging of subsystems by properly defining input and output vectors is highlighted to describe the dynamics of the HPP during various operating states. Eigenvalue (EV) and participation factor (PF) analyses demonstrate the necessity of assessing small-signal stability over a wide range of operational scenarios. A sensitivity analysis shows the impact of relevant system parameters on critical EVs and enables one to finally design and tune the central HPP controller (HPPC). The rapid control prototyping and control verification stages are accomplished by means of discrete-time domain models being used in both off-line simulation studies and real-time hardware-in-the-loop (RT-HIL) testing. The outcome of this paper is targeted at off-grid HPP operators seeking to achieve a proof-of-concept on stable voltage and frequency regulation. Nonetheless, the overall methodology is applicable to on-grid HPPs, too.

Keywords: hybrid power plant; off-grid electricity systems; model-based design; state-space model; voltage stability; frequency stability; small-signal analysis; control tuning; controller validation

1. Introduction

Off-grid electricity systems have attracted significant attention in emerging and frontier markets in order to conduct rural and island electrification and to supply remote industrial sites (e.g., mining areas) [1–3]. Such isolated grids—commonly associated with the microgrid (MG) concept—are characterized by increasing hybridization of the involved distributed energy resources (DERs). Traditional fossil-fueled production systems (e.g., diesel generators) are replaced or augmented by renewable generation (e.g., wind power, solar photovoltaic (PV)) and energy storage due to environmental, economic and social reasons. During the design stage of an off-grid hybrid power plant (HPP), it becomes evident that the most cost effective approach is to retain a fossil-fueled generator with time-limited operation to supply the net load demand only whenever it is needed [4–7].

It is crucial to ensure power supply and balance stability and control system stability in the HPP [8]. In this context, small-signal stability is concerned with assessing the occurrence of voltage, frequency or power oscillation modes and sufficient stability margins in every technically feasible operating state of the HPP. Moreover, a robust control solution and adequate tuning guidelines are required to keep the frequency and voltages within the operational limits.

Several publications address the topic of small-signal modeling and voltage/frequency stability assessment in MGs [9–13]. In [9,10] state-space models of inverters, grid and loads were developed and the sensitivities of eigenvalues (EVs) to control parameters were evaluated. The stability analysis was targeted for a MG consisting of inverter based DERs that share the grid-forming task among each other. A similar type of analysis was performed in [11] for a system with synchronous generator (SG), asynchronous generator and battery energy storage system (BESS) in PQ control mode [14]. Finally, in [12,13] the parallel operation of SG and grid-forming inverter was investigated to study the dynamics of the MG and to adjust some control parameters of the individual DERs to ensure system stability. All above-mentioned publications do not further extend their scope beyond modeling and stability analysis. Further developments and studies on control design and tuning for voltage and frequency regulation are missing which are required to achieve a proof-of-concept on the HPP control system.

In this regard, various MG control architectures are proposed in the literature and can be classified as centralized or decentralized [15,16]. However, in these publications it remains unclear how to effectively design and tune decentralized controllers with the objective of keeping the frequency and voltages within the limits during any possible operating condition. The most common way is to utilize a central system controller that dispatches commands to the individual DERs to achieve a global control objective. Here, control design and tuning methods need to take into account the dynamics occurring within the HPP. In the present literature, no clear guidelines on the models applied and control methods are identified.

The novelty of this paper is to propose a model-based design (MBD) approach that includes all necessary building blocks to achieve stable voltage and frequency regulation in off-grid HPPs; i.e., the required set of models; a systematic and complete stability assessment; a design and tuning method of a hierarchical control system consisting of central HPP controller (HPPC) and DER controllers; and final verification and validation of the control system. The HPP in the scope of this study consists of wind turbine generator (WTG), PV, BESS and a fossil-fuel generator set (hereinafter called genset). An overview of the different stages from development to testing is given in Figure 1.

Section 2 describes the requirement specifications which are distinguished into system, functional and modeling requirements (Step I).

In Section 3, a modular approach for state-space modelling of each DER, and subsequently the entire HPP is presented (Step II). State-of-the-art inverter models and generic rules for control design are referenced. The state, input and output vectors of the state-space models are summarized. A representative model of the genset including speed governor and automatic voltage regulator (AVR) is summarized in equal manner. Here, it is highlighted how to utilize in and output variables of the model depending on whether grid-forming inverter and/or genset provide the grid frequency reference. A set of numerical simulation models is proposed to be used during the rapid control prototyping (RCP) and control verification stage.

In Section 4 a voltage and frequency stability assessment is shown (Step III). First, EV and participation factor (PF) analysis was conducted for two relevant test scenarios; i.e., genset in or out of service. The occurring dynamic modes and the associated state variables of each DER were clustered with respect to their eigenfrequencies. Such a clustering method provides insight into the dynamic modes and control parameters that require further attention with regard to absolute and relative system stability. Subsequently, the sensitivity of EVs to certain DER control parameters was assessed. This is a useful step to gain more confidence on specific control parameters (e.g., droop gains) which need to be parametrized in the context of the entire HPP.

Section 5 deals with the design and tuning of voltage and frequency controller (Step IV). State-space models are converted to transfer functions which are used to tune DER control parameters and design and tune the central HPPC.

Subsequently Section 6 shows discrete-time domain models being applied to test the control algorithms under various operating conditions to identify the robustness of the design (Step V). This stage is called RCP.

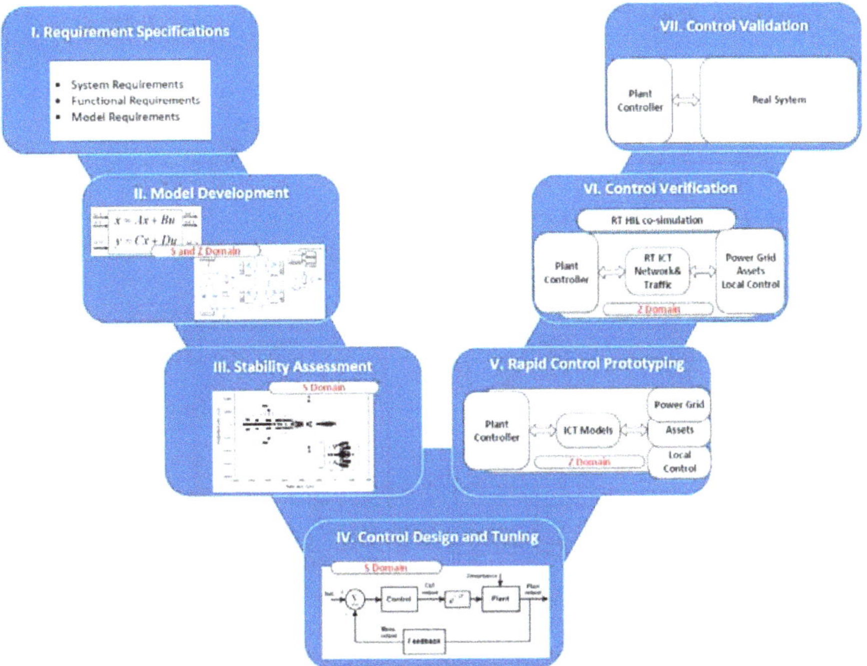

Figure 1. Proposed model-based design (MBD) approach for stability assessment, control tuning and verification in off-grid hybrid power plants (HPPs).

The final stage of proposed MBD approach is control verification and validation against system and functional requirements (Step VI and VII). In Section 7, an outlook is given for verifying the performance of HPPC platform by means of real-time hardware-in-the-loop (RT-HIL) testing. The developed control algorithms including physical implementation on target hardware are then ready for site testing as a final validation stage.

The mathematical formulations and procedures are demonstrated in general to enable studies in off-grid HPPs on either kW-scale or MW-scale and with modular expansion of the production subsystem.

2. Step I: Requirement Specifications

The single-line diagram (SLD) of the benchmark off-grid HPP investigated in this study is shown in Figure 2.

This paper uses the optimal configuration for an off-grid HPP, as derived in [3], since it constitutes a representative system optimized from a techno-economic perspective. The production subsystem consists of a full-scale converter WTG (80 kW) and PV (40 kW), a BESS (160 kWh/90 kVA) and a fossil-fuel genset (90 kVA). At least one grid-forming unit is required that provides the voltage and frequency reference in the HPP. Typically, SGs are responsible for this task. By allowing flexible HPP operation with partial shut-down of fossil-fuel generators it becomes evident that the BESS inverter system must implement grid-forming capabilities, too. Then, it is not desirable to switch between various control schemes, since it prevents smooth transition due to controller reset and eventual plant shut-down and restart. Hence, the optimum solution for the power management strategy is to use droop regulation. It ensures seamless transition between operation scenarios on the one hand, and on the other facilitates smoother integration of the HPP, if it becomes grid-connected.

The demand subsystem (90 kW peak) might comprise multiple low voltage (LV) feeders with residential, commercial and small industrial consumers. In this study, it is represented as an aggregated electrical load which is modelled as constant impedance (RL) load [17]. Production and demand subsystems are connected via the point of common coupling (PCC).

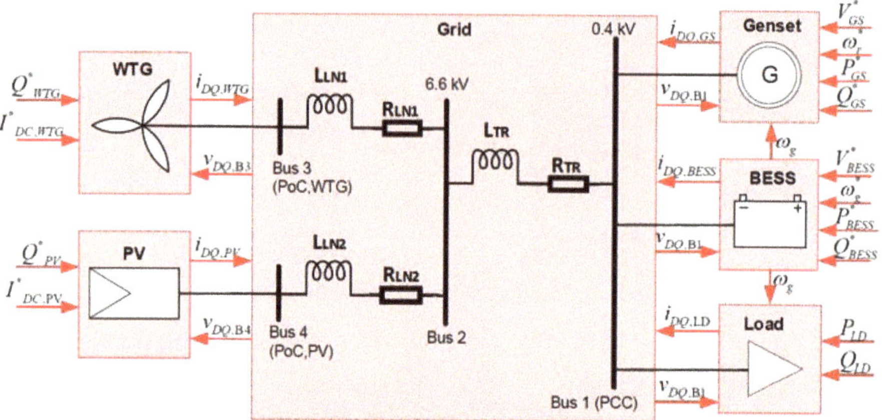

Figure 2. Single-line diagram of off-grid HPP (black) and functional diagram of state-space model (red).

2.1. Step Ia: Defining System Requirements

System requirements are related to the performance of the HPP system and are usually specified in so-called grid codes, technical regulations or guidelines.

First and foremost, it is essential in off-grid HPPs to ensure system stability for voltage and frequency regulation due to the characteristics of isolated MGs. The stability phenomena in MGs, which can be classified as control system stability and power supply and balance stability, are explained in [8] and summarized in [18]. Absolute and relative stability can be measured by means of an EV analysis. Absolute stability is ensured if all poles are located in the left half plane. Relative stability is associated with the damping ratio of the EVs. In order to avoid critically low damping of any voltage or frequency oscillations in the HPP, a reasonable target for the damping ratio is $\zeta \geq 0.05$ according to [19].

Then, operational requirements for off-grid HPPs are difficult to define by means of existing grid codes, since each MG size, the layout, the DERs involved, and hence, the related technical regulations, are unique [18]. However, some generic guidelines are found in [20]. A steady-state voltage and frequency profile of $f_g = \{48, 52\}$ Hz and $V = \{0.85, 1.15\}$ pu is suggested. Dynamic performance requirements for voltage and frequency regulation are not specified.

2.2. Step Ib: Defining Functional Requirements

Functional requirements refer to the necessary functions that are to be implemented in the HPP control system in order to satisfy certain system requirements. A comprehensive overview of the required control functions in MGs is provided in [21] and summarized in [18]. In this section, the specific elements of voltage and frequency control function are elaborated on.

As explained previously in this section one part of the overall power management strategy is to utilize parallel grid-forming DERs (i.e., BESS and genset) to regulate grid frequency and voltages within their nominal limits specified in Section 2.1. Moreover, the studies in [13] reveal that it is important to evaluate the transient power sharing performance between several grid-forming units—to avoid any unwanted oscillations within the HPP and to minimize the duration of unequal power sharing. This aspect is related to DER droop regulation (primary control) [13]. Primary control actions by DERs will leave steady-state errors in voltage and frequency. There are good reasons that these deviations from the nominal value should be eliminated, even if the voltages and frequency remain within the normal operating range. Firstly, it is not desired to operate the PCC voltage below nominal value, as it will lower the voltage levels in the demand subsystem as well. Depending on the feeder length and the connected loads, severe undervoltages can be expected that would lead to load disconnection. Secondly, it is preferred to avoid grid frequencies below nominal value for a longer time period, as they will lower the power consumption and efficiency of some frequency dependent loads; e.g., hydraulic pumps. Field measurements in an island power system with water supply pumps have revealed that the active load changes with approximately 8.5%/Hz [22]. Hence, steady-state voltage and frequency errors need to be compensated by secondary control actions of the central HPPC.

2.3. Step Ic: Defining Modeling Requirements

As part of the MBD process it is necessary to define certain requirements of the models being developed in Step II:

- Linearized models are required for stability assessment and control tuning in frequency domain.
- The model bandwidth shall be limited to a minimum value that enables the assessment of voltage and frequency stability and control.
- Numerical models are required that are applicable for RCP and control verification purposes. Computational effort of model execution shall be taken into account to achieve accelerated off-line simulation studies during RCP stage and to ensure real-time capability for RT-HIL testing.
- Simulation platforms must be carefully selected to reduce the modeling effort by re-using developed models throughout various MBD stages shown in Figure 1.

3. Step II: Modeling of Hybrid Power Plant

3.1. Step IIa: Modeling Plant Components in State-Space

The small-signal models were developed using the state-space approach, as it allows one to represent each plant component separately and subsequently merge according to the balance of plant (BoP) shown in Figure 2 [18]. With regard to stability analysis, any unstable system mode can be consistently assessed by means of frequency and damping ratio (EV analysis) and ascribed to the causative state variables (PF analysis). Furthermore, state-space models can be directly converted to transfer functions, and thus applied in the control tuning stage [18]. It is possible to develop state-space models in either S-domain or Z-domain. However, since the dynamics in the power system applications involve a wide range of time constants and various sampling times for the involved subsystems, a continuous time domain tuning is preferred. Some considerations on the required level of model details for assessing voltage and frequency stability are given in [18]. It should be noted that the models described do not facilitate harmonic stability assessment and are solely be used for voltage and frequency stability assessment and control design and verification.

A set of state-space equations describe dynamic states of the system. The linearized differential equations of each plant component model are obtained by linearizing around steady-state values with resulting matrices A, B and C; D linking state vector x; input vector u; and output vector y according to Equation (1) [23].

$$\Delta \tfrac{dx}{dt} = A\Delta x + B\Delta u$$
$$\Delta y = C\Delta x + D\Delta u. \tag{1}$$

Each plant component model presented in this section is represented in a local synchronous rotating reference frame (SRRF) with dq-variables. The global reference frame (DQ-variables) of the HPP model is defined on the grid rotating at angular frequency ω_g. Equations (2) and (3) describe the Park transformation of reference frames. The matrix in Equation (3) is valid for inverter based units where the local SRRF is aligned with the d-axis grid voltage v_d and grid voltage angle δ_g (Figure 3a). In case of the genset the local SRRF rotates at rotor speed ω_r and is aligned with the q-axis voltage v_q and rotor angle δ_r as expressed by Equation (4). and illustrated in Figure 3b.

$$\begin{bmatrix} v_d \\ v_q \end{bmatrix} = [T] \begin{bmatrix} v_D \\ v_Q \end{bmatrix} \tag{2}$$

$$T_{PES} = \begin{bmatrix} \cos \delta_g & \sin \delta_g \\ -\sin \delta_g & \cos \delta_g \end{bmatrix} \tag{3}$$

$$T_{GS} = \begin{bmatrix} \sin(\delta_g + \delta_r) & -\cos(\delta_g + \delta_r) \\ \cos(\delta_g + \delta_r) & \sin(\delta_g + \delta_r) \end{bmatrix} \tag{4}$$

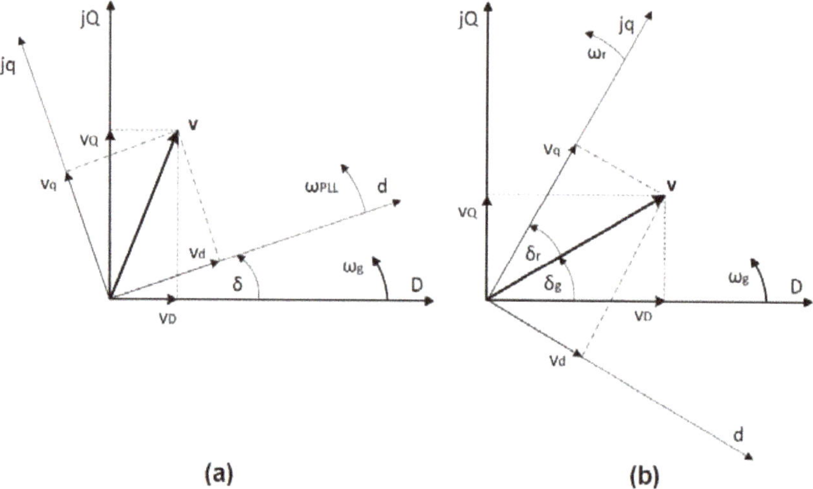

Figure 3. Relationship of voltage space vector components in local reference (qd) and global (DQ) reference frame (**a**) for inverter based units and (**b**) for genset.

The correct performance of linearized state-space models has been verified in MATLAB SimPowerSystems Toolbox by means of numerical models which are described in Section 3.3.

In the following subsections, the state-, input- and output vectors of each component state-space model are explained as they are of high relevance for the EV and PF analysis. The corresponding system matrices can be found in various references [17,24–27].

3.1.1. Grid-Forming Inverter

In Figure 4 the most typical structure of a grid-forming inverter with droop control mechanism and power-based synchronization is depicted. It is characterized by an ideal voltage source with low output impedance [14]. The grid-forming functionality can be part of the BESS, where the voltage is controlled by a DC/DC converter at the source side. The dynamic model of the grid-forming inverter is explained in [13] and [28] and serves as a basis for the subsequent state-space representation.

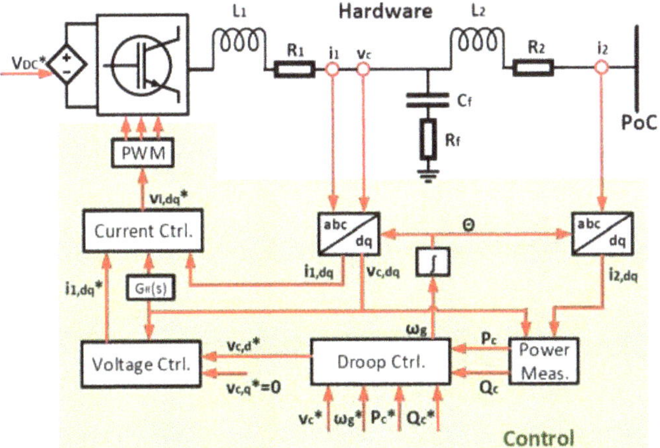

Figure 4. Schematic diagram of grid-forming inverter.

The state vector x_{FORM} of the small-signal model is given by Equation (5) where:

- i_{1d}, i_{1q}, i_{2d}, i_{2q}, v_{cd} and v_{cq} refer to the dynamic states of LCL filter. Note that it is necessary to represent the filter capacitance, as the voltage v_c is controlled in this control structure.
- φ_{id} and φ_{iq} are the state variables of the current controller (PI).
- φ_{vcd} and φ_{vcq} are the state variables of the outer voltage controller of type PI.
- $P_{c,avg}$ and $Q_{c,avg}$ refer to the dynamics of a low pass filter (LPF) for power measurements, leading to an average value for active and reactive power.

$$x_{FORM} = \begin{bmatrix} i_{1d} & i_{1q} & i_{2d} & i_{2q} & v_{cd} & v_{cq} & \varphi_{id} & \varphi_{iq} & \varphi_{vcd} & \varphi_{vcq} & P_{c,avg} & Q_{c,avg} \end{bmatrix} \tag{5}$$

The classic droop characteristics for active power sharing (ω_g/P_c) and reactive power sharing (v_c/Q_c) can be used, since the coupling impedance between grid-forming inverter and genset is mainly given by the grid-side inductor of the inverter ($X \gg R$). The essential components for the implementation of droop control are the LPF cut-off frequency $f_{LPF,PQ}$ applied for the power-based synchronization (Equation (6)) and the droop characteristics. They need to be parametrized in the context of the entire HPP to ensure stable operation in parallel to other grid-supporting DERs, as demonstrated later in Sections 4 and 5.

$$\begin{bmatrix} \frac{dP_{c,avg}}{dt} \\ \frac{dQ_{c,avg}}{dt} \end{bmatrix} = 2\pi \cdot f_{LPF,PQ} \cdot \left\langle \begin{bmatrix} P_c \\ Q_c \end{bmatrix} - \begin{bmatrix} P_{c,avg} \\ Q_{c,avg} \end{bmatrix} \right\rangle \tag{6}$$

It is defined that the grid-forming inverter provides the global reference frame to be used by the remaining HPP component models, in this way $d = D, q = Q$.

The input vector u_{FORM} is defined in Equation (7), where grid voltage variables v_{gD} and v_{gQ}, voltage reference V_c^*, frequency reference ω_g^*, active power reference P_c^* and reactive power reference Q_c^* act as input variables to the system.

$$u_{FORM} = \begin{bmatrix} v_{gD} & v_{gQ} & V_c^* & \omega_g^* & P_c^* & Q_c^* \end{bmatrix} \tag{7}$$

The output vector y_{FORM} provides the currents i_{2D} and i_{2Q} at the PoC and the grid frequency ω_g (Equation (8)).

$$y_{FORM} = \begin{bmatrix} i_{2D} & i_{2Q} & \omega_g \end{bmatrix} \tag{8}$$

Again, the power-electronic switches are not modeled explicitly, as their dynamic process is in the kHz range, and thus, not relevant for voltage and frequency control.

3.1.2. Grid-Feeding Inverter

Figure 5 shows a schematic diagram of a grid-feeding inverter which is characterized by a current-controlled source connected to the grid with high parallel impedance [14]. It is the most typical inverter control structure of grid-connected WTGs and PV systems [29]. The dynamics of the DER source side are not considered due to the decoupling effect of the inverter DC link. A comprehensive overview on the modeling and control design of a grid-feeding inverter is provided in [13] and [28] and was used in this study.

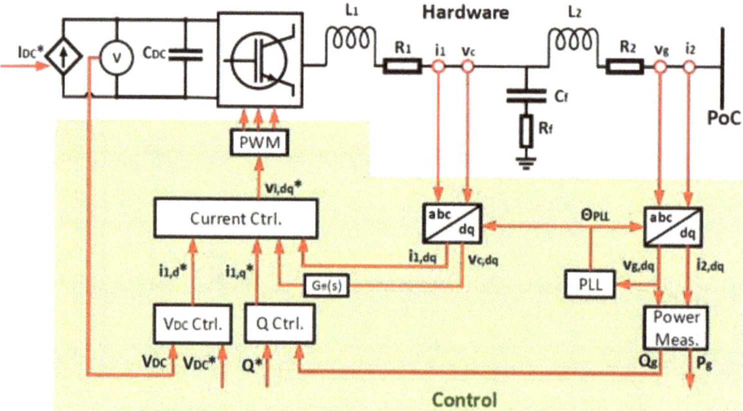

Figure 5. Schematic diagram of grid-feeding inverter.

Vector x_{FEED} (Equation (9)) describes the dynamic state variables of the system where:

- The dynamic states of LCL filter and current controller are the same as in grid-forming control mode. Note that it is not mandatory to represent the filter capacitance for the intended studies in this paper, as harmonics are not of concern. It is nevertheless retained to align the modeling representation of grid-forming and grid-feeding inverter control structure.
- φ_{vdc} and φ_{Qg} relate to the PI controllers of DC link voltage and reactive power respectively.
- V_{dc} corresponds to the dynamic of DC-link capacitor and $Q_{g,avg}$ to a LPF for reactive power measurement.
- φ_{PLL} belongs to a first-order filter of the phase-locked loop (PLL) and δ is the phase angle dynamic as a derivative of the grid frequency.

$$x_{FEED} = \begin{bmatrix} i_{1d} & i_{1q} & i_{2d} & i_{2q} & v_{cd} & v_{cq} & \varphi_{id} & \varphi_{iq} & \varphi_{vdc} & \varphi_{Qg} & V_{dc} & Q_{g,avg} & \varphi_{PLL} & \delta \end{bmatrix} \quad (9)$$

The input vector u_{FEED} is defined in Equation (10), where the grid voltage variables v_{gD} and v_{gQ}, the reactive power reference Q_g^* and the DC source current I_{dc}^* act as input variables to the system. I_{dc}^* represents an active power modulation of the WTG or PV system.

$$u_{FEED} = \begin{bmatrix} v_{gD} & v_{gQ} & Q_g^* & I_{dc}^* \end{bmatrix}. \quad (10)$$

The output vector y_{FEED} of the system (Equation (11)) provides the currents i_{2D} and i_{2Q} at the point of connection (PoC).

$$y_{FEED} = \begin{bmatrix} i_{2D} & i_{2Q} \end{bmatrix}. \quad (11)$$

It should be noted that an average model of the inverter is used to represent the pulse width modulation (PWM) of power-electronic switches, assuming that $v_{i,dq} = v^*_{i,dq}$ (Figure 5). This type of model preserves the average voltage dynamics over one fundamental period being necessary to design controls [8].

3.1.3. Generator Set

Figure 6 shows a schematic diagram of a genset with speed-governor droop and AVR droop function. The genset model consists of three main elements; i.e., electrically excited SG, speed governor and AVR. The SG is described by a 7th-order model including stator and rotor flux linkage dynamics and rotor dynamics [26]. The dynamics of prime mover and excitation system are modeled as a simple first-order time response [13,27]. Speed governor and AVR are implemented by a PID controller with design specifications given in [13].

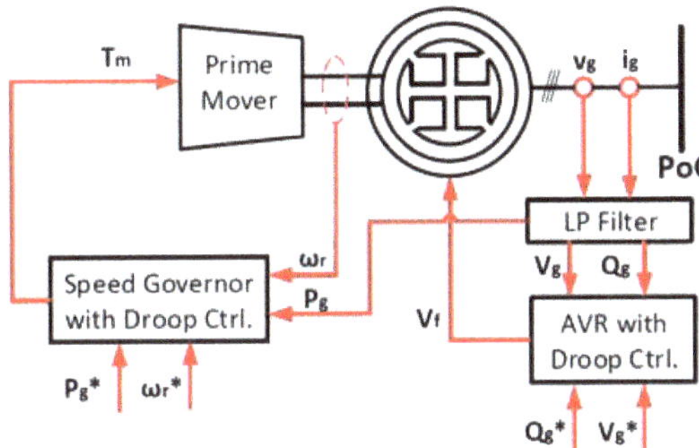

Figure 6. Schematic diagram of generator set.

The resulting state-space model of the entire genset is characterized by its state vector x_{GS} in Equation (12) where:

$$x_{GS} = \begin{bmatrix} i_{gd} & i_{fd} & i_{kd} & i_{gq} & i_{kq} & \omega_r & \delta_r & P_{g,avg} \\ Q_{g,avg} & v_{g,avg} & \varphi_{G1} & \varphi_{G2} & T_m & \varphi_{AVR1} & \varphi_{AVR2} & v_{fd} \end{bmatrix} \tag{12}$$

- The first five state variables result from the stator and rotor flux dynamics. i_{gd} and i_{gq} are the stator currents, i_{kd} and i_{kq} are currents in the damper winding and i_{fd} is the field winding current [26].
- The rotor dynamics are expressed by the swing equation (Equation (13)). It should be noted that $\Delta\omega_r$ is the rotor angular speed deviation from the grid angular frequency. H is the inertia time constant, D the damping factor coefficient and T_m and T_e the mechanical and electrical torques. An expression for the rotor angle displacement $\Delta\delta_r$ is given in Equation (14) [30].

$$H\frac{d\Delta\omega_r}{dt} = T_m - T_e - D \cdot \Delta\omega_r = T_m - T_e - D \cdot \left(\omega_r - \omega_g\right) \tag{13}$$

$$\frac{d\Delta\delta_r}{dt} = \Delta\omega_r = \omega_r - \omega_g \tag{14}$$

- $P_{g,avg}$, $Q_{g,avg}$ and $v_{g,avg}$ refer to the dynamics of a LPF for voltage and current measurement, leading to an average value for active and reactive power.
- φ_{G1} and φ_{G2} are the derivative and integral states of the governor PID controller, while T_m relates to the change in mechanical torque due to prime mover dynamics (fuel actuator and combustion engine).

- φ_{AVR1} and φ_{AVR2} are the derivative and integral states of the AVR, while v_{fd} corresponds to the dynamics of the excitation voltage.

Droop characteristics are implemented for both speed governor (ω_r/P_g) and AVR (v_g/Q_g).

The input vector $\mathbf{u}_{GS}$ is given in Equation (15), where grid voltage variables v_{gD} and v_{gQ}, voltage reference V_g^*, speed reference ω_r^*, active power reference P_g^* and reactive power reference Q_g^* act as input variables to the system. It should be noted that the grid frequency ω_g, imposed by the grid-forming inverter, is an input variable as well.

$$\mathbf{u}_{GS} = \begin{bmatrix} v_{gD} & v_{gQ} & \omega_r^* & P_g^* & V_g^* & Q_g^* & \omega_g \end{bmatrix} \tag{15}$$

The output vector $\mathbf{y}_{GS}$ (Equation (16)) involves the currents i_{gD} and i_{gQ} at the generator terminals and the rotor speed ω_r.

$$\mathbf{y}_{GS} = \begin{bmatrix} i_{gD} & i_{gQ} & \omega_r \end{bmatrix} \tag{16}$$

In cases when a grid-forming inverter is absent, ω_g is not an input variable any longer. In this case, the rotor speed becomes synchronized with the grid frequency ($\omega_r = \omega_g$).

3.2. Step IIb: Merging of State-Space Models

The remaining components of the HPP are distribution lines, transformers and loads. The corresponding state-space expressions are well-known and can be found in [17].

Subsequently, all grid components and the individual DER state-space models need to be connected according to BoP shown in Figure 2 (black colored). This is achieved by linking input and output variables of the models in the global SRRF, as indicated by the functional diagram in Figure 2 (red colored). As a result a multiple-input-multiple-output (MIMO) model with matrices A_{HPP}, B_{HPP}, C_{HPP} and D_{HPP} is obtained [18].

3.3. Step IIc: Developing Discrete-Time Domain Models for Control Verification

So far, linearized state space models have been developed in the S-domain for the purpose of small-signal analysis and controller tuning. At this stage, it was proposed to prepare in parallel numerical simulation models that, on the one hand, can validate the developed small-signal models, and on the other hand, can be used later on for RCP (Step V) and control verification (Step VI). Here, dynamic models of WTG, PV, BESS and genset were to be implemented by means of discrete-time domain models [31]. Thus, they can be used both for accelerated off-line simulation studies and for RT-HIL verification, as explained in Section 1.

The grid-forming inverter model described in Section 3.2 does sufficiently represent the dynamic behavior of BESSs. The transient response for voltage and frequency control is dominated by the inverter system, while the voltage/current dynamics of the battery cells can be neglected ($V_{DC}^* = const.$; see Figure 4) [31].

With regards to WTG and PV system, it is necessary to further extend the model capabilities beyond the inverter system in order to conduct realistic test scenarios for control assessment. More specifically, DC current variations are expected due to changes in power ($I_{DC}^* \neq const.$; see Figure 5). Some simple performance models of WTG and PV are proposed and validated by field data in [31]. They are capable of emulating power output variations according to wind speed v_w, solar irradiance G, temperature T and active power reference P_{WTG}^* or P_{PV}^*, respectively. Figure 7 shows how the inverter models described in Section 3.2 are coupled with the performance models proposed in [31]. P_{WTG} and P_{PV} are AC power output at the DER's PoC. Hence, the model coupling block accounts for inverter and possible transformer losses and calculates the DC current input to the inverter model by using the generated DC power $P_{DER,dc}$ of each DER, respectively (Equation (17)).

$$I_{DC}^* = \frac{P_{DER,dc}}{V_{dc}} \tag{17}$$

Input and output signals of the grid-feeding inverter model are as described in Figure 5.

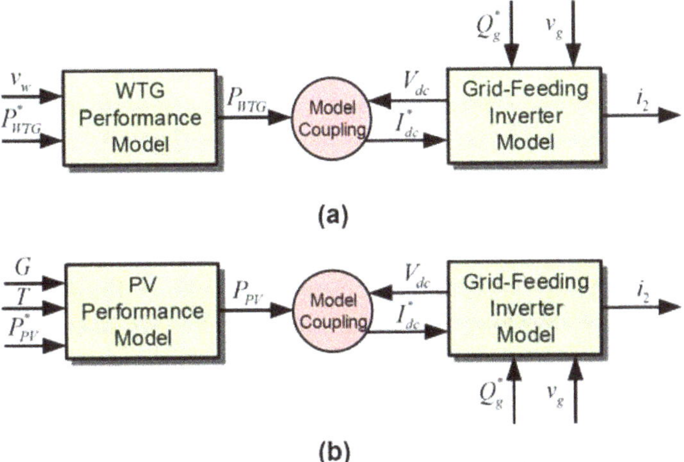

Figure 7. Schematic diagram of numerical models applicable to rapid control prototyping (RCP) and control verification: (**a**) wind turbine generator (WTG); (**b**) photovoltaic (PV).

4. Voltage and Frequency Stability Assessment

This section deals with the assessment of voltage and frequency stability by using the state-space models described in Section 3.1 and 3.2.

4.1. Step IIIa: Defining Relevant Operating Sccenarios

The EVs of state matrix A_{HPP} were determined for two essential operational scenarios of the HPP [3]: (1) only grid-forming inverters (i.e., BESS) and grid-feeding inverters (i.e., WTG, PV) are in operation; (2) all inverter based DERs and genset are in operation. For each scenario, matrix A_{HPP} is updated by model linearization around steady-state values, which are within the range of normal operating conditions; i.e., $V_g = \{0.85, 1.15\}$pu, $f_g = \{48, 52\}$ Hz, $P_{LD} = \{0, 1\}$pu, $P_{WTG} = \{0, 1\}$pu, $P_{PV} = \{0, 1\}$pu. In this way, a thorough stability assessment is assured as the system behavior might depend on its actual operating state, i.e., permitted voltage and frequency deviations [20], and partial or full loading of DERs.

The EVs represent dynamic modes of the HPP system and are characterized by its frequency and damping ratio. Then, in order to identify the dominant state variables participating in a particular dynamic mode, it is necessary to determine the participation matrix P_i. Its PFs p_{ki} are obtained by Equation (18), where the right (ϕ_{ki}) and left (ψ_{ki}) eigenvectors of the system matrix A_{HPP} are used. The magnitudes of p_{ki} provide a measure of the relative participation of the kth state variable in the ith mode and vice versa. [23]

$$P_i = \begin{bmatrix} p_{1i} \\ p_{2i} \\ \vdots \\ p_{ni} \end{bmatrix} = \begin{bmatrix} \phi_{1i} & \psi_{1i} \\ \phi_{2i} & \psi_{i2} \\ \vdots & \vdots \\ \phi_{ni} & \psi_{in} \end{bmatrix} \tag{18}$$

In both test scenarios, the cut-off frequency for the grid-forming inverter's power measurement filter is per default $f_{LPF,PQ} = 5$ Hz according to [13]. The droop characteristics are set to 5% for both frequency/active power control and voltage/reactive power control.

4.2. Step IIIb: Clustering of Eigenvalues

Figure 8a shows the EV map for HPP operation with only grid-forming and grid-feeding inverters. It was ascertained that all poles were located in the left hand side of the complex plane, indicating that the plant model is stable. The distribution of EVs demonstrates that their location depends on the linearization points. A detailed analysis of EV movement was not attempted due to the vast number of dynamic modes. Instead, the focus of this analysis was to determine absolute stability by identifying EVs in the right-half plane and relative stability by observing the EVs with lowest damping ratio.

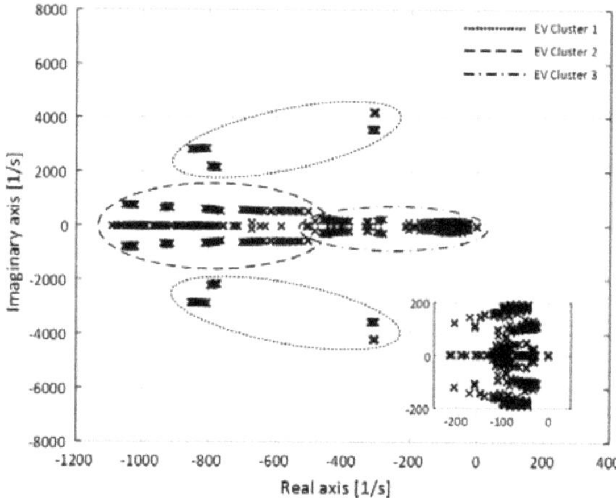

Figure 8. Eigenvalue map of A_{HPP}—Test scenario 1: only inverter based distributed energy resources (DERs) in operation.

In order to illustrate the extensive results of EV and PF analysis more effectively, a clustering of EVs was attempted. Table 1 presents several clusters of EVs according to their eigenfrequencies and associated state variables. The identified EV clusters are encircled in Figure 8.

Table 1. Clustering of eigenvalues for test scenario 1.

EV Cluster	Eigenfrequency f_n [Hz]	Damping Ratio ζ [-]	Associated Dominant State Variables
1	{185, 20 × 10³}	{0.03, 0.79}	Plant inductances and capacitances
2	{130, 160}	1	Inverter inner current control loops
3	{5, 30}	{0.39, 1}	Inverter outer control loops

Cluster #1 involves the time constants of physical plant parameters; i.e., inductances and capacitances of inverter LCL filters and DC link, plant transformers and distribution lines. Some EVs are highly underdamped ($\zeta = 0.03$). However, their sensitivity to a plant's operating condition is negligible and their damping ratio does not become negative which otherwise would lead to system instability. Too much attention should not be paid to EV cluster #1, as it concerns the high frequency range relevant for harmonic stability, which is not in scope of this study. Small-signal models of the inverter switching phenomena are required for accurate harmonic analysis.

EV cluster #2 concerns the state variables related to the inverters' current control loops. Their dynamics are non-oscillatory ($\zeta = 1$).

All dynamic states of the inverters' outer control loops are assigned to EV cluster #3. They include PLL, DC link voltage and reactive power controllers of grid-feeding inverters and voltage controller and power measurement filter of the grid-forming inverter. The EVs in cluster #3 are highly relevant for voltage and frequency stability of the HPP. The critical EV $\lambda_{1,cr}$ in this test scenario is the one with lowest damping ratio ($\zeta = 0.39$). $\lambda_{1,cr}$ exhibits an eigenfrequency of $f_n \approx 30$ Hz and corresponds to the DC link voltage dynamics of grid-feeding inverters. The property of $\lambda_{1,cr}$ is solely determined by the design of DC link voltage controller whose parameters are not subject to any change during HPP operation.

The slowest dynamic mode in the HPP is related to the power measurement filter of grid-forming inverter where the LPF cut-off frequency is $f_{LPF,PQ} = 5\,\text{Hz}$. The associated EV exhibits a damping ratio of $= 1$. This is expected, as dynamic load changes are compensated by the grid-forming inverter only. Test scenario 2 will demonstrate that the EV properties change if power is shared by multiple droop controlled DERs.

4.3. Step IIIc: Applying Corrective Measures for System Stability

Figure 9a depicts the EV map for a scenario where all inverter based DERs and the genset are in operation. It can be seen that one EV pair is located in the right-half plane, indicating that the system is unstable. This EV has an eigenfrequency of $f_n \approx 18\,\text{Hz}$ and its damping ratio ranges between $\zeta = \{0.02, -0.1\}$.

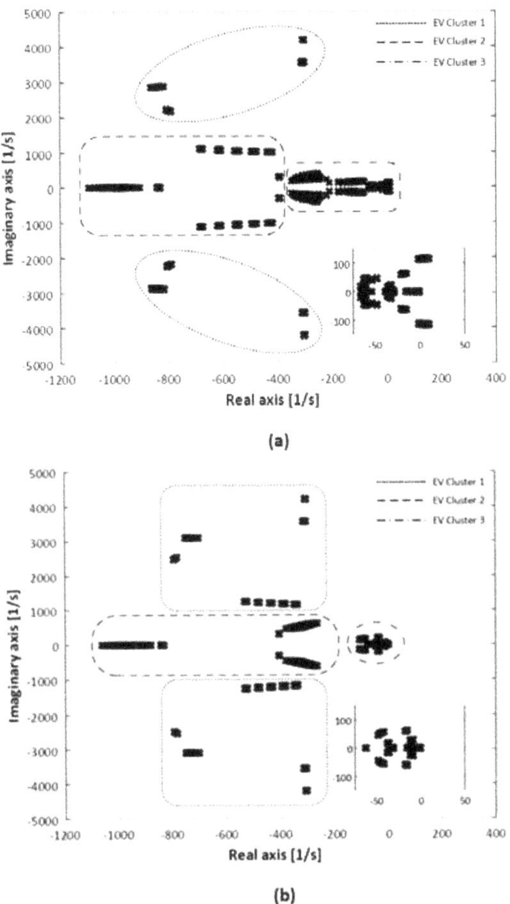

Figure 9. Eigenvalue maps of A_{HPP}—Test scenario 2: all DERs in operation: (**a**) unstable case; (**b**) stable case.

In fact, negative damping is ascertained only during operating points where the BESS is in charging mode. This emphasizes the need for assessing the entire range of scenarios for HPP operation.

The PFs reveal that this oscillatory mode is associated with state variables of SG stator and rotor flux (i_{gd}, i_{fd}, i_{kd}) and of grid-forming inverter (φ_{vcd}, i_{1d}). In order to stabilize the system the focus needs to be laid on these particular state variables. On the one hand, it is not feasible to modify the physical characteristics of the genset. On the other hand, adjustments in the inverter control loops can result in a stable system. The cascaded control structure for voltage and current regulation is designed according to the physical characteristics of the inverter (i.e., LCL filter, switching frequency) [13]. Hence, re-tuning of these control loops might not be desired. Another way is to adjust the voltage feed-forward filter $G_{ff}(s)$ (see schematic diagram in Figure 4). The voltage feed-forward term is used to minimize the initial transient effect of the current. Generally, a filter with very high cut-off frequency (e.g., 20 kHz) is applied, and thus, it was neglected so far [28]. However, it can be adjusted to lower values to prevent system stability issues. Thereby, two additional state variables $v_{cd,avg}$ and $v_{cq,avg}$ are introduced to the state-space model (Equation (19)). The cut-off frequency is tuned to $f_{LPF,vc} = 30$ Hz to yield a sufficient damping ratio of the EVs ($\zeta = 0.42$).

$$\begin{bmatrix} \frac{dv_{cd,avg}}{dt} \\ \frac{dv_{cq,avg}}{dt} \end{bmatrix} = 2\pi \cdot f_{LPF,vc} \cdot \left\langle \begin{bmatrix} v_{cd} \\ v_{cq} \end{bmatrix} - \begin{bmatrix} v_{cd,avg} \\ v_{cq,avg} \end{bmatrix} \right\rangle \tag{19}$$

The EV map in Figure 9b demonstrates that the HPP is now stable in every operating condition. A time-domain simulation using a numerical model of the HPP (see Section 3.3) is performed to verify the observed stability phenomena. Figure 10 shows the grid frequency during a time interval $t = \{0,5\}$s with $f_{LPF,vc} = 20$ kHz and during a time range of $t = \{5,10\}$s where the filter cut-off frequency is adjusted to $f_{LPF,vc} = 30$ Hz. Grid frequency oscillations with negative damping are observed for $t < 5$ s, while the system stabilizes after filter tuning at $t = 5$ s.

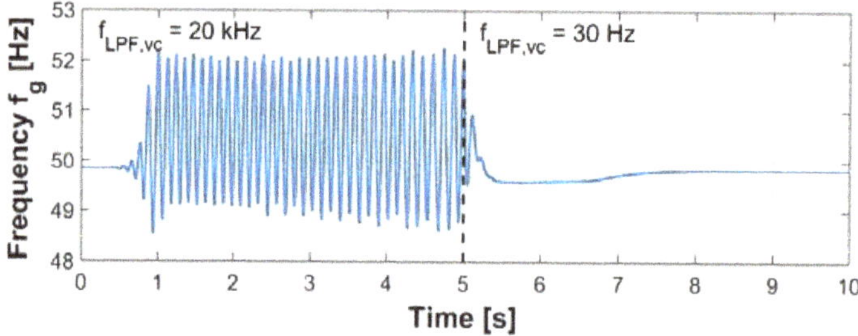

Figure 10. Verification of stability phenomena by time-domain simulations.

In the next step, the occurring EVs, shown in Figure 9b, are characterized by means of Table 2. The resulting EV clusters are encircled in Figure 9.

Table 2. Clustering of eigenvalues for test scenario 2 (stable case).

EV Cluster	Eigenfrequency f_n [Hz]	Damping Ratio ζ [-]	Associated Dominant State Variables
1	$\{225, 20 \times 10^3\}$	$\{0.03, 0.76\}$	Plant inductances and capacitances
2	$\{80, 160\}$	$\{0.61, 1\}$	Inverter inner current control loops, SG flux
3	$\{0.3, 34\}$	$\{0.19, 1\}$	Inverter outer control loops, SG flux, excitation system, genset mechanical time constants and control loops

The EVs in cluster #1 are associated with the same state variables as in test scenario 1.

The second cluster involves the dynamics of both inverters' current controller and SG stator and rotor flux. All EVs in cluster #2 are sufficiently damped ($\zeta \geq 0.61$).

The third EV group includes all dynamics within the frequency range of below 35 Hz. It is impossible to provide a more granular classification due to the vast number of state variables associated with each EV. The PF analysis revealed that the dynamic modes of cluster #3 are linked with inverters' outer control loops and all genset dynamic states; i.e., electrical time constants of SG, mechanical time constants (inertia, prime mover) and control loops (governor, AVR). It should be noted that the presence of the genset introduces some dynamics in the very low frequency range ($f_n < 2$ Hz), caused by speed governor and AVR. In this test scenario, the critical EV $\lambda_{2,cr}$ with the lowest damping ratio ($\zeta = 0.19$) exhibits an eigenfrequency of $f = 11$ Hz. The corresponding PFs indicate that state variables of grid-forming inverter ($P_{c,avg}$) and genset ($\omega_r, \delta_r, P_{g,avg}, T_m, \varphi_{G2}$) have a dominant impact on the EV. In conclusion, this dynamic mode is associated with the power sharing performance between grid-forming inverter and genset. Oscillatory behavior is caused by the inverter power measurement filter in combination with genset's inertia and active power control.

The major conclusion from this subsection is that stability issues are of much more concern in scenario 2 (parallel operation of grid-forming inverter and genset) than in scenario 1 (operation with only inverter based DERs). Additionally, it is demonstrated by means of Figure 9 that the EVs of cluster #3 are located nearest to the imaginary axis. Hence, it is most crucial to observe the low frequency dynamics (<35 Hz) during the small-signal analysis.

4.4. Step IIId: Performing Sensitivity Analysis

A sensitivity analysis was applied to investigate the impact of various parameters on system stability during test scenario 2. Of particular interest is $\lambda_{2,cr}$, as it exhibits relatively low damping. The criterion for absolute system stability is $\zeta > 0$. However, in order to avoid critically low damping of any voltage or frequency oscillations in the HPP, a reasonable target for the minimum damping ratio is $\zeta_{min} = 0.05$, as specified in Section 2.

4.4.1. Test Case 1: Power Measurement Filter of Grid-Forming Inverter

Power-based synchronization means that the grid-forming inverter is synchronized based on the power exchange between inverter and grid rather than measuring the voltage by means of PLL. The LPF of power measurements (see Figure 4) acts as a delay to any grid power variations, and hence affects the power sharing performance. Section 4.3 showed that the dynamic state of LPF ($P_{c,avg}$) is associated with $\lambda_{2,cr}$. Thus, the sensitivity of damping ratio to the filter cut-off frequency $f_{LPF,PQ}$ needs to be assessed. Up to now, a arbitrary value of $f_{LPF,PQ} = 5$Hz has been applied. In Figure 11 the damping ratio of $\lambda_{2,cr}$ is shown for various values within $f_{LPF,PQ} = \{1, 30\}$Hz and in steps of $\Delta f_{LPF,PQ} = 0.5$ Hz. It is observed that the damping ratio ζ hits bottom at $f_{LPF,PQ} = 5$ Hz and increases with rising values of $f_{LPF,PQ}$. However, it remains above $\zeta > 0.15$. Thus, relative system stability is not seriously influenced by the value for $f_{LPF,PQ}$.

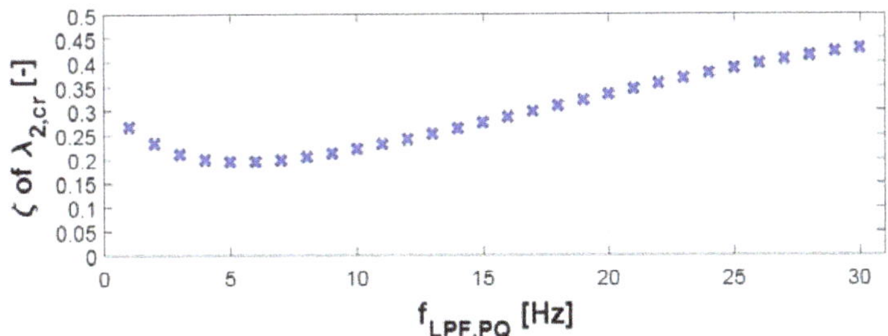

Figure 11. Sensitivity of damping ratio of critical eigenvalues (EV) $\lambda_{2,cr}$ to low pass filter (LPF) cut-off frequency $f_{LPF,PQ}$.

4.4.2. Test Case 2: Frequency/Active Power Droop Characteristic

The dynamic performance of power sharing between various DERs is dependent on the droop characteristic selected [13]. The correlation between frequency (or speed) and active power in steady-state is described by Equation (20), where m_P is the so-called droop gain and ω^* and P^* are frequency (or speed) reference and active power reference, respectively.

$$\omega - \omega^* = -m_P \cdot (P - P^*) \tag{20}$$

Typical values for the droop gain are in the range of $m_P = \{2, 12\}\%$ [32]; however, an optimum range is yet to be ascertained.

It is well known that the composite frequency/power characteristic of a power system contributes to the overall system damping [23]. In fact, there is an inverse relationship between system damping and droop gain m_P. Since EV $\lambda_{2,cr}$ is associated with state variables of both droop controlled DERs in the HPP, the sensitivity of its damping ratio to droop gain m_P was investigated. The results are presented in Figure 12 where the droop gain was incremented with $\Delta m_P = 1\%$. As expected, the damping ratio decreases significantly for increasing droop gains. It is not recommended to apply larger droop gains than $m_P > 10\%$, as the damping ratio falls below the limit of $\zeta_{min} = 0.05$.

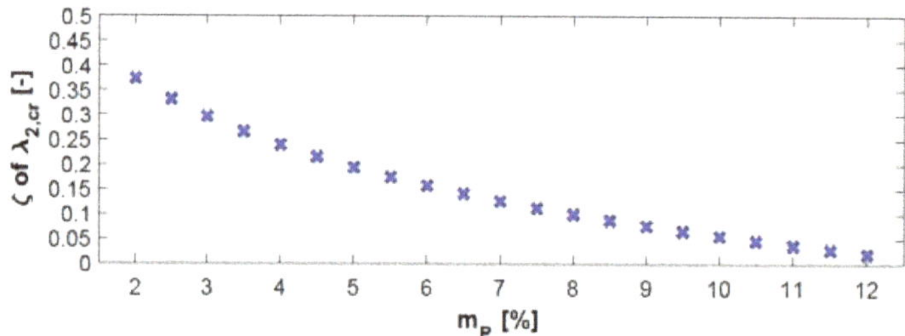

Figure 12. Sensitivity of damping ratio of critical EV $\lambda_{2,cr}$ to f/P droop gain m_P.

The major conclusion from this section is that selecting feasible values for the droop gain m_P are of much higher concern for system stability than tuning the power measurement filter of the grid-forming inverter due to large sensitivity of EV $\lambda_{2,cr}$ to droop gain m_P.

5. Step IV: Design and Tuning of Hybrid Power Plant Controller

This section deals with the design and subsequent tuning of the central HPPC. As explained in Section 2, primary control actions by DERs will leave steady-state errors in voltage and frequency which need to be compensated by secondary control actions of the central HPPC.

5.1. Step IVa: Designing Hybrid Power Plant Controller

Figure 13 depicts a system representation of the HPPC control architecture which is valid for both frequency and voltage control, yet being decoupled from each other. The plant system consists of DERs with primary voltage and frequency control (i.e., BESS and genset), the HPP internal grid consisting of the remaining DERs (i.e., WTG and PV), cables and transformers and a load which represents a disturbance to the system (ΔP_{LD} and ΔQ_{LD}). The measurement block contains an LPF for voltage and frequency measurements, respectively. The time delay block e^{-sT} reflects the communication (T_{com}) and sampling delay ($T_{s,HPPC}$) between HPPC and DERs when collecting all feedback signals and sending reference signals to the individual DERs. According to [33] an aggregated time delay describing the entire process can be defined by Equation (21).

$$T = 0.5 \cdot T_{s,HPPC} + T_{com} \tag{21}$$

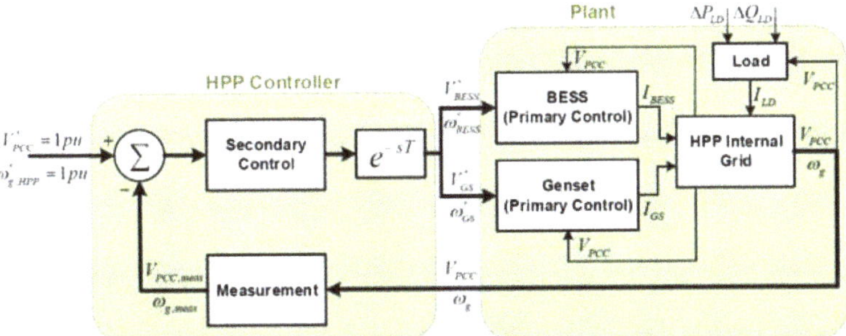

Figure 13. System representation for hybrid power plant voltage and frequency controller.

The secondary control loop should be slower than the primary controller. Instead of a standard PI controller, a simple integral (I) controller with transfer function $G_{SC}(s)$ was chosen (Equation (22)), since control inputs and outputs have equal units, and thus, a proportional gain is not required.

$$G_{SC}(s) = \frac{1}{\tau_{SC} \cdot s} \tag{22}$$

The control time response τ_{SC} shall regard the minimum bandwidth of the plant system which can be described by means of transfer functions using the respective input and output signals (u, y) of the MIMO state-space model with A_{HPP}, B_{HPP}, C_{HPP}, D_{HPP} and Equation (23) [23].

$$G_{u-y}(s) = \frac{\Delta y(s)}{\Delta u(s)} = C_{HPP}(sI - A_{HPP})^{-1}B_{HPP} + D_{HPP} \tag{23}$$

The secondary frequency controller was designed using $G_{\omega^*_{BESS}-\omega_g}$ and $G_{\omega^*_{GS}-\omega_g}$ The bandwidth of $G_{\omega^*_{BESS}-\omega_g}$ is infinite as the grid-forming inverter can regulate its frequency output almost instantenously. In case of the genset, the time response is given by speed governor and prime mover. Hence, the bandwidth ω_{bw} of $G_{\omega^*_{GS}-\omega_g}$ is applied to calculate the time response $\tau_{SC,f}$ of secondary frequency controller (Equation (24)).

$$\tau_{SC,f} = \frac{1}{\omega_{bw}\left(G_{\omega^*_{GS}-\omega_g}\right)} \tag{24}$$

The secondary voltage controller is designed using $G_{V^*_{BESS}-V_{PCC}}$ and $G_{V^*_{GS}-V_{PCC}}$ The bandwidth of $G_{V^*_{BESS}-V_{PCC}}$ is around 20 times higher than of $G_{V^*_{GS}-V_{PCC}}$ due to relatively slow dynamics of AVR and excitation system. Hence, the bandwidth ω_{bw} of $G_{V^*_{GS}-V_{PCC}}$ is applied to calculate the time response $\tau_{SC,V}$ of secondary voltage controller (Equation (25)).

$$\tau_{SC,f} = \frac{1}{\omega_{bw}\left(G_{V^*_{GS}-V_{PCC}}\right)} \tag{25}$$

The control sampling rate $T_{s,HPPC}$ is to be chosen according to the smallest time constant according to Nyquist (Equation (26)).

$$T_{s,HPPC} < 0.5 \cdot \min\left[\ \tau_{SC,f}\quad \tau_{SC,V}\ \right] \tag{26}$$

5.2. Step IVb: Assessing Control Performance

In this subsection, the dynamic performance of the HPP voltage and frequency control is assessed by looking into step response characteristics in time-domain. This is accomplished by using the MIMO state-space model of the HPP. During the test cases it is assumed that no communication delays are present ($T_{com} = 0$).

5.2.1. Test Case 1: Frequency Control

An active power load change of $\Delta P_{LD} = 0.25$ pu was applied as an input to the MIMO system of the HPP (see Figure 13) in order to test the performance of frequency regulation and active power sharing between grid-forming inverter and genset. In Figure 14 the output signals Δf_g, ΔP_{BESS} and ΔP_{GS} of the MIMO system are shown for two different droop gains, $m_P = 2\%$ and $m_P = 7\%$, respectively.

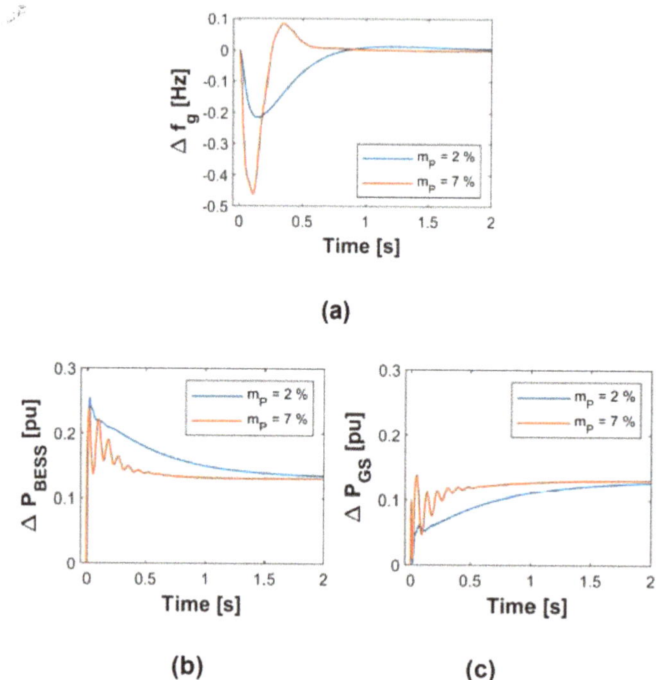

Figure 14. Transient load sharing performance between abattery energy storage system (BESS) and fossil fuel generator set (genset) for various f/P droop gains m_P; (**a**) frequency; (**b**) BESS active power; (**c**) genset active power.

As identified during the EV analysis (Section 4.2) the transient power oscillations are much more dominant for large droop gains due to decreased damping ratio of EV $\lambda_{2,cr}$. Larger values ($m_P > 7\%$) lead to undesired low damping, and thus, were not further considered.

By observing the settling time of active power sharing, it can be concluded that small droop gains lead to prolonged transient response of P_{BESS} and P_{GS}. While for $m_P = 7\%$ the steady-state value was reached at $t = 0.89$ s, a new steady-state value is obtained at $t = 2.45$ s for a droop gain of $m_P = 2\%$.

The transient response of grid frequency f_g is mainly affected with regard to its Nadir. A steep droop characteristic ($m_P = 7\%$) leads to larger initial frequency deviations than a flat droop characteristic ($m_P = 2\%$).

5.2.2. Test Case 2: Voltage Control

The aim of this test case is to assess the voltage control performance for various voltage/reactive power droop gains which typically lie within $m_Q = \{2,7\}\%$ [32]. The relation between voltage and reactive power is explained by Equation 27 where V^* and Q^* are voltage reference and reactive power reference, respectively.

$$V - V^* = -m_Q \cdot (Q - Q^*) \tag{27}$$

In order to assess the reactive power sharing performance between grid-forming inverter and genset, a reactive power load change of $\Delta Q_{LD} = 0.25$ pu was applied as an input to the MIMO system (see Figure 13). In Figure 15 the output signals ΔV_{PCC}, ΔQ_{BESS} and ΔQ_{GS} of the MIMO system are shown for the droop gains $m_Q = 2\%$ (a) and $m_Q = 7\%$ (b) respectively.

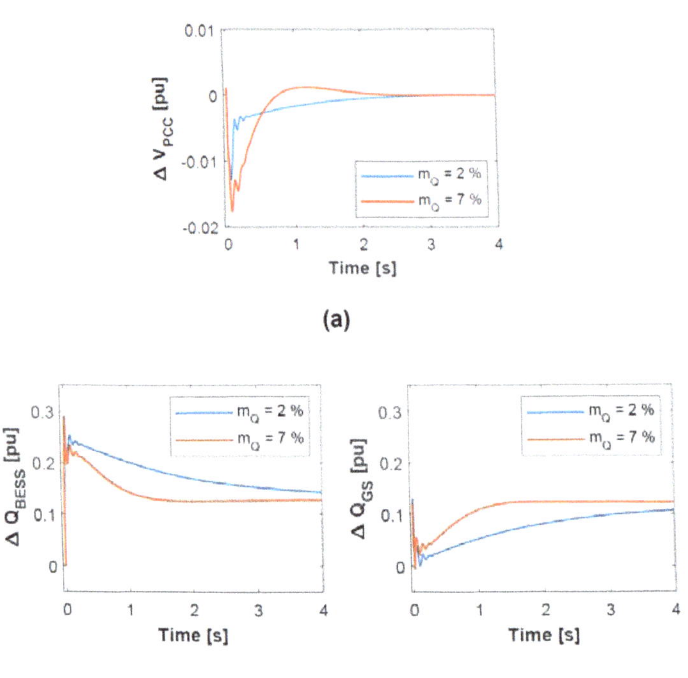

Figure 15. Transient load sharing performance between BESS and genset for various V/Q droop gains m_Q (**a**) PCC voltage; (**b**) BESS reactive power; (**c**) genset reactive power.

It was observed that the oscillatory behavior of voltage and reactive power was not affected by the droop gain.

With regard to settling time, similar conclusions to those in test case 1 were drawn: The steady-state value was reached at $t = 1.40$ s for a large droop gain $m_Q = 7\%$, while for a small droop gain of $m_Q = 2\%$ a new steady-state value was obtained much later at $t = 6.55$ s.

It is obvious from Figure 15 that the largest voltage drop occurs for a droop gain of $m_Q = 7\%$. However, in both cases the remaining voltage drop is $\Delta V_{PCC} \gg -2\%$ for a reactive power load change of $\Delta Q_{LD} = 0.25$ pu. Hence, it is not expected that the voltage limits are violated during a large-signal event (up to 100% change in reactive power) for any droop gain m_Q.

5.3. Step IVc: Control Tuning

The assessment studies have shown that the performance of voltage and frequency regulation is highly associated to the applied droop gains. In fact, the bandwidth values of the plant system, and hence the tuned parameters $\tau_{SC,f}$ and $\tau_{SC,V}$ of the secondary controller, depend on m_P and m_Q, respectively. Several aspects are to be considered for selecting optimum droop gains:

- Frequency requirements: Figure 14 has shown that the frequency Nadir is deteriorated for large droop gains. Under-frequency load shedding (UFLS) schemes might be triggered depending on the occurring event (e.g., loss of unit) and UFLS characteristic. Detailed numerical simulations are required in order to assess such requirements.

- Power sharing performance: Figure 16 summarizes in numbers how the settling time of active and reactive power increases with decreasing droop gains of BESS and genset. Proper droop gains shall be chosen according to the desired update rate of active and reactive power dispatch. Such a function is required, e.g., for BESS state-of-charge regulation and to avoid underloading/overloading of gensets according to dynamic fluctuations of renewable power. If active and reactive power setpoints are dispatched, e.g., every 3 s, the minimum droop gains shall be $m_P = 2\%$ and $m_Q = 4\%$ in order to allow settlement of the respective parameters (Figure 16).
- Control sampling rate: According to Equation (26) the required sampling rate $T_{s,HPPC}$ and hence the signal exchange between HPPC and DERs depends on the desired speed of secondary controller. In order to achieve the settling times depicted in Figure 16, the sampling rate must be $T_{s,HPPC} \leq 50$ ms for $m_P = 7\%$ and $T_{s,HPPC} \leq 250$ ms for $m_P = 2\%$. Hence, the computational power of the control platform and communication latency issues shall be well assessed prior to selecting a certain droop gain.

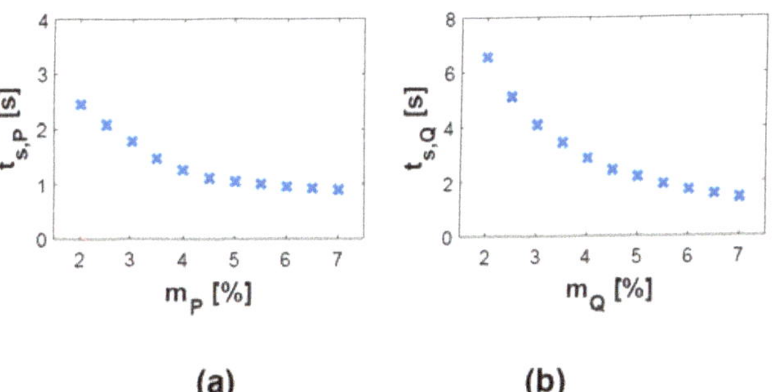

Figure 16. Settling time of (**a**) active power ($t_{s,P}$) and (**b**) reactive power ($t_{s,Q}$) of BESS and genset for various droop gains m_P and m_Q.

At this stage, it needs to be stressed that voltage and frequency control requirements for MGs/off-grid systems are usually not defined explicitly, and if so, are based on general technical guidelines only; e.g., as for rural electrification systems in [20]. However, the methodology for control design and tuning presented in this section enables to propose guidance on the practical implementation to the operators of such systems.

6. Step V: Rapid Control Prototyping

In this section, the performance of voltage and frequency control in the off-grid HPP shown in Figure 2 was evaluated by means of simulating representative test scenarios. While up to now all results were obtained by means of the linearized state-space model, in this section time-domain studies are described, which used discrete-time domain models (see Section 3.3) implemented in MATLAB SimPowerSytems Toolbox. In this way, the robustness of stability analysis and the HPPC design and tuning stage can be tested.

A set of wind speeds and solar irradiance measurements was utilized and fed into the performance models described in Section 3.3. A realistic aggregated load profile for the demand subsystem was applied where the load power factor was $\cos\varphi = 0.9$. The DER and HPPC settings during this test scenario are summarized in Table 3.

Table 3. Test settings for rapid control prototyping.

	BESS	**Genset**	**WTG**	**PV**	**HPPC**
P^* [kW]	−45	45	80	40	-
Q^* [kW]	0	0	20	10	-
$m_P = m_Q$ [%]	5	5	-	-	-
$T_{s,HPPC}$ [ms]	-	-	-	-	50

The following events are simulated in order to assess voltage and frequency control during various operating modes:

0. Operation with all DERs until $t < 10\ s$;
1. Disconnection of PV at $t = 10\ s$;
2. Disconnection of genset at $t = 15\ s$;
3. Disconnection of WTG at $t = 20\ s$.

Figure 17 presents the resulting grid frequency (a), voltage (b) and active and reactive power profiles (Figure 17c–g). The results show that grid frequency and PCC voltage are regulated to 1 pu after every single event. Power is shared evenly between BESS and genset after PV disconnection. The largest drop of frequency and voltage occurs after genset tripping, since it is highly loaded ($P_{GS}(t = 14.9\ \text{s}) \approx 60$ kW) and the BESS needs to solely compensate for the power deficit.

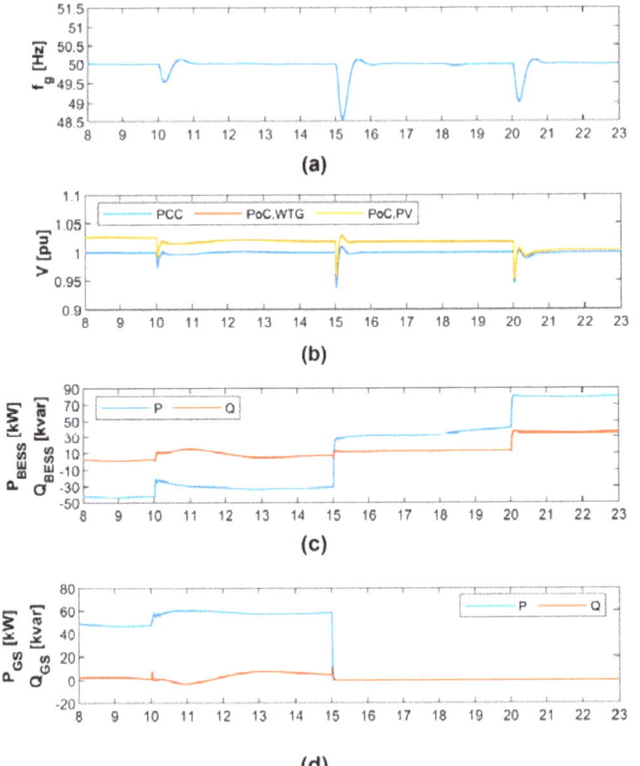

Figure 17. *Cont.*

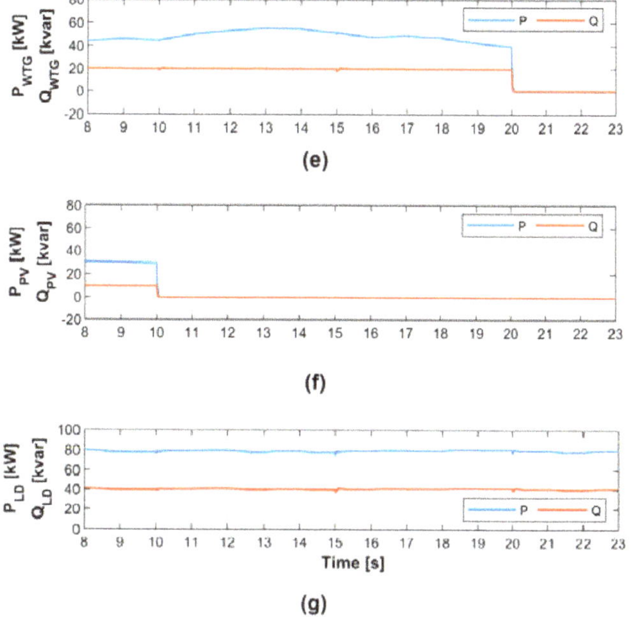

Figure 17. Results of control verification using discrete-time domain simulation models.

Usually, during the RCP stage iterations of control design and tuning are expected depending on the performed tests and their outcome. However, the presented simulation studies demonstrate the effectiveness of performing stability analysis and control design/tuning for voltage and frequency control by applying linearized state-space models which have been validated against numerical models. In this way, the entire MBD process is accelerated.

7. Step VI and VII: Control Verification and Validation

7.1. Step VI: Real-Time Hardware-in-the-Loop Verification

As a final step of MBD in HPPs, the control algorithms were implemented on a controller platform and tested. Verification of the HPPC was accomplished by connecting the controller platform to a RT model of the power system including DERs; i.e., WTGs, PV system, BESS, gensets. Moreover, for realistic testing a RT model of the communication networks was used. Thus, the controller platform including the developed algorithms could be assessed in realistic conditions. Grid events that cannot be measured in real life can also be replicated in a controlled environment while data traffic associated to specific communication network technologies are captured properly without actually involving the real technologies [34].

One might argue the verification stage might be dropped and on-site testing of off-grid HPPs can be realized immediately as being independent of any external grid parameters. However, it seems impracticable and cumbersome to deploy and test HPPCs on-site for remote applications, before gaining further confidence by extensive testing and verification in a controlled system environment.

The existing facilities in the Smart Energy Systems Laboratory at Aalborg University allowed all the above design and verification procedure. The architecture of this platform is shown in Figure 18 [35]. The information and communication technology (ICT) layer is the backbone for the setup and aims to emulate different technologies and topologies for the communication network. It is crucial to account for communication bandwidth, latency and potential packet losses for all signals being exchanged between HPPC and DERs. The RT digital simulator was based on OPAL-RT technology. Here, the numerical DER models developed in MATLAB SimPowerSystems and utilized during RCP stage can be re-used with minor adjustments. OPAL-RT's automatic code generation process enables one to easily compile models. The HPPC platform is based on Bachmann PLC technology. Here, a similar process allows the use of control algorithms in MATLAB Simulink and subsequent deployment of controller models to the PLC without manual coding.

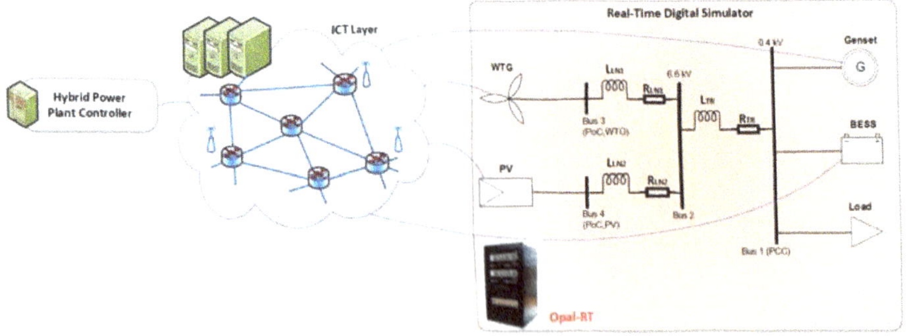

Figure 18. RT-HIL Co-Simulation Architecture in Smart Energy Systems Laboratory [35].

7.2. Step VII: On-Site Testing

The actual controller platform was tested on-site under operating conditions allowed by the physical power grid and assets. The testing campaign is typically limited in time and power system events in scope for the developed algorithms; e.g., large voltage and frequency excursions might not occur in the system during this period. Thus, an open loop approach is used. This means that the controller is fed with pseudo-measurements and the output of the plants is recorded. However, the actual impact on the power grid cannot be evaluated, nor can possible control interactions between assets. These recordings might be used to validate the numerical models developed and used in the previous stages. In fact, the performance models of WTG, PV and BESS applied during the RCP stage (Section 6) have been validated by field data and are presented in another publication [31].

8. Conclusions and Outlook

This paper proposes a detailed and practical guidance on applying MBD for voltage/frequency stability analysis, control tuning and verification in off-grid HPPs comprising both grid-forming and grid-feeding inverter units and synchronous generation. The different stages are summarized by means of Figure 1.

Initially, system and functional requirements for voltage and frequency regulation and the modeling requirements for assessment studies are specified in Section 2.

Then, a modular approach of setting up the state-space model is described by means of a benchmark HPP system used in this study (Section 3). Flexible merging of subsystems by properly defining input and output vectors is highlighted to describe the dynamics of the HPP during various operating states. The state-space models were used during the stability assessment and control tuning stage. Numerical simulation models were prepared in parallel based on the work in [31] and are applicable for small-signal model validation and the control verification stage.

EV and PF analyses were performed as part of the stability assessment (Section 4). The studies reveal that during particular load conditions instable dynamic modes occur which can be stabilized by tuning the inverter feed-forward filter. Furthermore, it is shown that clustering the vast number of system EVs enables one to identify critical dynamic modes with low damping ratio. A sensitivity analysis addresses the impact of relevant system parameters on these critical EVs. It is ascertained that false parametrization of active power/frequency characteristic by selecting large droop gains will r ove the system towards instability.

Subsequently, the control loops of the central HPP controller are designed with the purpose of frequency and voltage restoration (Section 5). It was demonstrated that the control performance largely depends on the applied droop gains of voltage and frequency regulating DERs. Some suggestions are provided for control tuning according to requirement specifications.

The RCP stage is accomplished by means of discrete-time domain models (Section 6). The off-line simulation studies confirm the effectiveness of performing stability analysis and control design/tuning for voltage and frequency control in the off-grid HPP.

An outlook is given for verifying the HPPC platform by means of RT-HIL testing as the final step of proof-of-concept (Section 7). The control algorithms developed, including physical implementation on target hardware, are then ready for site testing.

Overall, the proposed MBD approach fulfills the specified requirement specifications. It involves thorough stability assessment and control tuning stages which reduce iterations between various MBD stages significantly. The modeling effort is minimized by the use of validated discrete-time domain models of DERs in both RCP and control verification stages. Extensive RT-HIL testing in the described test setup [34] yields in high level of confidence where the need for control validation (step VII) can be significantly reduced.

The outcome of this paper is targeted at off-grid HPP operators seeking to achieve a proof-of-concept on stable voltage and frequency regulation. Nonetheless, the overall methodology is applicable to utility scale HPPs as well, where design and tuning criteria are given by the respective grid codes.

Author Contributions: Conceptualization, L.P.; methodology, L.P.; formal analysis, L.P.; investigation, L.P.; writing—original draft, L.P.; writing—review and editing, L.P.; supervision, F.I. and G.C.T. All authors have read and agreed to the published version of the manuscript.

Funding: This work was carried out as part of the PhD project "Proof-of-Concept on Next Generation Hybrid Power Plant Control." Innovation Fund Denmark is acknowledged for financial support through the Industrial PhD funding scheme. Additional funding by the Danish ForskEL-program through RemoteGRID project is appreciated.

Conflicts of Interest: The authors declare no conflict of interest.

References

1. Elkadragy, M.M.; Baumann, M.; Moore, N.; Weil, M.; Lemmertz, N. Contrastive Techno-Economic Analysis Concept for Off-Grid Hybrid Renewable Electricity Systems Based on comparative case studies within Canada and Uganda. In Proceedings of the 3rd International Hybrid Power Systems Workshop, Tenerife, Spain, 8–9 May 2018; Energynautics: Darmstadt, Germany.
2. Bitaraf, H.; Buchholz, B. Reducing energy costs and environmental impacts of off-grid mines. In Proceedings of the 3rd International Hybrid Power Systems Workshop, Tenerife, Spain, 8–9 May 2018; Energynautics: Darmstadt, Germany.
3. Petersen, L.; Iov, F.; Tarnowski, G.C.; Carrejo, C. Optimal and Modular Configuration of Wind Integrated Hybrid Power Plants for Off-Grid Systems. In Proceedings of the 3rd International Hybrid Power Systems Workshop, Tenerife, Spain, 8–9 May 2018; Energynautics: Darmstadt, Germany.
4. Kumar, R.; Gupta, R.A.; Bansal, A.K. Economic analysis and power management of a stand-alone wind/photovoltaic hybrid energy system using biogeography based optimization algorithm. *Swarm Evol. Comput.* **2013**, *8*, 33–43. [CrossRef]
5. Maitra, A.; Rogers, L.; Handa, R. *Program on Technology Innovation: Microgrid Implementations: Literature Review, Report*; Electric Power Research Institute: Palo Alto, CA, USA, 2016; Available online: epri.com (accessed on 30 November 2019).
6. Hafez, O.; Bhattacharya, K. Optimal planning and design of a renewable energy based supply system for microgrids. *Renew. Energy* **2012**, *45*, 7–15. [CrossRef]
7. Ghiani, E.; Vertuccio, C.; Pilo, F. Optimal sizing of multi-generation set for off-grid rural electrification. In Proceedings of the IEEE Power and Energy Society General Meeting, Boston, MA, USA, 17–21 July 2016; IEEE: Piscataway, NJ, USA, 2016; pp. 1–5.

8. Cañizares, C.A. *Microgrid Stability Definitions, Analysis and Modeling*; IEEE Power & Energy Society: Piscataway, NJ, USA, 2018.

9. Pogaku, N.; Member, S.; Prodanovic, M.; Green, T.C.; Member, S. Modeling, Analysis and Testing of Autonomous Operation of an Inverter-Based Microgrid. *IEEE Trans. Power Electron.* **2007**, *22*, 613–625. [CrossRef]

10. Etemadi, A.H.; Iravani, R. Eigenvalue and Robustness Analysis of a Decentralized Voltage Control Scheme for an Islanded Multi-DER Microgrid. In Proceedings of the IEEE Power and Energy Society General Meeting, San Diego, CA, USA, 22–26 July 2012; IEEE: Piscataway, NJ, USA, 2012.

11. Tang, X.; Deng, W.; Qi, Z. Investigation of the dynamic stability of microgrid. *IEEE Trans. Power Syst.* **2014**, *29*, 698–706. [CrossRef]

12. Katiraei, F.; Iravani, M.R.; Lehn, P.W. Small-signal dynamic model of a micro-grid including conventional and electronically interfaced distributed resources. *IET Gener. Transm. Distrib.* **2007**, *1*, 324. [CrossRef]

13. Gkountaras, A. Modeling Techniques and Control Strategies for Inverter Dominated Microgrids. Ph.D. Thesis, Technical University of Berlin, Berlin, Germany, 2017. Available online: depositonce.tu-berlin.de (accessed on 30 November 2019).

14. Rocabert, J.; Luna, A.; Blaabjerg, F. Control of Power Converters in AC Microgrids. *IEEE Trans. Power Electron.* **2012**, *27*, 4734–4749. [CrossRef]

15. Guerrero, J.M.; Vasquez, J.C.; Matas, J.; De Vicuna, L.G.; Castilla, M. Hierarchical Control of Droop-Controlled AC and DC Microgrids—A General Approach Toward Standardization. *IEEE Trans. Ind. Electron.* **2011**, *58*, 158–172. [CrossRef]

16. Guerrero, J.M.; Matas, J.; Garcia de Vicuna, L.; Castilla, M.; Miret, J. Decentralized Control for Parallel Operation of Distributed Generation Inverters Using Resistive Output Impedance. *IEEE Trans. Ind. Electron.* **2007**, *54*, 994–1004. [CrossRef]

17. Hadisupadmo, S.; Hadiputro, A.N.; Widyotriatmo, A. A Small Signal State Space Model of Inverter-Based Microgrid Control on Single Phase AC Power Network. *Internetworking Indones. J.* **2016**, *8*, 71–76.

18. Petersen, L.; Iov, F.; Tarnowski, G.C.; Raghuchandra, K.B. Methodological Framework for Stability Analysis, Control Design and Verification in Hybrid Power Plants. In Proceedings of the 4th International Hybrid Power Systems Workshop, Crete, Greece, 22–23 May 2019; Energynautics: Darmstadt, Germany, 2019.

19. EirGrid. *Operating Security Standards*; EirGrid: Dublin, Ireland, 2011.

20. IEC 62257. *Recommendations for Renewable Energy and Hybrid Systems for Rural Electrification—Part 2: From Requirements to a Range of Electrification Systems*; International Electrotechnical Commission: Geneva, Switzerland, 2016.

21. IEEE Standards Department. *P2030.7/D11 Draft Standard for Specification of Microgrid Controllers*; IEEE: Piscataway, NJ, USA, 2017.

22. Steinhart, C.J.; Finkel, M.; Gratza, M.; Witzmann, R.; Kerber, G.; Verteilnetz, L.E.W.; Germany, G. Determination of Load Frequency Dependence in Island Power Supply. In Proceedings of the 24th International Conference on Electricity Distribution, Glasgow, UK, 12–15 June 2017; pp. 12–15.

23. Kundur, P. *Power System Stability and Control*; McGraw-Hill, Inc.: New York, NY, USA, 1993.

24. Petersen, L.; Kryezi, F.; Iov, F. Design and tuning of wind power plant voltage controller with embedded application of wind turbines and STATCOMs. *IET Renew. Power Gener.* **2017**, *11*, 216–225. [CrossRef]

25. Huang, H.; Mao, C.; Lu, J.; Wang, D. Small-signal modelling and analysis of wind turbine with direct drive permanent magnet synchronous generator connected to power grid. *IET Renew. Power Gener.* **2012**, *6*, 48–58. [CrossRef]

26. Anderson, P.M.; Fouad, A.A. *Power System Control and Stability*; Wiley-IEEE Press: Hoboken, NJ, USA, 1995; ISBN 007035958X.

27. Knudsen, J. Modeling, Control and Optimization for Diesel-Driven Generator Sets. Ph.D. Thesis, Aalborg University, Aalborg, Denmark, 2017. Available online: vbn.aau.dk (accessed on 30 November 2019).

28. Yazdani, A.; Iravani, R. *Voltage-Sourced Converters in Power Systems: Modeling, Control and Applications*; IEEE Press: Piscataway, NJ, USA; John Wiley: Hoboken, NJ, USA, 2010; ISBN 9780470521564.

29. Teodorescu, R.; Liserre, M.; Rodríguez, P. *Grid Converters for Photovoltaic and Wind Power Systems*; John Wiley & Sons, Ltd.: Hoboken, NJ, USA, 2010; ISBN 9780470057513.

30. Sabor, O. Small-Signal Modelling and Stability Analysis of a Traditional Generation Unit and a Virtual Synchronous Machine in Grid-Connected Operation. Master's Thesis, Norwegian University of Science and Technology, Trondheim, Norway, 2015. Available online: repository.tudelft.nl (accessed on 30 November 2019).
31. Petersen, L.; Iov, F.; Tarnowski, G.C.; Gevorgian, V.; Koralewicz, P.; Stroe, D.-I. Validating Performance Models for Hybrid Power Plant Control Assessment. *Energies* **2019**, *12*, 4330. [CrossRef]
32. European Commission. *Network Code on Requirements for Grid Connection of Generators*; European Network of Transmission System Operators for Electricity: Brussels, Belgium, 2016.
33. Garcia, J.M. Voltage Control in Wind Power Plants with Doubly Fed Generators. Ph.D. Thesis, Aalborg University, Aalborg, Denmark, 2010. Available online: vbn.aau.dk (accessed on 30 November 2019).
34. Iov, F.; Shahid, K.; Petersen, L.; Olsen, L.R. RePlan Project: D5.1—Verification of Ancillary Services in Large Scale Power System, Project Report, Aalborg, Denmark. 2018. Available online: www.replanproject.dk (accessed on 30 November 2019).
35. Smart Energy Systems Laboratory at Aalborg University. Available online: www.et.aau.dk/laboratories/power-systems-laboratories/smart-energy-systems (accessed on 30 November 2019).

MDPI

St. Alban-Anlage 66

4052 Basel

Switzerland

Tel. +41 61 683 77 34

Fax +41 61 302 89 18

www.mdpi.com

Energies Editorial Office

E-mail: energies@mdpi.com

www.mdpi.com/journal/energies